OPENING THE GREAT DEPTHS

OPENING THE GREAT DEPTHS

THE BATHYSCAPH *TRIESTE* and Pioneers of Undersea Exploration

Norman Polmar and Lee J. Mathers

Naval Institute Press
Annapolis, Maryland

Naval Institute Press
291 Wood Road
Annapolis, MD 21402

Library of Congress Cataloging-in-Publication Data

Names: Polmar, Norman, author. | Mathers, Lee J., author.
Title: Opening the great depths : the bathyscaph Trieste and pioneers of undersea exploration / Norman Polmar and Lee J. Mathers.
Description: Annapolis, Maryland : Naval Institute Press, [2021] | Includes bibliographical references and index.
Identifiers: LCCN 2020050159 (print) | LCCN 2020050160 (ebook) | ISBN 9781682475911 (hardcover) | ISBN 9781682475928 (epub) | ISBN 9781682475928 (pdf)
Subjects: LCSH: Piccard, Jacques. | Walsh, Don, 1931- | Trieste (Bathyscaphe)—History. | Bathyscaphe—United States—History—20th century. | Underwater exploration—United States—History—20th century. | Explorers—United States—Biography.
Classification: LCC VM989 .P65 2021 (print) | LCC VM989 (ebook) | DDC 359.9/383—dc23
LC record available at https://lccn.loc.gov/2020050159
LC ebook record available at https://lccn.loc.gov/2020050160

∞ Print editions meet the requirements of ANSI/NISO z39.48-1992 (Permanence of Paper).
Printed in the United States of America.

29 28 27 26 25 24 23 22 21 9 8 7 6 5 4 3 2 1
First printing

This book is dedicated to the adventurers and scientists, both military and civilian, who in frail craft challenge the Earth's most inhospitable environment—the deep ocean.

Contents

Perspective

The bathyscaph *Trieste* was a scientific marvel that accomplished unprecedented scientific, technical, and military feats in the ocean depths. France and the United States both acquired and subsequently developed improved versions of the original Piccard bathyscaph. The French concentrated on the bathyscaph as a tool for deep-ocean scientific investigation, and their vehicles—the *FNRS 3* and the *Archimède*—were valuable tools for that research.

The U.S. Navy employed the three *Trieste* variants both as scientific tools and as operational military platforms. The *Trieste* program provided the Navy a nascent capability to put a manned deep submersible almost anywhere on the ocean floor for search and limited recovery operations. Indeed, the bathyscaph proved to be of unequaled value in locating and examining the remains of the sunken nuclear-propelled submarines *Thresher* and *Scorpion*, as well as other objects "lost" on the ocean floor.

Significantly, the *Trieste*—an "inner space" vehicle—was conceived and developed by Auguste Piccard, a pioneer and record holder in manned balloon flight. This volume recounts the "story" of how Piccard pioneered the development of atmospheric research employing manned balloons and then, with his son Jacques, employed the research balloon concept to develop the bathyscaph. His initial efforts had limited success, but the *Trieste* proved his design concept.

From the *Trieste*'s initial dives, the participants in the program realized that they were making history, diving to previously unexplored and unexploited depths, developing new capabilities, and opening a new frontier. Comparisons with developments in space and the space race between the United States and the Soviet Union were constantly made in official speeches and documents, and in conversations concerning the *Trieste* program and contemporary developments in undersea technology and capabilities.

The *Trieste*—actually three distinct vehicles carrying the same name—opened the entire three-dimensional volume of the oceans to exploration, exploitation, and operation. The bathyscaph was a first-generation system, a "Model-T" that spawned an entirely new industry and encouraged new concepts for naval operations. These new deep-sea technologies lacked the "gee-whiz" factor of the contemporaneous space race, but were just as significant in the development of new technology, new knowledge, and—especially—new military capabilities.

The "official" unclassified dive logs for all three versions of the *Trieste* are in the collection of the U.S. Naval Undersea Museum, in Keyport, Washington, and can be accessed at https://bathyscaphtrieste.org/contents/divelog/logt2.html. The museum's logs do not include the dives for the *Trieste II* dives between 1963 and 1967.

The many persons who assisted in researching and writing this book are listed in the Acknowledgments. However, it must be noted that one of the authors has had a close and long-lasting relationship with Dr. Don Walsh, who took the original *Trieste* to the bottom of the Mariana Trench in January 1960, since a few weeks after the "deep dive." Walsh's friendship and assistance made this book possible. Also, that author, as an employee of the Northrop Corporation, worked full-time for four years in the Navy's Deep Submergence Systems Project, where he had an association with some of the events described in this book.

Early *Trieste* "commanders" Larry Shumaker and Donald Keach also provided early and extensive assistance to this project. Both of these men passed away before we completed our research. The authors also have benefited from the assistance of Victor L. Vescovo, whose 2018–2019 dives into the world's six deepest trenches established a new world record for the deepest manned dive in history—besting the *Trieste*'s record, which had stood for almost 60 years.

Also, Pelham Boyer, formerly of the Naval War College, was most helpful in copy editing this volume. Janis Jorgensen of the Naval Institute provided assistance with illustrations, while Richard Russell, Glenn Griffith, and Rachel Crawford of the Naval Institute Press brought this project to reality.

Notes on style and usage:

- The spellings *bathyscaphe* and *bathyscaph* are both in common use. We have used the latter spelling as the accepted American style in contemporaneous official documentation and newspaper usage.
- Despite the widespread adoption of metric measurements, the American readership still prefers the English system of feet, yards, ounces, pounds, fathoms, etc.
- All times are presented as local time, except in reference to events in "space," where Zulu time (Greenwich Mean Time) is annotated: for example, 1600Z for 4 p.m. at the Greenwich meridian.

Norman Polmar
Lee J. Mathers

Foreword by Dr. Don Walsh

It has been more than 60 years since I first encountered the bathyscaph *Trieste*. Little did I realize at the time how much our fates would be intertwined in the decades since then. Over the years the story of this unique and historical submersible would be told in many different ways, but never comprehensively.

Thus, this book is long overdue. It remained for Norman Polmar and Lee Mathers to write this definitive story. They tell the little-known history of a quarter century of the U.S. Navy's use of three *Trieste* bathyscaphs to support in a unique way research and work into the deepest parts of the oceans. Equally important is the story of their highly classified missions of national importance. And this book does not stop with *Trieste* stories.

It was January 1958, when the Office of Naval Research (ONR) purchased the bathyscaph from Auguste and Jacques Piccard with the intention that it would be a research platform for oceanographers. That summer the *Trieste* was shipped from Italy to its new home, the Navy Electronics Laboratory (NEL) in San Diego, California. One of nearly a dozen Navy laboratories, NEL was chosen for its extensive oceanographic scientific programs and because deep-water operating areas were not far offshore.

At NEL the project was organized so that civilian scientists would determine the science missions while a Navy team of two officers and a few enlisted personnel would operate and maintain the submersible. The program manager and chief scientist was NEL's Dr. Andreas B. Rechnitzer, a newly minted marine biologist recently graduated from the Scripps Institution of Oceanography.

"Andy" had been involved with *Trieste* since the summer of 1957, when ONR leased it for a series of test dives near Capri. At the conclusion of that program he and other oceanographers who had sampled the craft's capabilities recommended that the Navy purchase it. And, he was instrumental in convincing ONR that it should be brought to his workplace, NEL.

With respect to military involvement, it was Rechnitzer's view that the Navy's submarine force would be the best source of personnel to operate and maintain the bathyscaph. The call for volunteers went out in the fall of 1958. I volunteered immediately. Surprisingly, I was one of only two naval officers who volunteered. Lieutenant Dick Davey was the other. Both of us were selected, but Dick had to step down due to a temporary health problem shortly after joining NEL.

Prior to joining NEL, I was serving in the submarine *Rasher* (SSR 269). A modified, diesel-electric "fleet boat," she had a maximum permitted diving depth of 300 feet. Little did I realize at the time that only one year after joining the *Trieste* I would be on the ocean floor at a depth of nearly 36,000 feet. That gave a new meaning to the term "maximum permitted depth"!

Once at NEL I was made *Trieste*'s officer-in-charge. I soon found out that there was little relationship between my submarine experience and being the pilot of a deep-diving submersible. They were as different as elephants are from dogs. The bathyscaph was my first command in what would become a 24-year Navy career. And it was certainly the most unusual of my Navy assignments!

Fortunately, I was able to get Lieutenant Larry Shumaker assigned as my assistant officer-in-charge. An Annapolis classmate (Class of 1954), he had served with me on the *Rasher*. I could not have asked for better colleague.

Andy, Larry, and I made a strong leadership team, and both remained my friends until the end of their lives. We led a group that rarely exceeded 14 permanently assigned people, but we got the job done. It was a great crew of Navy civilian employees and uniformed personnel. We were doing exciting things for the first time, and our team spirit was always high even in the longest days and weeks during our months of intensive testing at Guam.

We were certainly pioneers as the *Trieste* was one of only two such vehicles in the world—the French Navy's bathyscaph *FNRS 3* was the other. It meant that we had to "write the book" about deep submergence operations in terms of techniques and technologies. We learned by doing and by failures, although very few of the latter were serious. For example, if we needed a piece of equipment we had to design it and make it. There were no commercial vendors who catered to deep submergence technology requirements. The market was far too small.

Although I was with *Trieste* project for only three and one-half years, every day was exciting. We were fortunate that our seniors at NEL and in the Office of Naval Research did not get too involved with or too curious about the project. This let our group of relatively young people operate without intrusive oversight from our masters. Still, our funding was generous, and we rarely had to endure budget difficulties. Looking back, all this probably was because "they" did not want to know if we screwed up. It was "plausible denial." As Abraham Lincoln said, "Success has a thousand fathers, failure is an orphan."

I left the program in the summer of 1962 to return to sea duty in submarines. But that was not the end of my association with deep-sea exploration. Indeed, for the past 60 years I have had a professionally rewarding association with undersea technologies and operations. Particularly rewarding today is to see our "*Trieste* DNA" in current deep-ocean engineering devices and systems. It is found in such devices as remotely operated manipulators (arms), lighting systems, cameras, instrumentation, and sampling devices. We were the first to develop those items for seagoing applications.

Almost all of my colleagues from those early days of *Trieste* are now gone. But I fondly remember them all and the great things that our little team accomplished. It has been a great run for me, and I am happy that I had a role in exploring the deepest ocean.

My sincere thanks to Lee Mathers and to my longtime friend Norman Polmar for telling our story.

Don Walsh, Captain, U.S. Navy (Ret.), PhD
U.S. NAVY SUBMERSIBLE PILOT NO. 1

Acknowledgments

The authors wish to thank these individuals, who volunteered their time, memories, and expertise to knit this history together: Richard Abbott, assistant OINC, *Trieste*; Jon Anderson, executive officer, *Point Loma*; Barney Bakara, diver; R. D. Baker, OINC *Trieste*; Malcolm Bartels, OINC *Trieste*; Edward Beron, physician and diver; Charles Bishop, staff, ONR and CO Naval Undersea Center, San Diego; Tony Boegeman, engineer, Scripps Marine Physical Laboratory; George F. Bond, senior medical officer, DSSP; Norman Branchflower, CO *Point Loma*; Troy Brown, hospital corpsman, *White Sands*; David Byrnes, pilot, *Trieste*; Eugene Cash, staff officer, ComSubLant; Jerry Carroll, lead engineer, Naval Oceanographic Office's Teleprobe Program; James Cathro, assistant director, NRTSC; Joseph A. Cestone, head, sensors and ship control branch, DSSP; Frank Charlton, crew, *Trieste*; Chuck Copeland, crew, *Trieste*; John P. Craven, director and chief scientist, Deep Submergence Systems Project; Dennis E. Curtis, staff officer, *Thresher* search; Celia Donnelly, French linguist; Ron Doyle, chief staff officer, SubDevGru 1; Tony Dunn, assistant OINC *Trieste*; Millard Firebaugh, staff, SubDevGru 1; Bob Fishback, techrep, Westinghouse; Roger Flood, oceanographer; Frederick Forst, assistant OINC *Trieste*; Roy Fort, crew, *Trieste*; Ray Freeman-Lynde, oceanographer; Robert Frosch, Assistant Secretary of the Navy for Research and Development; Lorenzo Hagerty, executive officer, *Apache*; Mort W. Hans, field engineer, Sperry Corp.; David F. Helms, crew, *Trieste*; Philip L. Hendricks, physician and diver; E. E. Henifin, OINC *Trieste*; Al Hodgdon, techrep, Sperry Corp.; John Howland, assistant OINC *Trieste*; Robert M. Hynosky, son of a crewman on escort *Lewis*; Alan Jamieson, PhD, oceanographer; Daniel A. Johnson, sonar engineer, Straza Industries; Marty Klein, field engineer, EG&G; Melissa Knapp, daughter of Roger Whitaker; James Kuczkowski, crew, *Trieste*; Patrick Lahey, president, Triton Submarines; Bud Lester, pilot, *Trieste*; Larry

Lonnon, CO *Apache*; Jon Losee, physicist/field engineer, Navy Underwater Center San Diego; George Martin, assistant OINC *Trieste*; Rob McCollum, expedition leader, Five Deeps 2019; Dick Meaux, staff, DSSP; E. John Michel, crew, *Trieste*; Mitch Mitchell, docking officer, *Point Loma*; Richard Molasky, field engineer, Perkin-Elmer Corp.; Brad Mooney, OINC *Trieste*; Ed Moore, CO *Halibut*; Becky Muckenthaler, daughter of Roger Whitaker; Beau Myers, executive officer and OINC *White Sands*; Bob Nevin, OINC *Trieste*; Mary Hollis Newell, wife of OINC *Trieste*; Les Parsons, OINC *Trieste*; Bertrand Piccard, explorer; Steve Pope, crew, *Trieste*; Larry Porter, pilot, *Trieste*; Phil Pressel, engineer, Perkin-Elmer Corp.; Dix Preston, O-System officer, AFTAC; Patrick Raetzman, staff, OpNav; Dan Rean, assistant OINC *Trieste*; Gary Rees, chief staff officer, SubDevGru 1; Kim Riemenschneider, stepdaughter of Roger Whitaker; Stan Reinhold, crew, *Trieste*; Carl Romney, director, Seismology Center, AFTAC; R. Bruce Rule, head acoustic analyst, SOSUS/NAVSTIC; Mary C. Ryan, Naval Undersea Museum, Keyport, Washington; Clark Sachse, staff, DSSP; Robin Salser, crew, *Trieste*; Ross Saxon, pilot, *Trieste*; Elizabeth Noble Shor, historian, Scripps Oceanography Institution; Glen Sjoblom, assistant director, Navy Nuclear Propulsion Program; Mary Liz Spiess DeJong, daughter of Noel Spiess; David S. Smyth Sr., son of "Junior" Smith, Mare Island NSY; Mike Staehle, pilot, *Trieste*; Harris Steinke, submarine escape expert; Ray Stoetzer, staff officer, 1964 *Thresher* search; Tony Taromino, NavSea representative *Trieste* overhauls; Dick Taylor, pilot, *Trieste* and OINC *DSRV 2*; Burt Tharp, OINC *Trieste*; Duane Tollaksen, crew, *Trieste*; John B. VanVoorhis, techrep, Sperry Corp.; George Verd, staff, OpNav and CO *Dolphin* (AGSS 555); Victor L. Vescovo, explorer, Five Deeps Expedition; Tom Vetter, assistant OINC *Trieste*; Heinfield Voss, historian, Krupp; David Waltrop, historian, CIA; Frances Whitaker, wife of OINC *Trieste*; Wilson Whitmire, CO *Point Loma*; Jim Worthington, crew, *Trieste* and executive officer, *Point Loma*; and James A. Youse, Special Advisory Group, *Scorpion* search 1969.

The authors also are especially indebted to the *Trieste* Alumni Association for its aid in locating *Trieste* veterans and its help in establishing contact with those who had first-person stories to tell.

Abbreviations

ANGUS	Acoustically Navigated Geological Underwater Survey
ARIS	Acoustic Range Interrogation System
ATNAV	Advanced Technology Navigation
AUTEC	Atlantic Undersea Test and Evaluation Center
AUV	Autonomous Underwater Vehicle
avgas	aviation gasoline
BuPers	Bureau of Naval Personnel
BuShips	Bureau of Ships (now Naval Sea Systems Command)
CASREP	Casualty Report
CNO	Chief of Naval Operations
CO	Commanding Officer
ComSubLant	Commander, Submarine Force Atlantic
ComSubPac	Commander, Submarine Force Pacific
CTFM	Continuous Wave, Frequency Modulated (sonar)
CURV	Cable-controlled Underwater Research Vehicle
DOT	Deep Ocean Transponder
DSSP	Deep Submergence Systems Project
DSSRG	Deep Submergence Systems Review Group
EG&G	Edgerton, Germeshausen and Grier
IOU	Integral Operating Unit
LORAN	Long-Range Navigation
MIT	Massachusetts Institute of Technology
NASA	National Aeronautics and Space Administration
NavShips	Naval Ship Systems Command (Washington, D.C.; until 1974)
NEL	Navy Electronics Laboratory (San Diego, Calif.)
NRL	Naval Research Laboratory (Washington, D.C.)
NRTSC	Naval Reconnaissance and Technical Support Center (Suitland, Md.)

NURO	National Underwater Reconnaissance Office (Washington, D.C.)
OINC	Officer-in-Charge
ONR	Office of Naval Research (Arlington, Va.)
OpNav	Office of the Chief of Naval Operations
ROV	Remotely Operated Vehicle
RV	Re-entry Vehicle
SATNAV	Satellite Navigation
scuba	self-contained underwater breathing apparatus
SOSUS	Sound Surveillance System
SubDevGru	Submarine Development Group
SubRon	Submarine Squadron
techrep	technical representative
UQC	type of underwater acoustic telephone
UTE	Underwater Tracking Equipment

U.S. Navy Ship/Submarine/Submersible Designations

AC	collier
AD	destroyer tender
AG	miscellaneous auxiliary
AGDS	deep submergence support ship
AGER	intelligence collection ship
AGOR	oceanographic research ship
AGS	surveying ship
AGSS	research submarine
AK	cargo ship
AKD	cargo ship—docking well
AKS	stores-issue ship
APD	high-speed transport
ARD	auxiliary repair dock
ARD(BS)	auxiliary repair dock (bathyscaph support)
ARS	salvage ship
AS	submarine tender
ASR	submarine rescue ship
ATA	auxiliary tug
ATF	fleet tug
CGN	guided missile cruiser (nuclear propelled)
DE	escort (formerly destroyer escort)
DER	radar picket escort
DSRV	deep submergence rescue vehicle
DSSV	deep submergence search vehicle
DSV	deep submergence vehicle
LCM	landing craft mechanized
LCVP	landing craft vehicle and personnel
LSD	dock landing ship

PF	frigate
SS	submarine
SSB	ballistic missile submarine
SSGN	guided missile submarine (nuclear propelled)
SSK	hunter-killer submarine
SSN	submarine (nuclear propelled)
SSR	radar picket submarine
SST	training submarine
YTB	large harbor tug

Note: The *T-* prefix indicates a civilian-manned ship operated by the Navy's Military Sealift Command.

Columbus of the Stratosphere

My objective is not to break records and especially not to be a record holder, but to pave the way for new scientific research and aerial navigation.

—Auguste Piccard

Amelia Earhart—aviation pioneer and a cultural icon—hosted a dinner on Monday evening, 23 January 1933, at her home in Rye, New York. It was a small, private affair. Attending were the hostess and her husband-publisher George Palmer Putnam, aviator Charles A. Lindbergh, explorer Roy Chapman Andrews (who had just been awarded the coveted Explorer's Club Medal for 1932), and balloonist Auguste Piccard.[1]

Earlier, on 13 June 1932, during her victory lap through Europe following her solo flight across the Atlantic the previous month, Earhart and her husband arrived in Brussels, Belgium. There, Amelia received the title of Knight in the Order of Leopold, the same honor earlier presented to Charles Lindbergh.[2] She then was invited to the laboratory of Auguste Piccard, the new holder of the record for highest altitude manned flight, a Swiss physicist, and fellow Knight of the Order of Leopold. In conversations with Amelia and her husband, Piccard revealed that one of the reasons for his upcoming trip to America was that he hoped to meet his hero: Charles Lindbergh. As publisher of Lindbergh's autobiography *We*, Putnam was positioned to make that happen.

Piccard had arrived in New York on 12 January 1933, on the French liner *Champlain*. Following a lecture at the Franklin Institute in Washington, D.C., and a few days in the area of Marshalltown, Delaware, home of his twin brother, Jean Felix Piccard, Auguste returned to New York City.

At Earhart's dinner party on the 23rd, Piccard told Lindbergh in English, with a thick Swiss accent, "You are a great aeronaut, and I am a mere balloonist studying penetrating radiations. Our territories almost touch, but they are not the same." After appropriately polite demurs from Lindbergh, Piccard tossed a conceptual grenade into the conversation: Piccard related that before World War I he had drawn up plans to go to the bottom of the ocean in a "balloon." Lindbergh was astounded: "In a balloon! Please tell me about it."

"My gondola," related Piccard, "would have been very much like the one I used for exploring the stratosphere, but the balloon would have been inflated with oil instead of hydrogen, helium, or other gases. This, being lighter than water, would have served to lift us and we would have carried ballast just as we do in ordinary balloons." Piccard continued, "I subsequently decided to go up rather than down in a balloon because observations of cosmic rays were to be made best in the upper regions, where they are assumed to be more numerous. I really have quite as much in common with submarine navigation as with aeronautical navigation." He proceeded to describe water pressures and means of artificial respiration that would enable two men to work equally well in something approximating normal atmospheric conditions whether on the bottom of the ocean or high in the skies.

As the dinner progressed, the table was amused to witness the professor go into his lecture mode, reaching into his white vest to arm himself with a slide rule, which he used to convert metric measurements to English miles, tons, and Fahrenheit degrees as he described the conditions that would meet an underwater balloon on its descent. Throughout the meal Piccard alternated between eating utensils and the slide rule.[3]

Two days later, at a New York Town Hall lecture, Piccard was introduced to William Beebe, who held the record for the deepest manned dive into the ocean—a historic meeting of the two men responsible for the highest and deepest explorations of humankind, and perhaps a portent of Piccard's future efforts.

A Family of Scientists

Auguste Antoine Piccard and his twin brother Jean Felix were born on 28 January 1884, in Basel, Switzerland, the sons of Professor Jules Piccard, head of the Chemistry Department at the University of Basel. Jules had married Jeanne Christiane Hélène Haltenhoff in 1873. The twins were the babies of the family, with an older brother, Paul Georges, and an older sister, Marie Louise. Jules Piccard was internationally prominent in organic chemistry. Jules' brother, Paul Piccard, was an engineer and automotive pioneer with the Piccard-Pictet Company, producing some of the finest motorcars in Europe from 1906 to 1920.

Twin Jean Felix was born right-handed and Auguste left-handed. Auguste learned to use his right with equal facility and became renowned for simultaneous ambidextrous drawings on his classroom chalkboard. Francophones at home, Auguste became fluent in German at school, and eventually comfortable in English and Italian. Both brothers displayed an early interest in science, attending university at the Swiss Federal Institute of Technology at Zurich.

Auguste earned his bachelor's degree in mechanical engineering in 1910 and showed early promise as a scientist, and in 1904, while still an undergraduate he published his first paper "New Findings on Geotropic Sensitivity of Root Extremities." Upon graduation he took a position as a teaching assistant while continuing his studies. A year later he coauthored a paper with his thesis advisor, Peter Weiss, on limits in the magnetism that accompanies radioactive sources. In 1912, he developed a precision manometer, demonstrating his skill in the design and fabrication of scientific instruments.

In 1914, he defended his dissertation before a panel of professors of physics that included Albert Einstein and was awarded his doctorate for an investigation titled "The Magnetization Coefficient of Water and Oxygen." Also that year, Piccard published his first paper with the Paris Academy of Sciences. In 1915 Auguste was appointed a university lecturer in Zurich, an unpaid post or one paid for by student subscriptions or attendance.

In 1917 Piccard isolated a substance he called "actinouranium," which has since been identified as the isotope of uranium that we know as Uranium-235. That year he was appointed full professor at the Institute. During his years at Zurich he was a leading student of Einstein and collaborated with him to devise instruments to measure radioactivity. Piccard developed several of the instruments used by Einstein in these early years (1914–1922).[4]

Both Piccard twins were fascinated with ballooning. On 19 June 1912, Auguste made his first flight in a balloon, sponsored by the Swiss Aero Club as part of a series of flights to investigate the distribution of temperatures of the gasses within the balloon envelope. In 1913 the twins together flew a balloon from Zurich across Germany into France, for 16 hours recording air densities and temperatures. During World War I both men served in the Swiss Army's lighter-than-air service. By 1931, Auguste had 13 balloon flights under his belt and was a licensed balloon pilot.

(As a balloonist using hydrogen for lift, it is hardly surprising that Auguste did not smoke, but more to the point, he claimed to become physically ill if anyone exposed him to tobacco smoke.[5] He also avoided alcohol.)

In 1920 Auguste married one of his former students, Marianne Denis, the daughter of a French historian teaching at the Sorbonne. In 1921 the couple's first daughter of an eventual four was born; their only son, Jacques Ernest Jean Piccard, was born in Brussels the next year.

The Free University of Brussels

Auguste Piccard accepted the newly established chair of applied physics at the Free University of Brussels in 1922 and was swept up in the excitement and turmoil that physics was then experiencing. He attended the prestigious Solvay Congress sessions held in the 1920s and 1930s. The conference of October 1927, was the most famous; 17 of the 29 invitees were current or subsequent Nobel Prize winners. A junior professor, Piccard rubbed elbows with such physics notables as Albert Einstein, Marie Curie, Max Born, Hendrik Lorenz, and Niels Bohr. At the 1927 conference, Einstein uttered his famous remark "God does not play dice," to which Werner Heisenberg, a leading German theoretical physicist, reportedly responded, "Einstein, stop telling God what to do."

Early in his career, Auguste devised a new form of bifocals with two lenses that flipped up onto the forehead for distant vision or down over a second pair of lenses for close vision. He was proud of this innovation and wore his hinged wire-rim bifocals for much of his career. They never caught on commercially, but were to be a unique and recognizable signature for Piccard, becoming part of his image and reinforcing his "scientific" persona.

His son Jacques described how his father worked: "He is constantly thinking, and when he has any kind of important problem, he will think on it for days and days, even arguing with himself. Very often, he goes walking in a field or the forest, and people do not understand that he is working. He never accepts the first solution to a problem. He would say that the big danger is to find *one* solution. My father would say, you have to think and not find *a* solution, but find *the* solution, which is quite different."[6] The physicist Auguste Piccard always sought both elegance and simplicity in a solution to a problem, whether a scientific issue or an engineering one.

The Stratospheric Vehicle

In Brussels in the late 1920s, Piccard had been conducting experiments on the properties of gamma rays from radium when his interest shifted to cosmic rays. He felt that better measurements of cosmic-ray ionization were only to be had by ascending above 90 percent of Earth's atmosphere, thus eliminating most of the secondary scattering ionization and thereby investigating primary cosmic rays in situ. Accordingly, he designed a balloon system that could take him into the stratosphere—some 50,000 feet, more than 10,000 feet higher than any man had flown before.[7] But the first hurdle for such a project was the funding.

In his famous *Discours de Seraing* on 1 October 1927, marking the 110th anniversary of a Belgian steelmaking firm, speaking in its factory at Seraing, in Liège—King Albert I of Belgium, emphasized the central position held by "pure" science in the health and economic development of modern societies.[8] In response to the king's speech, a group of scientists and industrialists created the Fonds National de la Recherché Scientifique (FNRS) on 27 April 1928. In just three months private individuals, FNRS patrons, and the Solvay family endowed the Belgian Fonds National with more than 100 million Belgian francs to support scientific investigations. The FNRS was a model later emulated in France, Germany, and Switzerland.

King Albert was not just a proponent of basic science and an admirer of physicists, but also an aviation enthusiast. Piccard, perceiving a possible sponsor for his plans, presented himself as a "native son" from the Free University of Brussels, a physicist of sufficient stature to be invited to the tri-annual Solvay Congress, and an experienced balloonist. He was convincing and articulate, offering FNRS an opportunity to support pure science and to do so with a project that could bring worldwide publicity to Belgium's new encouragement of basic science. The FNRS agreed to finance the construction of a balloon capable of taking an entirely closed, hermetically sealed aluminum gondola with two men into the stratosphere. It would take two years to design and construct such an innovative aeronautical vehicle.

Piccard's design involved a balloon ten times the size of an average racing balloon, capable of lifting the 200-pound aluminum gondola with crew and equipment. The sealed cabin would have an oxygen generator capable of delivering eight hours of oxygen; Piccard prudently provided a duplicate of the device. This decision was to save the lives of the aeronauts on the first flight, which unexpectedly stretched beyond 16 hours. Piccard christened the balloon with its sealed gondola the *FNRS* in recognition of his funding agency.

Prisoners of the Air

The largest balloon to be fabricated to that time lay draped across acres of meadow adjacent to the Riedinger Balloon Factory outside of Augsburg, Germany. It was 26 May 1931, and the press was on hand to film what would be an attempt to break the 69-year-old manned balloon altitude record. That record had been set when British balloonists Henry Coxwell and James Glaisher had reached 39,000 feet in 1862 in a coal gas-fired hot-air balloon.[9]

Piccard and his assistant, physicist Dr. Paul Kipfer, were aiming for an altitude of ten miles—52,800 feet, well into the stratosphere—to obtain new data on cosmic-ray ionization. The delicate choreography required to inflate such a gigantic balloon commenced at 2300 under artificial light. Any crossing of lines, trampling on the balloon skin, or miscommunication with line handlers could result in fatal consequences. Two parachutes, wicker stools, thermos bottles of liquid oxygen, and one small bottle of drinking water were loaded into the gondola. The weather was perfect, and the press was advised to expect a return to Earth, at whatever location to which the untethered and unpowered balloon had traveled, about noon on the 27th.

When the balloon lifted off Piccard and Kipfer found themselves almost immediately in peril, the beginning of a flight that would be a literal fight for survival. Unbeknownst to them, an unauthorized rope had been attached to the balloon to help line handlers cope with heavy winds prior to liftoff, and that rope had become entangled with the control line that released hydrogen to achieve a controlled descent. Also unbeknownst to Piccard and his assistant, a penetration of the airtight aluminum skin of the cabin in which a valve was fitted had been torn. Piccard faced a pressure leak that might force him to abort at low altitude, not knowing that the means for aborting—the release of hydrogen—had been disabled. Piccard's log contained the entry, "We are prisoners of the air!"[10]

At noon on 27 May, the projected landing time, those following the flight on the ground became concerned. There was no response from the balloon to signals from the ground. Seven hours after the advertised landing time with the balloon still at high altitude, the press notified the world that "oxygen [was] believed gone." Actually, Piccard and Kipfer had patched the air leak and were sitting quietly to conserve their oxygen supply. Piccard knew balloon physics and therefore that sunset would cool the hydrogen, bringing the balloon to lower altitudes. With two eight-hour oxygen generators and a couple of thermos bottles full of liquid oxygen, they had more than 16 hours of safe flight. They needed only to wait.

By sunset, normal evening cooling had brought the aeronauts down to 15,000 feet, where they could open the top hatch and air out the cabin. They were descending into the high mountain and glacier country between Bavaria and Italy. Piccard dared not attempt to control the landing by releasing reserve ballast as the balloon might then regain altitude. Piccard's caution was well-advised; the gondola hit a snowbank on the Gurgl Glacier at about 2100.

The huge balloon had settled near the Bavarian village of Ober-Gurgl. Certain that Piccard and his assistant had died from asphyxia during the flight, the local schoolmaster organized a party to depart at dawn of the 28th to recover the bodies. As the party ascended the mountain it saw a pair of presumed tourists coming down off the glacier. After mutual long-distance hand-waving, the Ober-Gurgl party continued upward. At the balloon they found signs that the aeronauts had survived and realized that the "tourists" they had seen had been the balloonists making their way down.

It was not until noon that the rescue team finally caught up with the aeronauts, now about halfway to Ober-Gurgl. The balloonists were tired and hungry, but unharmed. Piccard asked that someone return to the village at top speed to send a telegram to his wife in Lucerne, Switzerland, telling her that all was well. From Ober-Gurgl the world was told: "Piccard alive and well."[11]

Dozens of newspaper and radio reporters were immediately en route to the small alpine village. Queries from the public and news organizations clogged telephone and telegraph lines, shattering the normally bucolic tranquility of Bavaria and Austria. One Munich police official, tired of the constantly buzzing telephone, groused, "I don't believe there is a single Bavarian who has not called up and asked about that cursed professor."

The press had created a frenzy.

Piccard was truly flummoxed by the sudden notoriety, the fame and unrelenting press attention that the successful completion of his problem-filled flight had attracted. He stated in an interview at Ober-Gurgl, that the idea that publicity might result "seemed nonsensical when I started. My laboratory life did not accustom me to such things. Now I begin to realize that strangers are quite interested."

The king of Belgium honored Piccard by making him a Knight in the Order of Leopold (the same honor given to Charles A. Lindbergh and Amelia Earhart). Kipfer was presented with a Knight's Cross. The Belgian government commissioned two commemorative bronze medallions by the famed Belgian sculptor Victor Demanet, one showing both Piccard and Kipfer in silhouette, and other the face of Piccard alone. Also, the German medalist Karl Goetz cast for Piccard and Kipfer two silver medals to commemorate the first stratospheric flight.[12]

Second Flight: Second World Record

By November 1931, Piccard was announcing his next balloon launch to investigate cosmic rays. Kipfer would not make the flight; his family would not agree to another flight where he would be risking his life.[13] The Fonds National de la Recherché Scientifique was financing this second flight as well, and King Albert I reportedly planned on attending the launch incognito. Before the flight the king inspected the gondola and even crawled inside of the cabin for a briefing.

On the evening of 17 August 1932, the massive balloon—now with a new, radio-equipped gondola—was once again spread out on the ground at the Dubendorf aerodrome near Zurich. Three hundred Swiss soldiers were on hand to control a crowd estimated at 16,000—many brought by special trains scheduled for the spectacle. The release was at 0507 on 18 August. Just before release, Piccard briefed the press to expect a ten-hour flight to terminate at about 1500. Piccard and his assistant, Dr. Max Cosyns, were attempting to better the 1931 altitude record, hoping to reach 65,000 feet.

The flight went flawlessly. Piccard was in nearly constant radio contact with the ground, while two aircraft and four automobiles followed the balloon's route. Although the maximum altitude officially recognized was, at 53,153 feet, far short of 65,000 feet, it did exceed the 1931 flight by several hundred feet and broke through the "magic" ten-mile ceiling. Reaching maximum altitude at about 1000, the two scientists conducted cosmic-ray ionization experiments until the early afternoon, then decided to descend.[14] The landing was at 1700 near Volta-Mantua, Italy. The bumpy landing dented the gondola and shook up its crew, but inflicted no significant damage or injuries.

With this second flight Piccard proved that he was more than a lucky survivor, more than a gifted amateur, and more than simply an academic. He had the intellect and discipline of a world-class physicist and the skills of a talented mechanical engineer. Enjoying the resources of a university and support from a national funding agency, and with the self-confidence to put his life on the line to achieve a scientific breakthrough, Piccard was set on his path.[15]

Piccard's cosmic-ray studies never yielded information that would justify the effort or the cost of his stratospheric experiments, although claims were made that these flights helped prove one of the predictions of Einstein's Theory of Relativity. Professor Erich Regener and his group in Germany were at the same time, sending unmanned cosmic-ray experiments into the stratosphere, perfecting self-recording electroscopes to be carried by balloons to extreme altitudes, much higher than the Piccard flights. A cost-benefit analysis would show that Piccard was on the wrong side of the equation.

On 28 January 1932, Professor Regener launched an unmanned instrumented balloon to 65,000 feet, and in August 1932, just a few days prior to Piccard's 1932 flight, to 92,000 feet. Regener's August 1932 flight carried, for the first time photographic plates to record charged-particle events. In meeting with reporters, Regener stated that he believed his results "will prove far more complete and exact than those obtained by Professor Auguste Piccard on his spectacular stratosphere flights." He was right.

Piccard's flights *were* a spectacle, albeit unintentionally. As he came to realize that he was not producing innovative results in cosmic-ray research, Piccard looked for other applications for what he had achieved. He almost immediately recognized that he had created the technological concept that would open the stratosphere and beyond to human endeavor and exploitation. The potential uses of his hermetically sealed capsule for future long-range travel were manifest. In 1933, he predicted, "New York would be a short day's journey from Europe. One could leave in the morning and attend a concert in

New York in the evening. This is a dream of the future but I'll end my comments by expressing not only my hope but my conviction that what I've told you won't remain a dream for long but will soon become a reality."[16] His brother Jean Felix somewhat confusingly proposed that Auguste's gondola could potentially take men to the moon—a long-term prediction of shirtsleeve environments for space travel.[17] In fact, Auguste had developed and flown the first airtight atmospheric-pressure cabin *after* the U.S. government had considered the idea and rejected it as too difficult for the available technology. Piccard had proved otherwise.

Auguste had twice set world high-altitude records for manned flight: on 27 May 1931, to 51,790 feet with Paul Kipfer, and on 18 August 1932, to 53,000 feet with Max Cosyns. Both flights were well-planned assaults on the stratosphere to attain useful scientific data. Others attempting altitude records soon were employing many of Piccard's innovations in balloon technology. His record balloon flights garnered reputation, recognition, and acclaim.

His awards in Belgium—the higher grade of Commander of the Order of Leopold was bestowed in 1932—were followed by equivalent honors from the French government, which granted Piccard a Commandeur de l'ordre national de la Légion d'honneur in 1932. Belgium granted Piccard the very rare honor of issuing three postage stamps to commemorate his balloon achievements. On the stamps, in addition to a photo of the *FNRS* ascending on the 1932 flight, there appears in small print the dates of the two flights and the name "Prof. A. Piccard." The stamps, intended to generate funds for the Fonds National, were sold out to collectors within a week. The Belgian sculptor Demanet produced bronze busts of Piccard, Kipfer, and Cosyns, the "explorers of the stratosphere."[18]

The press was enamored of Piccard and promoted him for the 1933 Nobel Prize in physics, as documented in *The New York Times* on 29 October 1933: "Piccard Expected to Get Nobel Prize; Professor Might Receive 1933 Award for Physics." There were candidates more qualified for the honor, and ultimately it went to P. A.M. Dirac and Erwin Schrödinger, jointly.

The press had made Piccard a cultural icon with the accolade "The Columbus of the Stratosphere." However, Auguste Piccard would eventually create his true legacy not in the stratosphere but in the ocean depths.

A Balloon into the Depths

It must be possible, I said to myself, to build a watertight cabin, resisting submarine pressure and furnished with portholes, to allow an observer to admire a new world.

—Auguste Piccard

On 5 September 1948, at the port of Antwerp, Auguste Piccard, the "Columbus of the Stratosphere," called together the international press corps. He revealed that a few days hence the Piccard-Cosyns expedition would lift anchor and sail for West Africa, where the two men would descend into the abyssal ocean depths to 13,200 feet—2½ miles down—bettering the world manned-dive record of 3,028 feet set by Americans William Beebe and Otis Barton in their tethered device in 1934. The Belgian Fonds National de la Recherché Scientifique was sponsoring the expedition. Piccard's "underwater balloon"—named the *FNRS 2*—had been quietly designed and fabricated over the previous two years.[1] The underwater balloon was now being publicly announced as a reality, and Piccard was making the most significant error of his public career.[2]

An Underwater Balloon

In later years, Piccard would wax nostalgic about a concept that had occurred to him as early as 1905, while he was still an undergraduate:

> I was a first-year student at the Zurich Polytechnic School when by chance I read the fine book of Carl Chun recounting the oceanographic expedition of the *Valdivia*.[3] Nets let down to considerably over a thousand fathoms brought back submarine fauna to the deck of the ship. They worked day and night. When a net was brought up in complete darkness,

> the oceanographers, leaning over the rails, were struck by the multitude of phosphorescent animals which the net contained in its seine. Certain fish were endowed with veritable headlights. . . . To observe these fish in their natural setting, there is only one means, to go down ourselves to the deepest part of the ocean.

Thus, the idea of a bathyscaph did not originate with Piccard's fascination with balloons and his ambition to reach the stratosphere, but rather the reverse: his concept of an airtight gondola for his stratospheric balloon came from an earlier idea of an underwater balloon taking men into the ocean depths in a sealed cabin. His "lecture" on the subject to Amelia Earhart, Charles Lindbergh, and others at the private dinner in January 1933 led to the first appearance of his idea in the press.

By 1937, in search of his next "big thing," Piccard had resurrected his idea of an "underwater balloon" and took his proposal to the Fonds National. His son Jacques related what happened:

> An official reception one morning brought the matter to a crux. King Leopold of Belgium attended; his father, King Albert, had been extremely interested in the stratospheric balloon flights. Approaching my father, King Leopold enquired, "Professor, tell me about your current research plans." My father was about to tell of some of his latest cosmic ray results but somehow these did not seem especially interesting. He said instead, "Your Majesty, I am planning to build a deep-sea submarine . . . for diving to the very bottom of the sea." King Leopold was fascinated. He did not pass on to pleasantries; he pressed for more precise information. Returning to his laboratory the next day, Professor Piccard called his assistants together. "I told the King yesterday that we are going to build a deep sea submarine. We have no choice now but to do it."[4]

When Piccard approached the Fonds National for financing, its director, Jean Willems, was most enthusiastic. In a feat of bureaucratic adroitness, Willems managed to fast-talk his way to approval of a grant of one million Belgian francs for the project. Piccard soon was touting it to the international press.[5]

Piccard next established a laboratory at the Free University of Brussels to investigate high-pressure physics, necessary to develop the materials and techniques to send an "underwater balloon" to a depth of 6,500 feet. Testing included one-tenth-size models of a watertight sphere. One sphere was a hollow ball made of an alloy of magnesium and aluminum that was tested to pressures simulating a five-mile ocean depth; the second hollow sphere, this one of steel, was tested to a seven-mile depth.[6] Piccard considered a Duralumin sphere, but was concerned that the art of casting the new alloy might not be sufficiently advanced to meet his requirements.

One of the most difficult challenges was to develop portholes through which to view the ocean bottom that could withstand extreme pressures. Glass, quartz, even tiny diamonds proved unable to survive without cracking, splitting, or shattering. In a coincidence of history, in 1933 the German chemist Otto Röhm patented the polymerization process that turns methyl methacrylate into polymethyl methacrylate, and in 1936 the resulting commercial "Plexiglas" became available. The Plexiglas could be fabricated into a truncated cone and fit into a smooth socket; the pressure of the ocean would compress it against the sphere's steel to form a leakproof seal.

Piccard was contemplating the use of a mix of solid paraffin and an as-yet-unspecified petroleum liquid for buoyancy, and had perfected a ballasting system using iron pellets controlled by an electromagnetic valve. The sphere was designed with a 1.38-inch (3.5-centimeter) wall thickness for dives down to 6,500 feet, more than twice the depth reached by Beebe and Barton in 1934. A 1938 diagram of Piccard's submersible showed two maneuvering thrusters and an external light. The principle of a "guide rope" was proposed. It came from ballooning—that is, ballast on a dangling rope would, when it touched ground, instantaneously remove weight from the vehicle. For safety, his planning provided for the use of electromagnets that would dump iron pellets, drop the guide rope, and even jettison whole ballast tub(s) should electricity to the magnets be interrupted.[7]

Piccard also gave considerable thought to how support ships would locate the deep-diving craft when it surfaced. He discussed mounting a telescoping radio antenna to transmit a radio-finding signal that would lead the ships and aircraft to the submersible. He also spoke of pyrotechnics, a water dye, and even acoustic underwater signaling.[8]

A Full-Scale Test Bed

On 30 July 1939, Piccard announced plans to conduct the first diving tests of his underwater balloon in the spring of 1940, suggesting that contracts had been let for the construction of both the sphere and the float. But the German invasion of Poland on 1 September 1939 initiated World War II in Europe, and the next five years saw the greatest destruction and devastation ever inflicted on and by humankind. In Piccard's words, "The disaster that we all deplore naturally put an end to all these beautiful projects."[9]

During the war Piccard returned to Switzerland with his family and obtained employment as a mechanical engineer at the Ateliers de Constructions Mécaniques de Vevey (Mechanical Constructions Workshops, Vevey). Despite his scientific detachment from politics and complete rejection of war as a method for resolving society's conflicts, he took great pride in the fact that his son Jacques joined the French Army in 1944 and received the Croix de Guerre while serving as a sergeant.[10]

After the war, upon returning to the Free University of Brussels, Piccard found that Belgium was bankrupt, its infrastructure damaged, and starvation a real concern in the cities. The U.S. Marshall Plan had yet to make its appearance, and there was little appetite for scientific innovation while basic survival was at issue.

Still, as early as December 1945, Piccard, seeking sponsorship for his underwater balloon concept, approached the Fonds National for reactivation of his research funding. The Fonds National was willing to support Piccard but no longer solely as a scientific program. For the first time Piccard had to accept political and spending constraints: not only was he to "buy Belgian," but the funds were available only with the stipulation that a Belgian national be named as co-director of the project with the same roles as Piccard himself. Accordingly, Max Cosyns, a Belgian national and Piccard's assistant on the 1932 record-breaking balloon flight, and Piccard again were teamed. Cosyns had joined the Belgian *résistance* during the war, been captured by the Germans, and interned at Dachau.

Piccard's plans, long in abeyance, were set into motion, By 1947 the redesigned underwater balloon was being called a "bathyscaph," from the Greek *bathos* for deep and *scaphos* for boat. Changes were undertaken to exploit new technologies developed during the war. But Piccard was now on a budget and forced to make uncomfortable compromises, notably on the weight, towability, and seaworthiness of the bathyscaph's float.

Nevertheless, a major upgrade was implemented to strengthen the sphere and expand the float to achieve a depth increase from the 6,500 feet intended in 1939 to a much more ambitious 13,200-foot capability for the planned submersible *FNRS 2.* The thickness of the sphere walls was increased from the 1.38 inches (3.5 centimeters) of 1939 to 3.5 inches (9 centimeters) in 1946, with concomitant increases to the size and buoyancy of the float. Piccard selected cast steel for the sphere. With 3.5-inch wall thickness, increasing to 6 inches around the two portholes, the *FNRS 2* sphere was designed for a maximum depth of 13,200 feet *with a 400 percent safety factor.*

Rather than the combination of solid paraffin and oil envisioned in 1938, Piccard now settled on gasoline to provide the buoyancy for the *FNRS 2* float. Six aluminum tanks containing 1,059 cubic feet of gasoline were enclosed in a 22¾-foot-long float fairing ten feet, five inches wide and covered with 20-gauge sheet iron.

By far the most sensitive aspect of the project was the 78-inch-inside-diameter sphere. It had to be fabricated as two matched hemispheres with a perfect joint that would be leakproof at all possible depths. As it would be a sphere, there would no forces acting to pry the hemispheres apart or cause them to slide away from each other as long as the joint was horizontal or nearly so. However, joint leakage was possible if the hemispheres were not machined perfectly. Piccard designed a special rubber gasket around the outside perimeter of the sphere's joint and a steel clamp to protect the gasket. The joint would never leak.

Piccard arranged to borrow from the Union of Mines of Upper Katanga in the Belgian Congo one gram of radium to conduct rudimentary X-ray testing of the casting. The bathyscaph team placed the radium at the center of the sphere and 18 square yards of photographic film around the external surface. After a 24-hour exposure the film was developed; it showed one small void. This defect was drilled out, and a tapered pin made of the same material as the sphere inserted from the outside. Piccard's calculations convinced him that this repair would not affect the sphere's strength at depth, but he decided that preliminary unmanned dives to one and one-half times the depth of each manned dive would be programmed to test sphere integrity prior to risking lives.

In 1938 ballast for the bathyscaph had been envisioned as a hopper filled with soft iron shot (BB-size round pellets), held in place directly beneath the sphere by an electromagnet. Other magnets would automatically drop the guide rope, the battery pack, and the entire ballast hopper if electrical power was interrupted. In 1948, Piccard's *FNRS 2* carried those forms of ballast plus 12 large boxes of scrap iron on the side of the sphere that could be individually released or dropped automatically in the event of loss of electricity to the magnets.

By September 1948, preparations were complete. Piccard and Cosyns had managed to "buy Belgian," stay within budget, and still produce a craft that would, they hoped, prove both the bathyscaph principle and establish a manned deep-diving record of 13,200 feet. Piccard personally was less interested in establishing new records than in proving his concept; from the sponsor's point of view Belgian's national pride and political considerations of the deep dive were very high priorities.

Is There Competition?

Meanwhile, in the foothills of the Italian Alps, another inventor and innovator was developing a deep-diving submersible. Italian mechanical engineer Pietro Vassena of Malgrate came out of the war with a dream to create the world's deepest-diving submarine. Working with submarine designer William Premuda, Vassena had his deep-diving midget submarine constructed at the Società Anonima Elettrificazione in Lecco, Italy.

The Vassena craft—named the *C-3*—was ready for testing on 19 February 1948, in Italy's Lake Como, more than 1,300 feet deep. His first test dive was to 55 feet, manned by himself and ex-submariner Nino Turati. On 21 February, he sent his submersible—unmanned—down to 770 feet for an integrity test. On 10 March 1948, another unmanned test took the *C-3*, tethered by a cable, to the bottom of the lake off Argegno, a depth of 1,328 feet. Finally, on 12 March, the *C-3* went to 1,371 feet with Vassena and Turati on board, establishing a new manned depth record for a free-swimming submersible.

However, still standing as the deepest manned dive was Beebe and Barton's 1934 record of 3,028 feet—in a tethered device.

In October 1948, Vassena took the *C-3* to Capri, off the coast of Naples, for further dives. On 8 October 1948, while being towed with an open deck hatch, the *C-3* foundered and sank, almost taking Vassena with it. The Italian Navy raised the midget sub four days later. During preparations for its 281st dive on 16 November 1948, the *C-3* flooded and sank for a second time while being handled by the Italian naval tug *Tenace*. It remained (and does today) in 2,000 feet of water near Capri.

Piccard was keeping an eye on Vassena's progress and had met with Vassena at his home in Sierre, Switzerland, early in 1948 for consultations. On 1 May 1952, Piccard would visit Vassena in Lecco.

On the West Coast of the United State another deep diver, Otis Barton, had constructed a larger version of the Beebe-Barton diving device; he called it the *Benthoscope* and had designed it for a maximum depth of 5,280 feet. Barton was a gentleman-adventurer with the personal funds to finance his ambition—to be the deepest diver on the planet. On 12 August 1949, Barton sent the tethered *Benthoscope* down in an unmanned dip to 6,000 feet for an integrity test. When it came up dry he was ready for his assault on the deep-dive record. On 16 August, Barton took the *Benthoscope* to 4,488 feet off the coast of southern California, setting the manned tethered-dive record.

The *FNRS 2* Expedition

The Belgian steamship *Scaldis* sailed from Antwerp on 15 September 1948, with the *FNRS 2* in her main hold. The ship's other holds contained cargo for delivery to West African ports. The Belgian government had provided the *Scaldis* to the expedition gratis for a limited number of days. In addition to the political and financial constraints already experienced, Piccard and Cosyns were now faced with strict time constraints on the availability of the freighter to support dives.

The *Scaldis* had been selected for this expedition because she could accommodate the *FNRS 2* in her hold and had deck cranes capable of lifting the bathyscaph into and from the water. In addition to Captain Laforce of the *Scaldis* and his crew of 50 men, the Piccard-Cosyns expedition numbered 12:

Auguste Piccard, co-director
Max Cosyns, co-director
Madame Cosyns, chemist (wife of Max)
Jacques Piccard, secretary (son of Auguste)

Georges Marlier, biologist, Free University of Brussels
Jean Pierre Van den Eeckhoudt, biologist, Free University of Brussels
Theodore A. Monod, Museum of Natural History, Paris
Claude Francis-Boeuf, oceanographer
Dr. Daniel Bouchet, physician
Louis Ockum, mechanic/engineer
Raphael Algoet, photographer
Henri Ghysels, Belgian government publicity officer.

Upon arrival at Dakar it proved necessary to dry-dock the *Scaldis* because of damage to her propeller. While repairs were under way, Piccard and Cosyns had the opportunity to become acquainted with their hosts at the French naval base under the command of Vice Admiral Alphonse A. J. Sol. The admiral provided every courtesy to the Belgians and assigned the French Navy's oceanographic research ship *Elie-Monnier*, two seaplanes, and two frigates, the *Croix-de-Lorraine* and *Leverrier* (the latter the former USS *Emporia* [PF 28]), to escort and assist the expedition.

In Dakar harbor the *FNRS 2* was wetted for the very first time. Once back in the water herself, the *Scaldis* pumped 7,040 gallons of gasoline into the float and loaded the ballast onto the submersible. Piccard confirmed slight positive buoyancy at full load. The bathyscaph was then de-fueled and returned to the hold. (The craft could not be lifted with the fuel loaded.)

On 19 October, the *Scaldis* and her escorts sailed for Boa Vista Island in the Cape Verdes. Captain Philippe Tailliez of the *Elie-Monnier* and his diving unit, led by Jacques Cousteau and Frédéric Dumas, had for more than a week been taking soundings to find an area with a gradually sloping bottom. Cousteau had become quite excited by Piccard's new concept, telling him, "Professor, your invention is the most wonderful of this century!" (Cousteau later was instrumental in convincing the French Navy to sponsor the development of a successor to the *FNRS 2* and, reportedly, was involved with negotiations between the Belgian Fonds National and the French Navy.)[11]

Portuguese authorities had granted permission for French naval units to operate within Portuguese territorial waters in support of the Piccard expedition. The *Scaldis* arrived in the area on 21 October, and the first dive commenced on 26 October, planned for 84 feet to test the controls and systems, with Piccard piloting and Professor Monod, chosen by lot, as observer.

A description of the process required for this dive makes clear that the *FNRS 2* was never conceived as an operational vehicle but rather as a full-scale test bed for the bathyscaph concept. To begin with, the fully ballasted *FNRS 2* was lifted out of the *Scaldis*' main hold and swung into position over the main deck, allowing the pilot and observer to enter the sphere. The sphere's hatch was bolted and the bathyscaph swung outboard and lowered into the water. Divers attached then fuel hoses and filled the float with gasoline. After what seemed like an interminable delay, the *FNRS 2* was released to descend the bottom at 84 feet. Time on the bottom was about 15 minutes, after which Piccard released scrap-iron ballast and returned to the surface. There the process was reversed: first de-fueling the float, then lifting the bathyscaph up over the main deck, where a special trolley was rolled up to take the weight of the hatch, after which it could be opened and the pilot and observer could exit before the craft was struck below. The total time that the crew was inside the sphere (on this occasion without communications due to an inoperative telephone connection) was almost 12 hours, of which the total time submerged was less than 20 minutes. After this first dive Piccard advised, "The apparatus will have to undergo a major overhaul and it will need certain simplifications as well before it is absolutely right."[12]

Interestingly, in Piccard's 1938 conceptual drawings the crew entered the sphere at the top by way of a ladder in a "tunnel" through the float, occupying the sphere after fueling and disembarking prior to de-fueling. The *FNRS 2* was designed ten years later, but after entering the crew was essentially trapped inside the sphere without hope of a timely exit or rescue. At sea it could take more than four hours for the ship to approach the surfaced bathyscaph, attach a crane cable and de-fueling hoses, de-fuel the *FNRS 2*, lift it up to the main deck, and unbolt the hatch. If sea conditions made it impossible to lift the *FNRS 2* onto the *Scaldis*' deck, the crew could be trapped for days with only a snorkel tube to provide fresh air.

Test Dive to Depth

Preparations finally were under way the morning of 3 November, for an unmanned dive to 4,620 feet off São Thiago Island. A timer was set inside the sphere for an automatic de-ballasting at 1640. A comedy of errors, problems, confusion, and equipment casualties caused delays of the actual descent until 1600, and there was great concern that the bathyscaph would not reach the bottom before the timer released the ballast. However, at 1629 the *FNRS 2* was sighted back on the surface, indicating that ballast release had been caused by contact with the bottom or by water entering the cabin.

As the *Scaldis* maneuvered to recover the bathyscaph, winds and seas picked up so quickly that divers were unable at first to attach hoses for de-fueling. The bathyscaph was eventually de-fueled, but the weather made it impossible for the *Scaldis* to hoist the *FNRS 2* on board. Towing the *FNRS 2* to Santa Clara Bay was the only alternative, but, as Piccard related, "we could only go very slowly," the float riding high and taking a severe pounding. By dawn on 4 November, the *Scaldis* had reached the quiet waters of the bay and recovered the battered *FNRS 2* with no further problems.

Piccard had been faced with a stark choice between the expedition and the bathyscaph, and he had chosen to protect the bathyscaph. "If the bathyscaphe got any heavier, it would end by sinking. We therefore had to take a quick decision: the petrol had to be replaced not by water but by carbon dioxide. . . . That would put an end to our diving and to the whole expedition, but we decided that it was better to sacrifice the petrol than to risk the loss of the bathyscaphe itself."[13] This decision could have ended his career—and nearly did.

Upon opening the sphere and examining the instruments, Piccard announced with some satisfaction that the bathyscaph had reached a depth of 4,592 feet and returned to the surface on autopilot without major damage to the sphere. Piccard noted that, if manned, this would have marked a new world depth record. But as it was, with the loss of the 7,000 gallons of gasoline and extensive damage to the float, the expedition was *fini*. The Belgians had neither funds to repair the float or buy additional fuel, nor authority to extend the loan of the *Scaldis*. They did not even have money to get the expedition members back to Europe. Piccard was facing a complete disintegration of his project.

The French Navy stepped in and generously made arrangements for all 12 members of the expedition to fly out of Dakar the evening of 12 November on an Air France flight to Paris via Casablanca. The *FNRS 2* was off-loaded by the *Scaldis* in Dakar and dismantled by the French Navy. The float was evaluated as a total loss, but the sphere, which had been fabricated by the Belgian firm of Émile Henricot, was salvaged and a few months later was transported by the French Navy to the naval base at Toulon.

Piccard was greatly disappointed that a full program of test dives could not be completed but took solace in the facts that the sphere had proved to be capable of dives to 4,592 feet and that his instrumentation and automatic de-ballasting systems had passed a real-world test. Always the optimist, he looked forward to repairing the float and returning to West Africa for the next series of tests. He had yet to realize the magnitude of the disaster.

Debacle at Dakar

But there was no disguising that the bathyscaph had totally failed on its first, highly publicized, at-sea trials. *The New York Times* opined, "A dream of forty years fizzled and died today when Profs Auguste Piccard and Max Cosyns decided to abandon their attempt to ride the weird underwater balloon they had created down to ocean depths of perhaps two and a half miles."[14]

Piccard could not reveal in his defense that his designs had been eviscerated by financial constraints nor acknowledge that he had erred in failing to conduct initial sea trials quietly and closer to home. He seemed oblivious to the fact that he had entered an area of design and innovation that had a long history of hard-won experience, of which he had none. When Piccard had stated in 1933 that "I really have quite as much in common with submarine navigation as with aeronautical navigation," he had implicitly misrepresented his knowledge of balloon design and engineering, which was in fact substantial, but properly characterized his knowledge of naval engineering, which was practically non-existent.

Within weeks Piccard was confronted by a Fonds National unwilling to continue funding the development of the bathyscaph concept, a shock that might have disabled a lesser man. The avalanche of disastrous consequences continued: the Fonds National asserted a claim to ownership of the *FNRS 2* and commenced an investigation into its disposition.

By far the most significant damage was to Piccard's reputation and esteem in the press, in the public's eye, and in the all-important opinion of the Fonds National. In 1939 and again in 1948, Piccard had stepped forward to claim to have solved the problems that denied human access to the deep oceans. In

September 1948, moreover, the Belgian publicity effort put Piccard in a position of seeming to have the temerity—some might say the hubris—to "call his shot," announcing that he was heading out to the Horn of Africa to dive two and one-half miles with his new bathyscaph. From the viewpoint of the press, Piccard had over-promised and under-delivered; his pre-dive promotional interviews and press releases had put newspapers and magazines worldwide in a position of hyping his effort, only to be embarrassed when the inventor did not "get it right."

Now the public's icon of scientific competence and mathematical certainty was shown to be over-confident and inadequately prepared. It mattered not that political and budgetary constraints had hobbled his program, nor that publicity organized by the Belgian government had overstated his technology's maturity. Piccard would have benefitted from any success; thus he was burdened with total responsibility for the failure.

At this point Piccard was without funding and facing the imminent prospect of losing all control of his bathyscaph. He was 63 years old, and for the past two decades the Fonds National had been his sponsor and had given him extensive support. Now Piccard had to find a new sponsor and raise funds for his bathyscaph project.

Admirals of the Abyss

To discover new countries,
To climb the highest peaks,
To travel through new areas of celestial space,
To turn our searchlights upon domains of eternal darkness,
That is what makes life worth living.

—Auguste Piccard

Jacques Piccard most probably saved his father's bathyscaph from becoming only a minor footnote in the history of oceanography. Jacques had been present at the Dakar debacle and had seen the new and promising technology fail due to the craft's unseaworthiness, not technological inadequacies. He recognized the bathyscaph's potential and knew how severe funding constraints had hobbled development of the *FNRS 2*. It had been a life-changing experience for the young Piccard.

Few other people knew or cared that financial limitations had caused trade-offs that had created an only marginally capable deep-diving craft. Even fewer knew or cared that the political decision to have a Belgian national as co-equal co-director had established a situation that made coordination and decision making difficult.

The Fonds National and the Belgian government wrestled with the situation. Had the failure off Dakar been a normal stumble in the development of a new technology, or had there been systemic problems? Was the bathyscaph concept fundamentally flawed? Was the lead scientist beyond his best years and not the man to direct the project in the future? Would it be better to terminate further *FNRS 2* efforts or to finance a redesign and rebuild of the *FNRS 2*? Was Piccard personally responsible for the disaster?

Committees would have to be formed and those issues discussed. Liaison would have to be established between the Fonds National and various government ministries. Time would be needed to gather facts. However, some early agreements could be reached. The first was that the damaged *FNRS 2* was not a total loss and the bathyscaph's sphere was an asset that could still be employed, both as a tool for oceanography and as a source of pride for Belgium and for the Fonds National. Second, it appeared that tight funding constraints contributed to the operational limitations of the bathyscaph. Third, designating Piccard and Cosyns as co-directors had had negative consequences. And, fourth, the French Navy was lavish in its praise of the bathyscaph concept and was indicating willingness to help make the *FNRS 2* an operational vehicle.

Who Will Rebuild the *FNRS 2*?

To its great credit, the Fonds National eventually came to a correct evaluation of the situation: the bathyscaph concept was worthy of further development. In April 1949, the Fonds National made a formal proposal to the French Navy, offering the *FNRS 2* for development and rebuilding, with French ownership and funding. *The participation of Piccard in his project was to be terminated.*

The French government was slow to respond. By late 1949, Jacques Piccard had offered an alternative program: that the Henricot sphere be shipped to Switzerland and there used to raise funds for reconstruction. In that way, neither the Fonds National nor the French Navy would have to finance it. During the winter of 1949–1950, even though the Henricot sphere remained at Toulon, the younger Piccard was able to obtain sufficient commitments from Swiss canton governments, universities, industry, and private individuals to fund the proposed project. In March 1950, he arranged with the industrial firm Sécheron in Geneva to prepare plans for a new float under his father's direct supervision. The French Navy agreed to support any reasonable plan—that is, one that did not cost its government a single sou. Neither decision nor guidance came from the Fonds National.

In May 1950, Jacques Piccard contacted the Italian Navy and received favorable responses; he notified the Fonds National that the Italians were an alternative to French support. Finally, in June, the French Navy agreed to ship the Henricot sphere to Switzerland, and Jacques Piccard traveled to Toulon to arrange transport.

Overnight, Jacques' plan unraveled. In Gallic fashion, probably during minister-to-minister conversations, telephone calls, or even a golf game, a decision was made that the Fonds National would provide nine million francs to the French Navy for the redesign and rebuild of the bathyscaph. On 9 October 1950, the Fonds National signed a formal agreement with its French equivalent, the *Centre National de la Recherché Scientifique*, and the French Navy. Among its provisions were:

1. The rebuilt bathyscaph would be named *FNRS 3*, honoring the Fonds National origins of the craft.
2. The Fonds National would pay nine million French francs to the Centre National.
3. After three successful deep dives the new bathyscaph would become the property of the French Navy.
4. Auguste Piccard would be invited to collaborate as a consultant.

Insult and Reaction

Those now involved with the French bathyscaph project almost immediately displayed aloofness toward Auguste Piccard. The object of the French effort was to make the *FNRS 3* an operational, seaworthy craft, and that goal was beyond Piccard's field of expertise. Further, the French Navy desired that operational characteristics were not made available to a Swiss scientist for security reasons.

The French Navy took a leisurely three years to develop an operational *FNRS 3*, with little if any input from Auguste Piccard. He was frustrated that his bathyscaph project had slipped from his control. However, it appears that as a scientist and having spent his life in the collegial halls of academia, he was concerned primarily with ensuring that his legacy as the founding father of the bathyscaph concept was secure and recognized. Accordingly, in 1950–1951 he forwarded all of his relevant notes, letters, and drawings to the French in the spirit of full cooperation, but they brought him no recognition for his initiatives. In contrast, Jacques Piccard had not been molded by decades of academic discourse and was personally insulted and outraged by the disregard and disdain shown to his father by the French. Also, unlike his father, Jacques had a demonstrated ability in fundraising, and he could see options. He could provide his father with an opportunity to wrest back the bathyscaph concept.[1]

During the winter of 1951–1952 Jacques traveled to the independent city of Trieste as part of the research for his doctorial dissertation (which he was never to complete) on the economic potentials of the Free Territory (see note 1). In Milan he managed to catch a ride for the lengthy drive to Trieste with Yolanda Versich, secretary and assistant to industrialist-entrepreneur Dr. Franz Kind, who with Dr. Georg Tugendhat was constructing state-of-the-art oil refineries in Britain, Belgium, Italy, and Trieste.[2] During their long drive, Piccard revealed the situation with his father's bathyscaph project. This discussion captured Yolanda's imagination.

When they arrived in Trieste she made a telephone call to Diego Guicciardini, director of the Aquila Petroleum Refinery—a Tugendhat/Kind project—and asked if he could accommodate the young scholar in the refinery's guesthouse. The refinery's hospitality ultimately was to extend for almost a year.

Trieste was abuzz with new projects and a major influx of American dollars through the U.S. government's Marshall Plan. The local shipyards had recently been revitalized and reopened, and the new Aquila Refinery was coming on line. Soon, with the imprimatur of Dr. Kind, Yolanda embarked on a major fundraising effort to help finance a new bathyscaph project. "But funding was found not solely in Trieste, the entire of postwar Italy was interested in being part of the project," related newspaper reporter Pietro Spirito.[3]

Jacques made the acquaintance of Diego de Henriquez, director of the War and History Museum in Trieste. Henriquez was a passionate historian and promoter of Trieste as an independent city. (Trieste became a part of Italy in 1954.) He wanted Trieste to become a great center of international culture and cooperation, and even hoped to make it the site of the world's first "spaceport" from which the first rocket to the moon would be launched. His energy and enthusiasm were infectious.

In conversations with Jacques, and apparently on a spur-of-the-moment inspiration, Henriquez proposed that Trieste be chosen to construct a new bathyscaph, as a vehicle to publicize the Free Territory's vitality and industrial capacity.[4] Jacques, bowled over by this man's energy and grand ideas, immediately responded with descriptions of his fundraising successes in Switzerland. A partnership was born, and within weeks, through the efforts of Henriquez and Yolanda, support from a few of the local industrialists for a bathyscaph initiative was in hand.

As the plans and drawings of the float prepared by the Swiss firm Sécheron had been complete and ready for months, it was possible to give reality to these ideas quickly. The drawings of the new float were passed to the Cantiere Riunite dell' Adriatico at Monfalcone in Trieste for evaluation. The firm had had extensive experience constructing submarines between the world wars, having launched 47 of the 100 Italian submarines built during that period.

The Terni Sphere

The bathyscaph's sphere would require the facilities of a world-class steel forge. Less than two hours north of Rome is the "steel town" of Terni. Rebuilt after more than 100 Allied bombing raids during the war, Terni now had the equipment, capacity, and skilled workers to take on the challenge of fabricating a forged (vice cast) steel, seven-foot-diameter sphere for the new Piccard bathyscaph. By the end of 1951 a contract had been let to the Societa per l'Industria e l'Eletricitá workshops in Terni for the sphere.

Just as the Henricot cast-steel sphere had been the most critical element of the *FNRS 2*, so too was the forged-steel Terni sphere of the new Piccard bathyscaph. Terni possessed one of the largest forges in the world, capable of forming a seven-foot sphere of wrought nickel-chromium-molybdenum steel. A forged-steel sphere would both be more resistant to the intermittent application of extreme pressures and allow greater operating depths than could a cast-steel sphere of equal weight. Auguste Piccard would be aiming for 20,000 feet, not the 13,200-foot limit imposed by the cast-steel Henricot sphere. To achieve such depths required forging a sphere of steel more malleable and less ductile than the Henricot sphere. The 20,000-foot depth would encompass some 98 percent of the world's oceans. Later, Auguste would write that supervising engineer at the *acciaierie* (workshop), Vincenzo Flagiello, "devoted all his energy and all his art to the making of this piece."

Flagiello specified the use of only the bottom two-thirds of each of the two 24-ton billets of steel the workshop began with to eliminate the majority of slag and other impurities. At the beginning of the forging process, each resulting billet weighed 16 tons and was about 3½ feet thick. The forging went on for several days. The billet then assumed the form of a round biscuit ten feet in diameter, almost a foot thick at the center and tapering to less than five inches at the rim. Next, a massive mechanical press forced the heated biscuit slowly into a hemispherical shape with an outside diameter of seven feet, five inches. Repeated re-heatings were required.

The hemispheres at this point each weighed almost 11 tons. Both were then machined down to a mirror-like finish and a thickness approximating the final thickness desired; the sphere now weighed just over five tons.

Non-destructive testing of the sphere halves was conducted, both radiographic examination and ultrasound tests, the latter only then making its way into commercial applications. The hemispheres were certified as nearly flawless. Final machining brought the sphere into the designed dimensions, and heat treatment to provide the desired hardness finished the process. Flagiello and the steelworkers at the Societa per l'Industria e l'Eletricitá Acciaierie in Terni had created an industrial gem seven feet in diameter.

The next step was to ensure that the joint between the two hemispheres was perfect. Unlike the Henricot sphere (and the later Krupp sphere), the Terni sphere had flanges on both hemispheres that allowed the two halves to be clamped together for a solid fit. The Terni sphere would not leak at any depth above its crush depth, which was calculated to be in excess of 50,000 feet, or two and one-half times the stipulated operating depth of 20,000 feet. Auguste and Jacques were more than satisfied.

The Monfalcone Float

While the work on the sphere was progressing at Terni, the Piccards were consulting with chief engineer Benvento Loser at the Cantiere Riunite. Loser, with 30 years of experience in submarine construction, was the head of the newly reestablished Naval Warship section at Monfalcone. He would provide the engineering expertise that had been lacking in the design of the *FNRS 2*.

Loser and his team considered the drawings and plans presented by the Piccards from the Swiss firm Sécheron wholly inadequate, and Auguste released the issues of the float's shape, hydrodynamics, towing characteristics, maneuverability, and safety into their care. From March 1951 through May 1952, the Loser team enthusiastically worked to turn Piccard's ideas into the plans, drawings, and specifications.

Reportedly, it also was through Loser's knowledge of petroleum products that super-refined, high-octane petroleum distillates (such as 100-octane aviation gasoline) were selected for the craft's floatation.[5] The float contained nine flotation compartments, overall measuring 44½ feet in length and 11½ feet in cross-sectional diameter. When empty the float weighed 43 tons; it would be filled with 67 tons of gasoline for floatation. The float was designed to provide positive buoyancy to the sphere even in the event that any two of the nine gasoline flotation tanks were punctured. The firm ESSO Standard Italia would generously contribute the aviation gasoline to fill the float. Unlike the *FNRS 2*, the crew of the new bathyscaph would access the sphere from the top, via a ladder in a tunnel down through the float, thus enabling them to enter after it was fueled and ballasted and to re-emerge before the float was de-fueled. The float and sphere of the bathyscaph would be trucked across Italy to the Navalmeccanica Castellammare di Stabia shipyard near Naples for fitting-out and the mounting of the Terni sphere.

In January 1952 Auguste Piccard joined Jacques in Italy to help coordinate and manage completion of the new bathyscaph. "Professor Auguste Piccard," Yolanda related, "arrived in Milan and in meeting me told me that out of gratitude, the bathyscaph under construction would be given my name. I was flattered. . . . I suggested however to name it the *Trieste* since the idea was born in Trieste."[6]

A key event occurred in the spring of 1952, when Auguste and Jacques were residing at the Savoia Excelsior Hotel in Trieste. Upon returning to the hotel one evening, Jacques was handed a letter from the mayor of Trieste. Jacques read it and exclaimed, "Oh, wonderful! The sphere is all right. The Mayor of Trieste has arranged with the Italian government for payment."[7] In May contracts were placed for the necessary instrumentation and controls with firms in Germany, Switzerland, and Italy.

Jacques, perceiving that all the pieces of the puzzle were falling into place, turned his back on academia forever. His dreams and his destiny were to find fulfillment in following his father's legacy to their logical conclusion—to the bottom of the deepest point of the world's oceans. Jacques was not to be a professor of economics; rather, he was to become, in his father's footsteps, an explorer, adventurer, and proponent of investigating the mysteries of the deep ocean.

The final assembly at the Navalmeccanica Castellammare di Stabia (another Fincantieri company) was no simple matter of bolting the sphere to the float. It involved more than 24,000 hours of labor, outfitting the sphere with instrumentation and controls, and fitting out the float. The shipyard work took seven months to complete.

On 1 August 1953, after 15 months of planning, construction, fitting-out, and dry testing, the bathyscaph *Trieste* was "launched" at Castellammare. There was no fanfare, no speeches, no music, and no champagne bottle smashed against the hull: a yard crane lifted the unusual watercraft from its cradle and deposited it gently into the water quayside. The *Trieste* was then loaded with nine tons of ballast and 25,000 gallons of aviation gasoline for a floatation test. Auguste Piccard's calculations proved correct: the *Trieste* rode low when fully loaded, with practically no freeboard.

On 9 August 1953, a very cautious program of shipyard testing commenced with the Italian naval tug *Tenace* towing the bathyscaph in the harbor of Castellammare. The *Tenace*, the former U.S. Army tug *LT 154,* was a 127-foot, oceangoing wooden tug—the same tug that had been instrumental in the loss of Vassena's deep submersible *C-3* on its 281st dive in 1948. The *Trieste* was put through five partial-submergence dives with shipyard engineers observing, gathering weight and trim data, and further test controls.[8]

The *Trieste*'s baptismal Dive #1 occurred on 11 August 1953, with the Piccards taking the bathyscaph down to 26 feet in Castellammare harbor. For the first four dives Auguste was recorded as a pilot, but Jacques was much more familiar with the controls than was his father, who was the theoretician not the operator. More accurately, however, it was a mutual effort, with both father and son learning a new discipline. From Dive #5 on 29 September 1953, up to the Mariana Trench deep dive in 1960, Jacques was the pilot for every operational dive.

On 13 August, with both Auguste and Jacques on board, the *Trieste* made a test dive in the harbor to 53 feet, remaining submerged for ten minutes in the presence of journalists on board the harbor tug *R-27*. The next day the *Trieste* was towed out to sea off Castellammare by the *Tenace* for Dive #3, to 140 feet.

The first deep dive was scheduled for 25 August, south of the island of Capri. The tow to Capri revealed that the *Trieste* had a tendency to "yaw," swerving as much as 45 degrees to port and then to starboard of the *Tenace*'s base course. (This situation later would be partially corrected by the addition of a keel plate to the bathyscaph.) The Italian Navy corvette *Fenice* rendezvoused with the *Tenace* and the *Trieste* to provide security in the dive area. The danger was that the *Trieste* could come to the surface and be rammed by a passing ship.

The first attempt to dive, early in the morning of 25 August, resulted in an automatic "hold" when current to the electromagnet holding the 770-pound guide rope was interrupted, automatically releasing that weight. Auguste decided to proceed without the guide rope and recommenced the dive, only to find that the forward ballast tub was dumping its load of iron shot because of an electrical interruption to the ballast release valve. As a result the *Trieste* went to only 60 feet before it returned to the surface automatically as the iron shot drained away.

Knowing and understanding their new craft intimately and fearing a press misreporting of an abortive dive—a nightmare from the Dakar experience—Auguste found, on the spot, a work-around to allow a dive the next day without returning to Castellammare for repairs. This involved sealing shut the forward ballast-tub valve and refilling the tub. It would allow a dive but leave only the rear tub available for incremental release of ballast for depth control and return to the surface. In an emergency the forward tub could be dropped as a complete package—the tub and full load of shot.[9]

With these modifications, the bathyscaph was ready to dive in the afternoon of the 26th. Dive #4 was planned to reach 3,540 feet, south of Capri. Still without a fathometer, lacking the guide rope, and unable to see the bottom until the last moment, the crew hit and buried the sphere 4½ feet into the bottom silt. After 15 minutes of checks, the *Trieste* regained the surface without incident, although badly out of trim; total dive time was 45 minutes.

A World Record

The bathyscaph could now be returned to Castellammare for repairs, including the new keel plate, without embarrassment. When they were complete, the *Trieste* was towed out to the island of Ponza, from which there was easy access to deep Mediterranean waters. Joining the *Tenace*, the bathyscaph, and a diving support boat from the shipyard, was the corvette *Fenice*, which made her rendezvous the morning of 29 September. Heavy seas that afternoon forced the little flotilla into the Bay of Ponza, and there Jacques made a shallow dive (Dive #5) with Armando Traetta, a professor at the L'Istituto Tecnico Industriale Leonardo Fea, to ascertain whether the tow through rough seas had caused any damage.

The next day, with a crew of shipyard workers assisting, all preparations were made for the first deep dive of the new Piccard bathyscaph. The dive commenced at 0819 on 30 September, with Auguste and Jacques on board. This would be the last time that Auguste would dive a bathyscaph and would mark the landmark event of one man claiming both the highest manned ascent and the deepest manned descent, albeit 22 years apart. During this dive to 10,300 feet the *Trieste* passed through the then-standing world depth record of 6,900 feet set by Georges Houot and Pierre Willm in the French *FNRS 3* just 32 days earlier.

With what seemed stunning ease and alacrity, Auguste Piccard's renown, fame, and place in history were made secure. He whispered to his son, "The credit is all yours." Jacques smiled and demurred. When back on board the *Fenice*, the exhausted Auguste made a brief statement to the press that was as simple as it was gracious: "It was very important, very lovely. And I must say that the chief merit of this undertaking goes to my son Jacques. It was he who controlled the *Trieste*."

Auguste was speaking from the heart, insisting that his son's energy, commitment, and drive be recognized as the primary forces that had rescued Auguste's reputation and legacy from relegation to the footnotes of history. Actually, both men had achieved places in history.

After a 36-month design-and-build gestation, the French Navy launched the *FNRS 3* in June 1953, making the craft's first test dive, to 42 feet, on 17 June 1953. Jacques Piccard had pushed the *Trieste* through construction in a mere 15 months starting in May 1952 and had made its first test dive, to 26 feet, in August 1953, only 45 days behind the first French dive.

Subsequently, the French put the *FNRS 3* through a carefully designed series of dives, each to greater depth, reaching 6,900 feet on 14 August 1953. Through 29 September 1953, as noted above, the *FNRS 3* held the world depth record. The *FNRS 3* regained the world depth record on 15 February 1954 off Dakar at 13,290 feet, which was not exceeded until 15 November 1959, by the *Trieste* off Guam (18,150 feet). The *FNRS 3* was retired in January 1960, after 93 dives over seven years of service. (It is on display at the Musée de la Marine at the naval base at Toulon.)

The Honors Given to an Admiral

The *Trieste* in Dive #6 descended to 10,300 feet off Ponza on 30 September 1953, surfacing at 1035 after two hours, 16 minutes submerged. Auguste and Jacques transferred to the *Fenice*, asked the captain to proceed to Ponza, and arranged for the *Tenace* to tow the *Trieste* there. The corvette prepared to "pipe" the Piccards off the ship upon arrival at Ponza. Vice Admiral Massimo Girosi, who was flying his flag on *Fenice*, gave the quarterdeck watch special instructions to which the officer of the deck reacted in surprise: "But those are the honors given to an Admiral!" To which he replied, "Admirals of the Abyss—they deserve it." The salute may have been the highest honor ever granted a Swiss mariner.[10]

All of Ponza was celebrating. Auguste recorded, "We reached our lodgings under a rain of flowers thrown from the windows." That evening the Piccards were invited to a dinner by the municipality and granted the rights of honorary Freemen of Ponza.[11]

The following day, 2 October, en route to Castellammare, the small flotilla paused south of the island of Ischia for Jacques to take the *Trieste* down to 2,150 feet with Victor Aldo De Sanctis, who had designed the *Trieste*'s external lighting system, to test the lights for motion picture filming.[12] Descent and tests on the bottom were without incident, but upon de-ballasting for ascent Jacques and De Sanctis smelled an acrid gas. Fearing chlorine gas from the *Trieste*'s batteries, Jacques immediately dumped the ballast, tubs, and the guide rope for an emergency surfacing. The odor proved to have come from the electric motor of the Cameflex camcorder carried by De Sanctis, but the episode proved that emergency surfacing procedures were flawless. After this incident the *Trieste* always carried emergency breathing apparatus. For the remainder of the return to Castellammare the *Tenace* was brought up to six knots to test the *Trieste*'s towing characteristics with her new keel; the bathyscaph towed easily at six knots.

The *Trieste* flotilla arrived at Castellammare after nightfall to find illuminated boats coming out to meet them and fireworks lighting the sky. The shipyard workers turned out en masse to celebrate the Piccards' achievement. The entire town was *en fête*—the *Trieste* had returned to its homeport in triumph. The next few days saw a parade and a reception wherein both Piccards were declared honorary Freemen of Castellammare di Stabia.

Upon their return to the village of Chexbres in Switzerland, the Piccards were celebrated yet again. "We were surprised to see our syndic [mayor], our pastor and the whole council of the commune, with a gendarme and a horticulturist in their company, assemble before our villa and present me with a beautiful blue cedar. We planted the tree. It bears a plaque in memory of 30 September 1953, a touching mark of esteem from our friends of Chexbres."[13]

Back at Castellammare, the *Trieste* was cradled for the season. The bathyscaph would not dive again for almost a year.

The press once more bestowed its favors on Auguste Piccard. He made for a great story: the rumpled Swiss scientist had bested the haughty French. To the Piccards, especially to Jacques, who had thrown his career into the effort, this was the sweetest moment. Auguste held a unique place in the history of technology, in the history of exploration, in the history of manned flight, and in the history of manned dives. He chose that moment to retire from deep diving.

His son, Jacques, would carry the family's legacy into the future and to even greater depths.

4 White Knight

The ocean's bottom is at least as important to us as the moon's behind.

—Gordon G. Lill, Oceanographer, Office of Naval Research

On the morning of Monday, 20 January 1958, the National Academy of Sciences hosted a meeting in Washington, D.C., with 55 of the most influential men—and one influential woman—in the field of oceanography. More than 40 of the participants were active-duty U.S. naval officers or civilians working for the Navy. The morning session discussed the bathyscaph as a scientific tool, the afternoon session the military implications of such a capability. The U.S. Navy was contemplating the acquisition of the Swiss-Italian deep-diving submersible *Trieste*.

To bring the participants current on the *Trieste* program, Dr. Columbus Iselin, director of the Woods Hole Oceanographic Institution, provided a summary:

> The Undersea Warfare Committee of the Academy [the National Academy of Sciences] has been interested in [the *Trieste*] and has been thinking about it for quite a few years. We have not thought that the advantage of being able to go to any depth was the overpowering thing. We have thought perhaps it would be nice to get down as deep as the axis of the main sound channel.
>
> I have probably been the worst enemy of this thing that it has ever had, and I am going to publicly admit that I was wrong.[1]

Dr. Andreas B. (Andy) Rechnitzer of the Navy Electronics Laboratory (NEL) in San Diego later noted that this statement by Iselin "changed the whole tenor of the meeting before we even got started." Attendees almost immediately followed the lead of this oceanographic emeritus, who endorsed bringing the *Trieste* to the United States as the basis for an American deep submergence program.[2]

A Long Courtship

As early as the winter 1949–1950, Auguste Piccard had initiated contact with the American embassy in Brussels, offering the U.S. Navy control of the bathyscaph as an operational deep-diving craft. In the immediate post–World War II world the United States was the only major economy not bankrupted or physically devastated. The United States had the economic resources, the scientific and industrial infrastructure, and an obvious requirement to emphasize oceanography as part of military preparedness against an aggressive Soviet adversary. Nevertheless, Piccard's plea in those years fell on deaf ears.

The French had the political will at that time to develop the bathyscaph, but required financial assistance from the Fonds National to do so. It appeared that the Belgians would have been happier not spending the nine million francs required by the French Navy, in fact quite content to allow the bathyscaph to go to the Americans. From April 1949 to September 1950, there was an opportunity for the U.S. Navy to step up, but the will was lacking. Several years would pass before a second opportunity arose.

In the spring of 1954, Jacques Piccard submitted an unsolicited proposal to the National Science Foundation in Washington. That agency was then in its infancy and had very modest sums for the support of scientific projects. Piccard's three-page proposal offered to collaborate with any group of American oceanographers and suggested an operating base in the Caribbean for dives into the Puerto Rico Trench. The proposal identified no U.S. partner and was not specific as to budget, purpose, or the scientific advantage for the United States. It stated simply, "It should be noted that for great depths, the bathyscaph is the only existing means of permitting man to make direct observations." It was amateurish in format and lacked specificity.

No U.S. response was forthcoming.[3]

In the fall of 1954, Jacques Piccard made eight *Trieste* dives in the Mediterranean. In those dives several Italian observers were taken to depths from 150 to 500 feet to conduct scientific investigations. When one of the scientific observers cancelled at the last moment, Jacques offered his wife

Dr. Andreas B. Rechnitzer

A native of southern California, Andreas B. (Andy) Rechnitzer was an avid diver and one of the first scientific scuba divers in the United States. He joined the Navy during World War II and remained a member of the Naval Reserve after the war. He attended graduate school under the veterans' program, ultimately attaining a PhD in biology from the Scripps Institution of Oceanography.

After earning his doctorate he joined the Navy Electronics Laboratory as an oceanographer. An invitation offered by the Office of Navy Research to participate in a series of deep dives in the Piccard bathyscaph led Rechnitzer to Italy where, after being introduced to the Piccard bathyscaph, he wrote the Navy's 1956 bathyscaph evaluation and final report. Rechnitzer recalled: "I wrote up a white paper that the Navy should buy it for the U.S. oceanographic community and they should deliver it to me in San Diego. Lo and behold, they did it."[1]

Don Walsh thought of Rechnitzer as his "recruiter and mentor—the driving force behind the *Trieste* program. He had great people skills—you could not help liking Andy." Rechnitzer had the personality, entrepreneurial drive, imagination, and people skills to establish the *Trieste* program from scratch. Walsh observed: "Since the whole bathyscaph concept was an unknown, Rechnitzer exploited the Navy's institutional ignorance to get things done. We operated in a vacuum, but a well-funded one."

Rechnitzer left NEL in 1963 for a position with North American Aviation's Autonetics division, heading its submersible program to design and build the *Beaver IV* deep submersible.[2] He remained with North American until the firm phased out its ocean systems program in 1970. He then accepted a dual-hatted position in the Pentagon: in his "white" role, Rechnitzer served from 1970 to 1973 as the Science and Technology Advisor to the OpNav Deep Submergence Systems Division, a title covering his significant and demanding "black" role as the Program Director, Program D (Advanced Technologies) for the National Underwater Reconnaissance Program. Program D was charged with identifying technologies and developments that could impact Navy underwater missions. Rechnitzer would be a knowledgeable voice in assisting Dr. Frosch in 1971–1972 managing the *Trieste*'s classified operations.

Upon the disestablishment of Program D in 1973, Rechnitzer served as the senior civilian science and technology advisor to four successive Oceanographers of the Navy, from 1974 until 1984. He held several other positions with the U.S. Navy: Head, International and Interagency Affairs Branch, Office of the Oceanographer of the Navy (1974–1978); Science and Technology Advisor in the Office of the Chief of Naval Operations, Naval Oceanography Division (1978–1984); and Coordinator of Polar Affairs at the Office of the Chief of Naval Operations. In 1975, Rechnitzer conducted exploratory SCUBA dives (at age 55) under seven feet of ice in the Antarctic.

Subsequently, Rechnitzer took the position of senior scientist at Science Applications International Corporation (SAIC) in La Jolla, California, remaining from 1985 to 1998. Dr. Andreas B. Rechnitzer passed away in 2005.

---

[1] Eric Hanauer, *Diving Pioneers: An Oral History of Diving in America* (San Diego, Calif.: Watersport, 1994), 101.

[2] "Scientist Urges Studies of Deep Ocean Rescue," *San Diego Union*, 12 August 1963; Hanauer, *Diving Pioneers*, 105.

the vacant seat to decrease her anxiety about his chosen profession. Marie Claude's dive to about 500 feet in the bay of Naples probably marked the deepest dive by a woman to that date, echoing her husband's aunt Jeannette Piccard's balloon ascent to 57,579 feet in October 1934, as the highest ascent by a woman to that date.[4]

There were no *Trieste* dives during 1955, due to problems with the batteries and insufficient funding to solve those problems. Still, modifications and improvements were made, including the provision of new floodlights and a fathometer. On 20 September 1956, the *Trieste* was put back into the water and during October she made a series of dives off Capri to depths increasing from 500 feet to 12,110 feet (Dives #18 through #22).

Jacques now was facing a classic problem. He needed to find "paying customers." Adventurers might be willing to pay for a ride, but scientists are a conservative and cautious lot, especially conservative with their budgets. The *Trieste* needed an "agent" to promote her as a new capability. Jacques decided that he would have to become his own agent and find entry into the closed world of oceanographic financing.

Piccard in London

Jacques flew into London in early April 1955, to be interviewed on television. Sir Robert H. Davis, an expert in undersea salvage, watched the show and invited Piccard to speak to a group of engineers under the aegis of the Royal Society of Arts. The younger Piccard gave a short presentation on the *Trieste*. After his talk, Jacques met Dr. Robert S. Dietz, an oceanographer in the London Office of the U.S. Navy's Office of Naval Research (ONR). Piccard later wrote:

> After my talk, several people came and spoke with me, requesting many details about the bathyscaphe, and several people presented their cards. One of the gentlemen who gave me his card asked to speak with me. I looked at his card and, as I was busy with other people at the same time, the only word on the card that I noticed was "attaché." The word *attaché* in French is used especially for the press, so I thought he was a reporter. As I had another engagement after the lecture, I told him, "If you would like to speak with me, come to my hotel at midnight. There is practically no time before then." He agreed to meet me there.

A few minutes later, I noticed that his card read "Attaché to the U.S. Embassy for the Office of Naval Research." I realized that he was not a reporter. On the contrary, this was an extremely important contact. I sought him out again to tell him that I would be very glad to see him before midnight. "If you would like to come just now we will have about half an hour to talk."

Dietz picked up the narrative: "There was a language barrier, my spoken French being worse than his English, but we managed."

Piccard continued: "He asked if I was interested in collaborating with the U.S. Navy. I said, 'Yes, of course. Would you like to visit the *Trieste* at Castellammare di Stabia in Italy?' "[5] This was the invitation for which Dietz had been angling.[6]

The Office of Naval Research

After World War II an informal group of scientifically trained U.S. naval officers sought to maintain the close relationship between the Navy and academia that had flourished during the war. Their efforts were rewarded in 1946 with the creation of the Office of Naval Research. From the start a funding agency of a different stripe, ONR was characterized by freewheeling and liberal support of research projects at universities, non-profit institutions, and industrial laboratories, including allocating a million dollars–plus over ten years to the Scripps and Woods Hole oceanographic institutions. The ONR support of nearly all major scientific fields and the generosity of its grants gave it the appearance of an office of *national* research before the founding of the National Science Foundation in 1950.[7] Thus, ONR was the ideal sponsor for the *Trieste* project.[8]

For Dietz, the meeting with Piccard was anything but unexpected. In 1953, Dietz received a Fulbright fellowship to the University of Tokyo, during which time he had an opportunity to make a dive in the tethered Japanese diving bell *Kuroshio I*. In 1954, he accepted a position with the Office of Naval Research to help its London office identify and investigate new European technologies and equipment that could be of interest to the U.S. Navy. Dietz later recalled, "I considered the *Trieste* to be an important breakthrough."

Dietz reported back to ONR in Washington about his investigation into the Piccard efforts, recommending that ONR show interest in this new capability to examine the ocean depths. Initially he ran into resistance, specifically from Gordon G. Lill, the head of the Geophysics Branch. Lill was reflecting the feeling of several ONR scientists that "we can study the seafloor by unmanned devices. We don't need to mess around with this funny old gas bag called *Trieste*."[9] Eventually Dietz, the first U.S. official to show an interest in the *Trieste*, was able to generate enthusiasm in ONR about acquiring the bathyscaph for the U.S. Navy.[10]

In June 1955, as invited, Dietz visited Castellammare and was so impressed by the *Trieste* that he promised to return as soon as possible with a senior representative from ONR. Dietz recorded: "Here was a privately owned submersible of radical design. A father-and-son team had built it, the physicist father providing the concepts and the son carrying the plan through to realization. It was refreshing to see a man alone with the 'effrontery' to own and operate a submarine. This was the business of navies. Modern technology is supposedly too sophisticated for back yard Edisons."[11]

The representative whom Dr. Dietz brought back to Castellammare was indeed senior: Dr. Thomas J. Killian, formerly the chief scientist of the Office of Naval Research (1947–1952) and in 1955 the most senior civilian at ONR, the deputy chief and the chief scientist. Killian's visit to Castellammare was auspicious; he returned to the United States "quite enthusiastic" about the *Trieste.* He called a conference within ONR to organize investigation of possible uses of the bathyscaph by the Navy. The result of this meeting was that Lill and seismologist Dr. John N. Adkins, Lill's immediate superior, spent a week in Naples with Jacques Piccard. Lill became an immediate convert and was to spearhead the *Trieste* effort for the next year. Adkins extended an invitation to Piccard to present a paper at the symposium on Aspects of Deep Sea Research in Washington on 1 March 1956.

Deep Sea Symposium 1956

At the symposium both Piccard and Dietz presented papers. Dietz's presentation included the statement:

> It may be that the Navy might see sufficient justification for the bathyscaph strictly from its military potential, even though this seems to be unlikely to me. Nevertheless, it seems that the principal justification for a bathyscaph program must be based upon its usefulness as a tool for deep sea research. Its future, then, depends upon the approval, or

> disapproval, of oceanographers such as those of you who are gathered here. *The time has arrived when oceanographers should cease having no opinion at all* [emphasis added]. With the recent successful assault on Mt. Everest, attention has naturally turned to conquering the world's greatest depth, the nearly 11,000-meter [36,000 feet] Challenger Deep off Guam. This will, no doubt, be accomplished *within the next few years*, because, from an engineering point of view, it is a comparatively straightforward project.[12] Of course, we must recognize such an effort for what it is, a prestige project [emphasis added].[13]

Subsequently, Dr. Willard Bascom of the National Academy of Sciences called for a motion to establish the will of the participants: "I think it's time for us to stand up and be counted."[14] A declaration was presented that, upon the change of one word, was passed unanimously by the 56 members of the symposium: "The careful design and repeated testing of the bathyscaph has clearly demonstrated the technical feasibility of operating manned vehicles safely at great depths of the ocean. The scientific implications of this capability are *far-reaching*. We, as individuals interested in the scientific exploration of the deep sea, wish to go on record as favoring the immediate initiation of a national program, aimed at obtaining for the United States, undersea vehicles capable of transporting men and their instruments to the great depths of the oceans."[15]

The resolution and news of the unanimity of its support were transmitted to the president of the National Academy of Sciences and to the Chief of Naval Research.[16] Lill, Killian, and Dietz had designed and executed a major bureaucratic coup.

Piccard's 100-Day Tour

Following the symposium, Killian arranged for Dietz to escort Jacques Piccard around the United States to visit oceanographic activities, institutions, and universities. Killian was the key man in these events and was to become a chief conspirator in a cabal of administrators, scientists, and naval officers who would shepherd the *Trieste* through the tangled web of bureaucratic inertia and resistance. Dietz later wrote, "For four years the *Trieste* sat at Castellammare, just twenty miles from Naples, a great U.S. base. But in all those years only one official came to see it. Beginning in 1955, I fought to obtain Navy support for the *Trieste*. This was finally forthcoming in the summer of 1957."[17] Dietz became Piccard's official escort during his 1956 tour of the United States.

While Piccard recognized some of the political facts surrounding his 100-day tour, he did not yet appreciate that ONR was precisely the right organization to "make things happen." Unbeknownst to Piccard, it was not solely the usefulness of the *Trieste* to potential U.S. scientific customers that was being promoted; Killian, Lill, Dietz, and others were using him to generate a groundswell of appreciation for what the *Trieste* could accomplish for the U.S. Navy. What ONR had decided to "make happen" was acquisition of the *Trieste* for the Navy. Dr. Dietz would play a key role in accomplishing that goal.

In early 1956, in Washington, the first, informal discussions took place between Dietz and Jacques Piccard concerning the possibility of the *Trieste* diving into the Challenger Deep in the Mariana Trench—the "deepest spot" in the world's oceans. Jacques suggested that it might be possible to dive the Challenger Deep with the existing Terni sphere but warned that the depth was very close to its 150 percent margin of safety. One or two unmanned test dives might be required to prove that the craft was safe for that depth. No further discussions on the subject occurred for almost a year, but for Piccard the hook was deeply set.[18] He later described his tour of the United States:

> Oceanography in America was, in one sense, a new science. Born of war's exigencies, it had become a part of the Navy. Oceanography was no longer a hobby to the Navy; it was an absolute necessity, the very key to survival.
>
> The United States would clearly be a fine future home for the *Trieste*. I knew already that the rank and file of oceanographers was enthusiastically behind the bathyscaphe. But in the upper echelon, her support appeared less strong. Some laboratory heads had misgivings. I sensed opposition in places where final monetary decisions are made.[19]

As part of his tour of U.S. oceanographic facilities, Piccard spent the period 6–21 March 1956, in San Diego visiting Scripps. At a lecture for the public at nearby Pacific Beach, Piccard drew an overflow audience of 300. In a letter dated 26 April, Scripps director Roger Revelle formally requested ONR to fund an effort by Scripps and the Navy Electronics Laboratory to bring the *Trieste* to San Diego for oceanographic and naval evaluation.[20]

With the unanimous resolution from the Aspects of Deep Sea Research symposium and the "unsolicited" request from Scripps, ONR had a foundation upon which to initiate action. Accordingly, it would sponsor a program of scientific dives in the summer of 1957 off Castellammare to evaluate the *Trieste*.

Investigations and Preparations

In the fall of 1956, Rear Admiral Rawson Bennett II, the Chief of Naval Research and head of the Office of Naval Research, sent Commander Charles B. Bishop of ONR on a special visit to Castellammare. He was to report on the potential of Piccard's bathyscaph from the point of view of the uniformed Navy. Bishop arrived during weather that prevented diving but spent three days on Capri getting an intensive briefing on the *Trieste* from Piccard. Bishop could see the potential—not only as a scientific research tool but also as a unique military operational capability. During an interview with Vice Admiral Massimo Girosi, Bishop was able to obtain an Italian commitment to support the upcoming dive program. He reported back to Admiral Bennett with a positive recommendation concerning the *Trieste*.[21]

After his American tour, Jacques attended a meeting of the Scientific Committee on Oceanic Research at Gothenburg, Sweden, in January 1957. There he again met with Gordon Lill and Robert Dietz, the leading proponents of acquiring the *Trieste* for the U.S. Navy. It was an opportunity to tighten up plans and the schedule for the proposed summer dives off Capri. Over coffee one evening, the magic name of the Challenger Deep arose:

"'An important reason for working with the U.S. Navy this summer is that it could open the door for an eventual assault on the deepest spot in the world,' Dietz suggested. 'You mean the Challenger Deep?' Piccard asked. Dietz nodded, 'Exactly. Who else but the Navy could easily support such an operation? They have the facilities, ships, and a base near at hand.'"[22]

Dietz was an expert salesman and appeared to know his man. Jacques *wanted* that dive.

In February 1957, Arthur E. Maxwell, the head of the ONR oceanography section and Lill's right-hand man, negotiated a contract with Piccard for

$40,000 to conduct a minimum of 15 deep dives in the Tyrrhenian Sea off Naples to allow American, Italian, Swiss, and other scientists to evaluate the *Trieste* as a tool for oceanography.[23] Because the money was considered sufficient for only the consumables and maintenance of the *Trieste* during the summer, not for amortization or indirect costs of the vehicle, and because the Italian Navy was providing extensive support, the program offered up to one-half of the dives to the Europeans. Piccard, however, was able to stretch this $40,000 to 26 dives for the season. The dives commenced in late June 1957, extended through October, and gave specialists in several disciplines opportunities to experience a foray into the depths.

The U.S. bathyscaph evaluation program began in April 1957. A group of American and European oceanographers met with Jacques Piccard at Castellammare to inspect the *Trieste.* The Americans were:

Dr. Russell V. Lewis, Underwater Sound Laboratory, New London, Conn. (acoustics)
Mr. Roberto Frassetto, Hudson Laboratory, Columbia University, New York, N.Y. (acoustics)
Dr. Andreas B. Rechnitzer, Navy Electronics Laboratory, San Diego, Calif. (biologics)
Mr. Morton R. Lomask, Hudson Laboratory, Columbia University (acoustics)
Mr. William V. Kielhorn, Woods Hole Oceanographic Institution, Falmouth, Mass. (magnetics)
Dr. Robert S. Dietz, Office of Naval Research, Washington, D.C. (oceanography)

In addition to scientific dives, several dives were made for demonstration or familiarization purposes; to the list of American riders was added Captain William G. Stearns II, U.S. Navy.[24] To handle the crowded schedule, Piccard hired, among others, the reliable and knowledgeable Giuseppe Buono, full-time. He had worked at the Navalmeccanica as a mechanic and had become obsessed with the bathyscaph. He learned everything about the *Trieste* and eventually became indispensable to Piccard as "topside engineer," a position that held the lives of the bathyscaph's crew in trust. In addition to the routine and seemingly unheralded assistance of the Navalmeccanica shipyard with divers, diving boats, crane services, dock space, and shop space the Italian Navy agreed to provide a tug and at-sea security.

During the summer of bathyscaph evaluations, Piccard and Dietz reopened informal discussions with Andy Rechnitzer on taking the *Trieste* to Guam to dive into the Mariana Trench. Nothing could be decided, but all agreed that the proposal had great merit.

The chairman of the National Science Foundation's influential Committee on Undersea Warfare, Harrison S. Brown, on a personal vacation in Europe, also visited Castellammare to see the *Trieste*.[25] His committee was enthusiastically "on board" about acquiring the *Trieste* for the U.S. Navy. Allyn Vine from Woods Hole also found a way to drop in for a few days. Vine was to become one of the strongest proponents of a manned deep submergence capability in the U.S. Navy.

Of the 26 dives conducted by the *Trieste* during the summer series, 8 were for acoustics studies, 4 for biologics, 3 for light penetration of the ocean depths, 1 for magnetism, 1 for gravity, and 1 for geology, plus 8 dives solely for demonstration, photography, or equipment checks. All of the acoustic and all but one of the biologic dives were for American scientists. The design of the tests and of the experiments revealed that the *Trieste* was being considered as an operational research asset for solving naval problems, not for pure science.[26]

On 4 October 1957, the Soviet Union launched *Sputnik*, the world's first artificial satellite into a low Earth orbit. It was visible with the naked eye, and amateur radio operators around the world could tune into its intriguing and ominous beeping sounds. The two-foot-diameter, 185-pound satellite stunned the world. With *Sputnik*, the Soviets were the first into space—the scientific and military implications were massive. The shock reverberated through American society, defense agencies, Congress, and the White House.

The Challenger Deep

In early January 1958, ONR again asked Jacques to discuss the *Trieste*, this time in Lill's office in Washington, D.C., in Building T-3 at the corner of 17th Street and Constitution Avenue, Northwest.[27] There, on 10 January, Lill and Piccard discussed with Maxwell the possibility of diving the bathyscaph into the Mariana Trench. On a large map on Lill's wall they determined that the Challenger Deep was about 180 miles southwest of Guam, where the U.S. Navy maintained a large base.

Jacques recalled the discussion:

> "You say the Navy still maintains a big base at Guam with cranes and tugboats? A mere two hundred miles from the Challenger Deep? . . . *Alors*, we can decide right now to make that dive. *Pourquois pas*?"
>
> "But can it be done with the *Trieste* as she is?" Lill asked.
>
> "Yes it could be done right now but with the Terni sphere, the margin of safety would be narrow. An unmanned test would have to be made first."
>
> "And if the sphere failed. What then? Scratch one bathyscaph?"
>
> "*Sans doute*," I agreed, "the float too would doubtless be shattered by the terrific implosion. . . . It's a foolhardy chance to take. It would surely be better to have a new thicker-skinned sphere compatible with the existing float. We could have one built in Europe."[28]

1958 Bathyscaph Conference

The schedule for the Bathyscaph Conference at the National Academy of Sciences in Washington on 20 January 1958 indicated that the participants would be asked to discuss whether to (1) follow up on the 1957 dives in the *Trieste* with additional investigations, (2) acquire the *Trieste*, (3) develop similar capabilities from scratch, or (4) reject the technology as undeserving of further development.

A statement made early in the conference indicated that ONR already was on course to acquire the *Trieste* for the United States. Lill stated, "We have high hopes of eventually modifying this craft so it has greater horizontal mobility."

The official Navy reactions to the 1957 diving program were positive, as reflected in the final report prepared by Dr. Rechnitzer, recommending that the Navy:

1. Continue deep sea research using a bathyscaph.
2. Encourage modification and further development of the bathyscaph and/or bathyscaph-type craft.
3. Promote the development of more versatile deep submersible research craft.

4. Evaluate the usefulness of bathyscaphs and deep submersible craft for military purposes.
5. Develop acoustic and oceanographic instruments for use on the bathyscaph.[29]

Dietz's presentation at the 1958 Bathyscaph Conference included a summation of the French *FNRS 3* program:

> The French are now building a new "super-bathyscaph," designed to go to 10,000 meters [32,800 feet], a three-man affair, very much along the lines of the older one, with bigger engines, bigger electric power, twice the amount of gasoline as the *FNRS 3* and also the windows will be made so that they come to a very small aperture inside, so it would be necessary to have an optical arrangement to view through this window. . . . [*T*]*hey presumably plan to dive in the deep trenches* because this factor of 10,000 meters is really only met there [emphasis added].[30] *Trieste* is neither a research ship nor a tool in being. It is something that has to be developed further. This is something we realized. We only wanted to demonstrate that it could be used and that it had great possibilities.[31]

Maxwell briefed the assembly with a candid and straightforward synopsis of ONR's views: "I would like to briefly state what ONR's position is at this time. . . . [I]t is ONR's feeling at this time that what we should do, as a minimum, is continue, perhaps, over in the Mediterranean, the same sort of program which we carried out this past year. However, we feel that we would be on much better ground if we could bring the bathyscaph over to the United States and have it more in our own operational control."[32]

Two weeks later, a Navy news release quoted Dr. Franz N. D. Kurie, technical director of NEL, as stating that the Office of Naval Research had negotiated an agreement with the Piccards to bring the *Trieste* to San Diego but that no contract had yet been signed.[33]

Finally, a Contract

With the favorable results of the 1958 Bathyscaph Conference in hand, ONR prepared a budget proposal for the acquisition of the *Trieste* and obtained congressional approval for its inclusion in the next budget cycle. Thus, although detailed negotiations with Piccard could be conducted early in 1958, it was not until 1 July that contracts were finalized.

ONR's Maxwell and the U.S. purchasing agent for Italy met with Jacques Piccard to discuss the contractual conditions by which the Navy would procure the bathyscaph. Initially Piccard objected to selling the *Trieste*, wanting instead to lease it to the U.S. Navy for three years. Maxwell later disclosed the difficulty of negotiating with Piccard:

> We sat down with Jacques Piccard, and we negotiated a deal on this. We talked not only about getting the *Trieste* as it was, but buying a new sphere that could go to the deepest part of the ocean; getting his engineer [Giuseppe Buono]; getting Piccard part time, etc. We would get this all settled at night and he would say, "I'll meet you in the morning." He'd go home and talk with his father, Auguste Piccard, and he'd come back and we'd start all over from scratch. We spent a week on that.[34]

Neither Jacques nor Auguste Piccard had any objection to transferring the *Trieste* to the United States, either on a long-term lease or as an outright sale. Both were now agreeable to withdrawing from the *Trieste* with enough money to pursue Auguste's next idea for a special-purpose submersible. The U.S. Navy insisted on an outright sale owing to its plans to modify the Piccard bathyscaph to dive to 36,000-plus feet and the potential for classified operations. Jacques Piccard would agree to an outright sale only if the *Trieste* was modified with a new sphere for the "Deep Dive"—that is, for an assault on the deepest place in the world, the Challenger Deep. This met with ONR agreement. Jacques agreed to the outright sale at a price that would reflect the fact that the *Trieste* was a "used" craft that would require expensive maintenance and modification by the Navy.[35] Jacques also added a stipulation: he personally would make the record dive into the Mariana Trench.

The agreement was signed, although the exact amount of the sale could not yet be determined until Piccard had time to develop the price of a new sphere from a European source. Lill later summarized the agreement: "They were not interested in making a huge profit out of the sale. In fact, they didn't. They sold it to us for $267,000, which probably just about recovered their costs. The only stipulation was that if we bought it, that Jacques wanted to take it and make the record dive into the Mariana Trench, which we agreed to. We wrote it into the contract."[36]

In early summer, Piccard met Dietz in London. In the middle of their discussions, Dietz suddenly exclaimed, "I've got a name!" Jacques, familiar as he was with English, was confused by this non sequitur. Dietz recognized Jacques' confusion and continued, "For the project. It is time for a name. A name will give it substance, reality. What do you think of Project Nekton?"

"Nekton" is the scientific term for motile (free swimming) sea animals, as opposed to drifting animals, such as jellyfish. Since the *Trieste* had a propeller and could "swim against the current," it was an appropriate name, and Jacques approved. Dietz later sent a letter to the Office of Naval Research, with a copy to Piccard, proposing "Nekton" as the name. It was approved as the official Navy designation for the dives into the Challenger Deep.[37]

The contract to buy the *Trieste* went into effect on 1 July 1958, and Piccard returned to Castellammare to clean and refurbish the bathyscaph prior to its crating and transport. However, events were to intervene. ONR advised that the worsening Lebanon crisis that year might interfere with transportation to San Diego and therefore pushed for an accelerated delivery: in the last week in July on the U.S. Navy cargo ship *Antares* (AK 258). Jacques had barely four weeks for cleaning, disassembly, and crating before the *Trieste* was loaded on board the *Antares* on 26 July for shipment to Norfolk, Virginia.[38] Once the *Trieste* was safely on board the *Antares*, Piccard could turn his attention to the fabrication of the new deep-diving sphere.

The Second Sphere

Piccard approached Terni to fabricate the new and stronger sphere in March 1958, but Terni declined; its existing capabilities were fully committed.[39] A friend recommended he contact Krupp in Essen, Germany. Piccard did so on 1 April 1958, and met with enthusiasm and willingness to help.[40] However, when Piccard produced exact drawings and specifications, Krupp threw up its collective hands in dismay. The problem was that its 15,000-ton forge had been dismantled and sent to Yugoslavia by the British as part of war reparations. The largest surviving forge Krupp had was just 3,000 tons and incapable of forging this new sphere as specified.[41]

Piccard asked, "What is the largest forging you can handle with existing Krupp equipment?" The reply was 24 to 28 inches in height and 78 inches in diameter. Piccard had an inspiration: What about forging the sphere in several rings to be assembled into a sphere? Krupp originally balked at the idea but upon further investigation found that it was possible in three segments, two caps and a center ring.[42] The next problem was how to join them. It had been relatively easy to join the two Terni sphere halves by clamped their flanges. Three pieces made using flanges more difficult.

Piccard's solution was to *glue* the three segments together with an epoxy resin. Krupp provided a price on 25 April 1958, and agreed to a five-month delivery schedule.[43] This information was forwarded to ONR. At the beginning of September the Navy advanced credits so that Piccard could proceed with the finalization of the contract for the new sphere, and Piccard signed the agreement with Krupp on the 16th.[44] Work on the new sphere started early the following month.[45]

For the "bargain basement" price of $267,000, the Navy was acquiring the Piccard bathyscaph. The financial package included the *Trieste* with two spheres, plus consulting contracts that would bring both Jacques Piccard and his indispensable topside engineer/mechanic Giuseppe Buono to the United States for one year.[46]

After more than two years of bureaucratic juggling and consensus building and later detailed planning for substantial upgrading and improvement of the *Trieste*, ONR had triumphed. The *Trieste* would fulfill the requirement implied for the U.S. Navy "by the responsibilities of its mission, to operate throughout the entire ocean environment."[47]

Lill recorded ONR's thinking:

> The fact that it [the *Trieste*] existed as a unique way of getting down to look at the seafloor was one thing that led us to this. The other was that we thought the Navy was by no means taking advantage of the depths

> of the ocean in its submarine and anti-submarine warfare activity. We thought this would be a way of interesting the Navy in going deeper to take advantage of the ocean depths—to hide submarines, to move submarines about. At that time, they had no requirement for very deep diving in the U.S. Navy. I argued with many admirals to try to convince them that they ought to use the ocean depths as well as the ocean surface. Buying the *Trieste* kicked off the whole thing.[48]

Maxwell later observed, "We got it for a steal. I think we paid about $1 million for it, but that also bought the new sphere; it bought Piccard's services and the *Trieste*. All of this was about $1 million and it got the Navy started in the deep-submersion business."[49] The bathyscaph clique had triumphed. The Navy would gain the capability to operate in the deepest ocean depths.

5 The New World

Such implosions can be disquieting.

—Jacques Piccard, of Dive #55

The U.S. Navy cargo ship *Antares* loaded the disassembled bathyscaph *Trieste* along with 30 crates of instruments and parts at Castellammare on 26 July 1958. Arriving at Norfolk after an uneventful Atlantic crossing, the *Antares* transshipped the cargo to the commercial steamship *P&T Leader* for transit via the Panama Canal to San Diego.[1] The *Leader* arrived in San Diego on 23 August. Offloading of the *Trieste*'s float, the Terni sphere, and associated crates onto a barge occurred at the B Street pier. The barge then was towed by tug to the nearby San Diego naval base.

A Navy press conference on the 23rd was held at the B Street pier with Captain John M. Phelps, commanding officer and director of the Navy Electronics Laboratory; Dr. Ralph J. Christensen, NEL associate technical director for research; Dr. Edwin L. Hamilton, chief of NEL's ocean floor studies; and Dr. Andreas B. Rechnitzer of NEL; and Arthur E. Maxwell of the Office of Naval Research.[2] They told the press that the *Trieste* had been purchased from the Piccards and would remain on the barge until Jacques Piccard arrived from Italy. The first dive could be expected "about mid-October."[3] The program announced for the *Trieste* was: (1) the *Trieste* would be "semi-permanently" based NEL; (2) the early deep dives would be made outside San Diego Bay, off Point Loma; and (3) over the first year of operation scientists would venture by stages to depths of two miles in "shakedown cruises" to determine the craft's capabilities.

Christensen revealed some of the underlying motivation for manned deep submergence operations: "We think of this undersea research as pretty much of a space program. There is as much threat under the seas as there is from the space above. We feel this is an area in which the United States should put more effort."[4]

At the arrival conference Rechnitzer was characterized as the NEL oceanographer "who will coordinate bathyscaph research with other scientific agencies." In reality, Rechnitzer was NEL's *Trieste* program director. On 3 October 1958, Andy Rechnitzer corralled Rear Admiral Elton W. Grenfell, Commander, Submarine Force Pacific, in San Diego and coaxed him into the *Trieste*'s entry tube for news photographers. Grenfell envisioned a dual use for *Trieste*-type craft: "While engaged in research it would still be available for emergencies."[5] Discussions between Rechnitzer and Grenfell centered on Submarine Force participation in the *Trieste* program, ideally providing the craft's operating crew. Grenfell was open to the idea, and he provided a favorable endorsement when NEL later made an official proposal to that effect.[6]

Jacques Piccard in San Diego

Late in the evening of 22 October 1958, Jacques Piccard and Giuseppe Buono arrived in San Diego to supervise the assembly of the *Trieste*. They were met at the airport by a contingent from NEL, including Rechnitzer, Christensen, and Commander J. A. McAllister, executive officer of NEL. Jacques told the press that he expected to have the *Trieste* ready for diving within a month or two. Once the craft was ready he would leave to bring his wife and infant son back to San Diego.[7] He later related, "When we arrived, the bathyscaph was practically untouched. The float and the sphere were still on the barge at the Naval Repair Facility [San Diego] where it had been unloaded from the ship. In other words, we had many things to do—clean the sphere, clean part of the float, make a careful inspection and so on."[8]

Buono and a "pickup team" of personnel from NEL spent November cleaning, priming, painting, and partially assembling the *Trieste* while awaiting the completion of a cradle being constructed to hold the *Trieste* when ashore. With the work progressing, Rechnitzer turned his attention to a longer-range problem: Who would maintain and operate the *Trieste*? After his meeting with Admiral Grenfell, Rechnitzer was able to arrange a working lunch the last week of October with Captain Ralph E. Styles, the commander of Submarine

Flotilla 1, on board his flagship, the submarine tender *Nereus* (AS 17), in San Diego Bay. The acting flotilla secretary, a lieutenant, scheduled the lunch and attended the meeting. The secretary related:

> Andy [Rechnitzer] brought Jacques Piccard with him. At lunch, they briefed the Commodore [Styles] on the immediate Navy plans for the bathyscaph, how it worked, its future importance to the Navy, etc. When Commodore Styles thanked Andy, he asked how the submarine force could help. There was not a bit of delay. "We need two submarine-qualified officers and about five enlisted men to maintain and operate *Trieste*, and we need them soon." NEL would provide technical support and program guidance but there needed to be a military crew to run the vehicle.[9]

The flotilla secretary was Lieutenant Don Walsh.

Commodore Styles agreed that Andy's request for two officers was reasonable and said he would try to get the Commander, Submarine Force Pacific and the Bureau of Personnel to agree. Meanwhile Captain Walter L. Small, Styles' chief staff officer, sent out a message to all the flotilla's submarines in the area asking for two officer volunteers, preferably submarine-qualified lieutenants. The response was unexpected: only one man volunteered: Lieutenant Richard (Dick) Davey. Walsh later recalled, "With some delicacy, I asked if I could put my name on the list. Captain Small reluctantly agreed, so Dick and I became the total volunteer pool." Captain Styles, with the minimum of paperwork, temporarily assigned Davey and Walsh to NEL. "Now the very thing that harpooned me into a desk job in the first place was working in my favor," Walsh would recall.[10] His desire to get out of the desk job in the commodore's office on board the tender *Nereus* and back into the operating Navy launched him on an unimaginable career trajectory.

Jacques returned to Europe in November 1958, to approve Krupp's manufacturing plans and to finalize the design and procurement for the interior instrumentation and controls required for the new sphere. He recalled, "When the work was started and everything was going well, when Buono started to understand a few words of English, I decided to return for about three or four weeks in Europe. I had to supervise the construction of the new sphere in Essen, Germany. I spent most of the month of November in Switzerland and in Germany and I came back at the beginning of December to San Diego."[11]

First Dives off San Diego

At this time Lieutenant Davey was formally named officer-in-charge of the *Trieste.* Jacques, upon his return to San Diego in early December 1958, found the new bathyscaph cradle ready and final bathyscaph assembly began. On the 16th Jacques and Buono ballasted the *Trieste* with iron shot. The bathyscaph was then floated and early on the 17th towed across the channel to the fuel piers at Naval Air Station North Island to fill the float. Jacques described the evolution: "We filled the bathyscaph with about 25,000 gallons of gasoline. Everything went well, the operation was extremely easy, and we made a first preliminary equilibrium test . . . which showed that the gas was 115 [octane]. From there we returned to NEL to establish the exact equilibrium of the bathyscaph, and add a bit more ballast to make it ready for the first dive."[12]

The *Trieste*'s harbor test dive on 17 December, to 70 feet, had Jacques piloting and Rechnitzer observing. Jacques would pilot all U.S. Navy *Trieste* dives through the next year, until the 1959 year-end holiday period.

Two days later, Dive #50 with John F. Light, an independent underwater cinematographer, was made to 860 feet in the Loma Sea Valley off San Diego. The *Trieste* was towed and supported by the NEL-based landing craft *YFU 45.*[13] Jacques described this first deep dive off San Diego:

> On the 19 December, the sea was favorable for this first dive with John Light. Everything was OK during the descent, the equilibrium was right, no trouble at all. After we were on the bottom about 20 minutes or half an hour, we heard a strange noise in the battery boxes and I discovered a battery was more or less burning. These batteries are extremely powerful and if they short circuit, it could be dangerous. So, I decided it was best to come up. The dive was relatively short and not very productive from a scientific point of view, but sufficient to show to the NEL people that the bathyscaph was ready to operate.[14]

Just before Christmas 1958, Jacques flew back to Switzerland, first to spend the holidays at home and then to monitor the construction of the new sphere in Germany and the fabrication of its controls and instruments. The inside equipment was being prepared just a few miles southeast of Jacques' home in Lausanne. Jacques explained, "I had to stay there [in Europe] for about three months until the sphere was completely finished, until the inside fittings [were] installed in the sphere and until we could ship the sphere from Germany to San Diego. . . . I observed every significant operation at Krupp; most of the small operations too. I was there almost continuously."[15]

Krupp was not merely reprising the Terni sphere with greater wall thicknesse, but fabricating a unique sphere in three sections to be glued together. Critical inspection points in this process included the casting, forging, machining, non-destructive ultrasonic and magnetic-particle inspections, micro-finishing and lapping of the joints, and gluing of the joints. Krupp's archives state that "careful control of the forging temperature and degrees of reduction ensured the highest quality of the work. It was possible to achieve the best direction of the flow lines and an extremely uniform consolidation of the metal."

The Piccard specifications called for the steel to meet a minimum yield of 113,755 pounds per square inch; Krupp managed to produce a yield 15 percent stronger.[16] Frequent non-destructive testing during the sphere's manufacture certified that all three sections were entirely free from defects. Jacques later stated, "Meticulous care was bestowed upon forging and machining it [i.e., the sphere]. We even went as far as to dispense with an unmanned test dive such as was carried out in the waters near Dakar with the very first sphere, so convinced we were that the security offered was amply sufficient, even when diving to extreme depths."[17]

The Krupp sphere's observation windows were thicker than the Terni sphere's window; the main Plexiglass viewports on the Krupp sphere were just over seven inches thick, with an inside diameter of two and one-third inches and an outside diameter of 15.75 inches. The windows were at opposite sides of the sphere to provide a 70 percent field of vision.[18]

In late February 1959, Jacques slipped through the Swiss export licensing bureaucracy in five minutes of fast talking and got the interior instruments and equipment on a train to Essen, Germany.[19] In Essen the sphere and the interior equipment were crated for shipment and trucked to Hamburg, Germany, for loading on 20 March onto the Danish-flag steamship *Panama* as deck cargo.[20]

San Diego, 1959

In January 1959, the only Navy personnel officially attached to the *Trieste* program were Lieutenant Dick Davey as officer-in-charge and Chief Machinery Repairman E. John Michel. Davey would soon be disabled and hospitalized. Lieutenant Don Walsh arrived as assistant officer-in-charge later in January and found himself temporarily in charge of the *Trieste*. The winter months would see only slow progress in readying the *Trieste* for the spring diving program.

Don Walsh, still listed as the assistant officer-in-charge, officially reported for duty to Rechnitzer on 3 March. On the 11th the *Trieste* was lifted into its cradle at the NEL waterfront area of Point Loma. This marked the start of an overhaul that ended on 22 April, shortly after the return of Jacques from Europe. This overhaul included detailed inspection of the craft, replacement of worn parts, and installation of a cathodic protection system to reduce metallic corrosion. And, the *Trieste* float was repainted.

During the second week of April 1959, Jacques arrived in San Diego; his wife and infant son followed a few weeks later. On 15 April, Prince Victor Emmanuel DeSaare of Italy, a personal friend of Jacques, arrived in San Diego to work as an unpaid deck hand on the *Trieste*.[21]

At least that was the public image. DeSaare, a college student, was indeed the son of the last king of Italy but was nevertheless a private citizen, as Italy had renounced its monarchy. He was now on summer vacation and would spend only a brief time with Jacques and the *Trieste*. The rest of his summer, like many college students, he spent skin diving and scuba diving off the California coast and partying.

Rechnitzer's efforts to establish the *Trieste* program at NEL were reaching early maturity by mid-April 1959. Dedicated facilities at the NEL pier area on Point Loma included a combined office/shop area of about 1,250 square feet in Building 149; an open pad for dry-docking the *Trieste* on its cradle; a 4,067 square-foot fenced concrete pad adjacent to Building 149; and a 17-foot line-handling boat. The boat was sufficient for harbor work but inadequate for open ocean operation.

After the five-week overhaul the *Trieste*—with its fresh coat of white paint with blue stripes—was lifted into the water by a floating crane.[22] The bathyscaph was weighed to determine its displacement after the addition of new equipment. Tons of iron pellets for ballast were then hand-loaded. The next day, 22 April, the bathyscaph was towed to nearby North Island, where its float was filled with 28,000 gallons of aviation gasoline.[23]

On 28 April the *Panama* arrived in San Diego with the 28,665-pound Krupp sphere.[24] Jacques observed: "We now had the capability for descending to the bottom of the Earth at hand. There was still no authorization from Washington to proceed with Project Nekton."[25] In the meantime, the Terni sphere remained in the *Trieste*. In the second week of May, Walsh formally was designated officer-in-charge of the *Trieste*, because of Davey's health issues.[26] Walsh obtained permission from Captain Styles to find a replacement as assistant and approached Lieutenant Lawrence (Larry) Shumaker, a shipmate from his previous assignment on the submarine *Rasher* (SSR 269). Shumaker agreed and reported on board on 29 July as the assistant officer-in-charge.

Spring Diving Program

On 12 May 1959, Walsh made a test dive with Jacques to 50 feet in San Diego Harbor, supported by a landing craft and the NEL line-handling boat. A few days later the Navy personnel permanently assigned to the *Trieste* program grew from two (Walsh and Michel—Shumaker not yet released from the *Rasher*) to three, with the arrival of Master Chief Electrician's Mate Harlan DeGood. Within a few weeks, DeGood was able to convince an old shipmate to volunteer, and Engineman 1st Class William McCartney joined the *Trieste*. Thus, until departure for Guam in October the uniformed element of the *Trieste* team numbered five Navy personnel—Walsh, Shumaker, DeGood, Michel, and McCartney—plus any men that they could beg, borrow, or steal from NEL on a temporary basis.

The first U.S. deep-diving season began on 19 May 1959, with Dive #52, in which Jacques and Rechnitzer went to 720 feet. The *Trieste* was towed and supported by the fleet tug *Mataco* (ATF 86), sailing for the Loma Sea Valley, 11 miles west of the entrance to San Diego Harbor. As a bit of scientific theater, Robert Dietz arranged to carry a half dozen fresh eggs on the outside of the craft

Dr. Don Walsh

Don Walsh[1] was born in Berkeley, California, and raised in the San Francisco Bay area, where he developed his love of the sea. He enlisted in the Naval Reserve in 1948 while still in high school. His first assignment in the reserves was to Attack Squadron (VA) 62A, then flying TBM-3 Avengers, at Naval Air Station Oakland. Walsh qualified as a "plane captain"—the enlisted maintenance mechanic responsible for a specified plane—before he had his automobile driver's license.

He recalled:

> I was not a scholar in high school. My grades were not good enough to get admitted to Cal Berkeley or Stanford. In those years, the only other option was San Francisco State University, and they did not have a program in aeronautical engineering. My goal was to be a Naval or Marine aviator. Then I became aware of the Naval Academy and how I could get an appointment through the reserves.
>
> At my graduation ceremony in June 1949 . . . my mother handed me a telegram. Congressman George P. Miller had appointed me to West Point and I had to be there the next week! I had requested Annapolis, *not* West Point. I turned it down even though I was a year away from competing for an Annapolis appointment.

Walsh took the Annapolis competitive examination and passed. But his dream of becoming a Navy carrier pilot was squashed during his last year at the Academy (1954) excluding him from aviation because of his eyesight. His second choice was the submarine service. Walsh also applied for nuclear power training, but was rejected due to his grades.

After the then-mandatory two years in the surface Navy, Walsh attended submarine school and was assigned to the radar picket submarine *Rasher* (SSR 269). Two years later, a submarine-qualified lieutenant, Walsh was "shanghaied" by the Commander, Submarine Flotilla 1 as his unofficial flag secretary. Although he felt "trapped" by it, that assignment led to his becoming the officer-in-charge—commander—of the bathyscaph *Trieste* and making the "deep dive" to the deepest point of the oceans in January 1960.

---

[1] Don Walsh's name is "Don Walsh," not Donald. He explained to the authors that his parents were poor and could not afford the additional letters "ald."

to test the permeability of eggshells. If calculations were correct, the eggshells would allow water to seep through to equalize the pressure and the eggs would be undamaged by the pressure at more than 700 feet. (Newspapers later reported that according to Rechnitzer, the six eggs survived without a crack.)[27]

From Walsh's perspective, "This operation made it abundantly clear that a great deal more thought should be given to the organization and planning of sea operations. The *Trieste* was not very seaworthy during surface operations in sea states much greater than sea state one. . . . [S]ea operations would continue to be extremely difficult until satisfactory support equipment was furnished for these operations and permanent personnel could be assigned to the project for continuity and safety."[28]

The second of the season's dives was on 22 May, to 4,100 feet, about 15 miles off Point Loma, with Jacques and Walsh in the sphere. Again, the *Trieste* was towed and supported by the *Mataco*.[29] The *Trieste*'s fathometer became degraded by the enormous amount of plankton in the water.[30] Walsh observed with surprise, and possibly dismay, that none of the controls or instruments in the sphere were labeled—Jacques, as the developer of the craft, knew exactly where everything was and what each control did. It would take extensive experience for others to learn how to operate this unique craft.

The third and fourth dives of the season took Kenneth V. Mackenzie, an NEL physicist, into the San Diego Trough for acoustic studies. Mackenzie's first dive, #54 on 28 May, was a harbor dive to 60 feet to prove the external instrumentation mounted by NEL. The following day the *Trieste* was towed about 20 miles off San Diego by the auxiliary tug *Koka*. This 29 May dive was the deepest of the season's descents; Mackenzie rode the submersible down to 4,200 feet to measure sound velocities. Piccard and Mackenzie submerged at 1344 and returned to the surface at 1738, having spent 2½ hours on the bottom.[31] Mackenzie reported: "This is the first device we have had to take precise measuring instruments to extreme depths. What we are trying to do is to give the Navy the best basis for calculating underwater sound speed that can be had." Jacques stopped the descent twice, at 500 and 1,200 feet, for sound velocity measurements.[32]

Jacques too described the dive:

> We had just finished hovering at 85 fathoms (510 feet) and were slowing, settling deeper, when we heard a sharp implosion, followed by

> bubbling sounds. Then, a little later, a second implosion startled us. Under the high sea pressure, this kind of collapse can be as violent as an explosion. Such implosions can be disquieting. The force could be sufficient to rupture the skin of the float with dire consequences. We tested our equilibrium. We were not losing gasoline, so we continued our dive to the bottom at 700 fathoms (4,200 feet). Back in port, we found the cause of the implosions. The underwater camera fitted beneath the float had failed. The manufacturer had assured us that the camera had been tested to 18,000 feet.[33]

The next *Trieste* dive was on 5 June to 770 feet, with Rechnitzer; the fleet tug *Tawasa* towed the craft to the dive spot.[34] The dive was routine, but the return to Point Loma was not. The *Tawasa* had both the bathyscaph and the 17-foot line-handling boat in tow. The small boat broke away. The bathyscaph's float rammed the boat, which was forced underwater and struck the sphere. Upon return to port the *Trieste* was thoroughly inspected; no damage to the sphere was apparent, but NEL officials decided that this would provide a good stopping point in the diving schedule.[35]

The *Trieste* was cradled at NEL on 18 June for the modifications that would be required prior to the deployment to Guam for Project Nekton. Work continued through September to lengthen the float, increase the size of the ballast tubs, and mount the new Krupp sphere.

Nekton Approved

Project Nekton was an Office of Naval Research initiative that had been in preparation for more than a year. Once the *Trieste* was under the operational control of NEL the details of Nekton were developed, with Rechnitzer and Walsh the primary drafters. NEL's work had elicited indications of interest at the highest levels of the Navy Department, but Project Nekton still had no formal approval from the Navy's leadership.

Walsh traveled to Washington to obtain that approval. The young lieutenant had not only organized the maintenance and operations of the bathyscaph, but worked closely with Rechnitzer to design a program to put the *Trieste* to its ultimate test: the 36,000-foot Mariana Trench. His tour of the halls of the Navy's leadership started at the Office of Naval Research, where he was passed up the chain of command, office by office, until he stood before the Chief of Naval Research, Rear Admiral Rawson Bennett. From there he was sent to the Bureau of Ships (BuShips), eventually reaching Rear Admiral Ralph K. James, its chief. From there he was sent on to OpNav, the office of the Chief of Naval Operations. A preliminary meeting at the Pentagon with representatives from ONR, BuShips, and the Undersea Warfare Development Division in OpNav, headed by Captain Charles B. (Swede) Momsen Jr., achieved agreement as to purpose, but no one was willing to make a final decision.[36] Walsh recalled:

> I was put on sort of a "conveyor belt"—the commanders passed me to the captains who in turn passed me to the admirals. And even within the ranks of the flag officers, I still moved upward rather rapidly without any real decisions being rendered. After not too much delay, I found myself in front of Admiral Arleigh Burke [the Chief of Naval Operations]. I knew I would get an answer here![37]
>
> Admiral Burke asked me who would be making the dive. I replied that I, as the Navy commander of the *Trieste*, and Dr. Rechnitzer our chief scientist, would be on board. Furthermore, I told him that Lt. Shumaker would be in charge of topside activities. Burke then said, "I want you to tell Shumaker that if the *Trieste* does not come back, then you, Walsh, are the lucky one because I will have Shumaker's balls."[38]
>
> He agreed to the project. However, he directed that our intentions not be publicized until we were successful. He did not want a high-visibility flop. I called NEL and gave them the good news.[39]

Admiral Burke's demand for secrecy was caused by the Navy's embarrassment over the failures of the first two Vanguard satellite launch attempts, on 6 December 1957 and 5 February 1958, which still were deeply felt at the Pentagon in view of the Soviet success in orbiting *Sputnik* in October 1957.

The new Krupp sphere attached during the overhaul of the *Trieste* added several tons of weight, which necessitated adding thousands of extra gallons of "avgas" for buoyancy and proportionately more ballast to compensate for the additional buoyancy. Professor Auguste Piccard arrived in San Diego on 5 July for a week of consultations on the modifications.

The expansion of the float was undertaken by the Naval Repair Facility in San Diego. The "fix" added two four-foot sections and removed two internal bulkheads to increase gasoline capacity by 5,000 gallons to 34,000 gallons. During this work all welds were inspected using X-ray and visual inspection. These inspections found that 80 percent of the welds in the float would not pass basic U.S. Navy standards and that about 20 percent were outright bad. The bad welds were cut out and rewelded to Navy standards.[40] Approximately 70 percent of the electrical wiring was replaced with new and improved wire and connectors.[41] With the larger float the ballast tubs needed greater capacity, and each was duly enlarged from 11 to 16 tons. Four tubs were constructed and shipped to Guam to allow replacement on site should there be problems or accidental release of tubs. The enlarged float then was sandblasted, inspected, and repainted. Meanwhile, the *Trieste* team was busy installing the new interior equipment in the Krupp sphere.[42] The Terni sphere was carefully set aside and protected as the jewel that it was.

Formal authorization for Project Nekton was in hand by the first week in September 1959. NEL had been coordinating informally with the Commander, Naval Forces Marianas since July, and with formal authorization the laboratory could finalize the services and support needed from the Guam facility for the upcoming operation.[43]

On 9 September 1959, the bathyscaph was "launched" at NEL. A much-abbreviated program to test the new sphere and float was arranged: a single harbor dive to test controls and a single deep-ocean dive to test integrity. Walsh noted, "Due to the short time remaining, it was decided that instead of proceeding as before and training operators for the *Trieste* before departing for Guam, that one test dive would be made at sea and then the craft would be taken out of the water, disassembled and sent to Guam."[44] The test dive in San Diego Harbor to 62 feet with Jacques and Shumaker took place on 11 September. No serious problems were encountered, and an ocean test dive was scheduled for 15 September, with Jacques and Rechnitzer (Dive #58).

This open-ocean dive, supported by the salvage ship *Gear* (ARS 34), descended only to 590 feet. The dive revealed a leaking hull connector. Jacques decided that it was a minor problem that could be fixed before the dive off Guam. It was "go" for disassembly and packaging the 150-ton deep submergence craft, its cradle, tons of iron-shot ballast, and the support equipment—a total of 350 tons of cargo.[45]

Departure for Guam

On 2 October 1959, the *San Diego Union* reported, "The *Trieste* will leave for deep-sea tests off the Marianas Islands on board the Grace Line freighter SS *Santa Mariana*, chartered by the American President Lines. . . . Dr. Andreas B. Rechnitzer, the NEL scientist in charge of the bathyscaph project, said the *Trieste* provides 'unique' study possibilities because it can carry observers directly to the research area." Three days later the *Santa Mariana* departed San Diego carrying the *Trieste,* the support equipment, and supplies. Two enlisted men from the *Trieste* crew were on board.

On 28 June 1959, a news item from Washington, D.C., reported that the research ship *Stranger* out of San Diego "was plying west across the Pacific for a two-year expedition in the Gulf of Thailand. On board were scientists from the Scripps Institution of Oceanography, equipped to probe deeper than ever before into the secrets of the seas. And the bathyscaph *Trieste*, after a day run at sea, was slung in NEL docks for further adjustments to permit exploration at greater depths than ever before."[46] There were other indications in the press as well that "something" was about to happen.

"Hints" such as these could be recognized by people with knowledge of the Navy's acquisition of the bathyscaph, and such individuals could guess about the probable intended use of the *Trieste*.

6 The Ocean's Deepest Hole

You have always believed in me and my destiny.
—Jacques Piccard

Not until 1952 was the deepest point in the ocean recognized as being within the arc of the Mariana Trench. The British research ship *Challenger* first recorded extreme depths in the vicinity in 1875, reporting 26,850 feet. Soundings in 1899 by the U.S. Navy collier *Nero* (AC 17) south and east of Guam recorded 31,694 feet, at a point that later became known as the Nero Deep. It was not until another British survey ship, HMS *Challenger II*, recorded 35,760 feet about 220 miles southwest of Guam on 14 June 1951 that the Challenger Deep was documented. The location was named for both *Challenger* survey expeditions.

Four deep-ocean trenches in the Pacific were at one time or another identified as having the deepest point of the world's oceans: the Kurile Trench (north and east of Japan); the Mindanao Trench (east of the Philippines); the Tonga Trench (east of the Tonga Islands); and the Mariana Trench (south of Guam). The Capricorn Expedition in 1952, conducted by the Scripps Institution of Oceanography's research ship *Horizon*, took precise echo-ranging soundings of the Mindanao, Tonga, and Mariana trenches and obtained Soviet soundings of the Kurile Trench. At that point the Mariana Trench was definitively identified at 35,810, plus or minus 30 feet—the deepest point in the world's oceans. Thus, the Challenger Deep was selected for the 1960 dive when the U.S. Navy decided to put the *Trieste* to its ultimate test.[1]

In subsequent years repeated surveys by more sophisticated methods and equipment have revealed the Challenger Deep to be a depression measuring about seven miles east to west and about one mile in width, with gradually sloping sides. In October 2010, the U.S. Navy survey ship *Sumner* (T-AGS 61) recorded a maximum depth of 36,070 feet.[2]

In 1952 Jacques Piccard, lecturing in Lausanne, Switzerland, stated that "man would someday assault the Mindanao Deep off the Philippines," which he then thought was the deepest point in the oceans. After the lecture a former instructor of his from the University of Geneva, Professor André Rey, approached him: "Ah, but now you must go to the Mariana Trench where the British oceanographic ship *Challenger II* has just discovered a spot nearly 36,000 feet deep."[3] Of course, Piccard had no idea that the man who would make that assault was himself. However, the idea was firmly planted.

Diving off Guam

The *Trieste* advance team, consisting of Lieutenant Walsh, Lieutenant Shumaker, and Chief Michel, arrived at the U.S. naval base on Guam by aircraft on 10 October 1959 to prepare for the arrival of the bathyscaph. Rear Admiral William L. Erdman, Commander, Naval Forces Marianas, extended full and enthusiastic cooperation to the Nekton team and opened the door for unlimited support from the ship repair facility on Guam.

On 22 October, the freighter *Santa Mariana* arrived and offloaded the *Trieste* and support equipment in Guam's Apra Harbor. The civilian members of the *Trieste* team arrived shortly thereafter and, working long hours for seven days a week, had the *Trieste* assembled and waterborne 13 days later. The first test dive was made on 4 November, to a depth of 70 feet inside of Apra Harbor.

Progressively deeper dives began on 10 November, three miles off the harbor entrance, with a descent to 4,900 feet with Piccard and Rechnitzer in the sphere. On 15 November, Piccard and Rechnitzer took the *Trieste* to a world-record depth of 18,150 feet some 30 miles southeast of Guam. Because of heavy seas, the tow to the site took almost 24 hours, and the dive was then further delayed by difficulty in uncoupling the tow from the auxiliary tug *Wandank* (ATA 204). With the escort ship *Lewis* (DE 535) present in support, the dive commenced at 1015 and gained the bottom at 1310.[4] Their ascent began after 12 minutes on the ocean floor. At 1548, when the bathyscaph was 50 feet beneath the surface and rising, "two violent explosions" occurred without warning.[5] Piccard characterized the noise as a one-two volley from a rifle. Before he or Rechnitzer could react the craft was on the surface. Piccard and Rechnitzer had reached a world-record depth on this dive, exceeding by almost a mile the record of 13,290 feet achieved on

15 February 1954, more than five years before, by Houot and Willm in the *FNRS 3* (see chapter 3).[6]

A Mission-Critical Casualty

Back in Apra Harbor a visual inspection revealed the problem: the center ring of the Krupp sphere had broken free of its epoxy bond, and the center section had shifted about one-eighth of an inch out of alignment. The sphere would have to be minutely inspected and repaired. Project Nekton had suffered its first mission-critical casualty.

This was not only the first critical casualty, but the first chink in Piccard's confidence in the bathyscaph's reliability and safety. The cause of the epoxy failure was basic physics. The surface temperature of the water being in excess of 80° Fahrenheit and the ocean bottom at about 35° Fahrenheit, the caps and rings expanded and contracted during descent and ascent but at different rates and at different times, stressing the epoxy glue ultimately to failure and allowing the center ring to slip.

On 18 November the *Trieste* was de-fueled and lifted onto its cradle at the Guam repair facility for inspection. Both of the center ring's joints had failed completely. Several solutions were considered, including airlifting the sphere back to Krupp in Germany, shipping it to Japan for a complete refurbishment, or handling the problem on Guam. Although it was impossible to determine without a complete disassembly if the joints had developed any corrosion, Piccard agreed to a Navy proposal that the sections be realigned with hydraulic jacks—without regluing the joints.

There is a tradition in the Navy that chief petty officers "can do anything," and Chief Machinery Repairman John Michel was about to reinforce that tradition. Michel recorded his work on the Krupp sphere:

> Plan One was to remove the windows in the door and front view port, replace these with aluminum cones that I machined at the Ship Repair Facility, put a spindle through each with nuts, washers, and a 60-ton hollow bore jack, in line like a shish-kebob.
>
> Jacques, although a sharp individual, is not an engineer. He would have had a fit or a heart attack if he could have seen what my plan was. I examined the front section of the sphere to locate the most out-of-place

> spot and marked it with a pen. This was some 3-mm out of position. I had [a civilian worker] practice lurching [a forklift-mounted wooden ram] and then had him line up with my X mark. When we were both confident, with him hitting and me with the pump on the jack, we had a go at it. After some repositioning and three or four hits, the sphere section was in almost perfect alignment once again.
>
> My next chore was to draw up a banding system to affect a mechanical method for holding the sphere sections together. This was fabricated by the [repair facility] and then installed by Giuseppe Buono.[7]

Michel located the only 60-ton hollow bore jack in the western Pacific at his previous duty station, the destroyer tender *Prairie* (AD 15), then docked at Yokosuka, Japan. The jack arrived at Guam by air within two days, a gift from his old shipmates.[8] A rubber gasket was affixed with Permatex to the exterior joints and held in place with steel bands, to reduce "weepage," and the sphere was ready to test.

The *Trieste* was readied, launched, fueled, ballasted, and made a 60-foot harbor dip on 14 December, with Piccard piloting and with Jack Cawley, an instrument designer with the General Atomic Division of General Dynamics, observing.[9] The sphere appeared to be tight, but weeping was noted at the joints.[10]

A World-Record Dive

On 18 December, Piccard and Walsh made a dive to 5,700 feet to test the new support bands and joint gaskets. Mild weeping continued, and evidence of a corrosion problem at the joint surfaces appeared. At that point a long-term problem could be foreseen, cumulative corrosion at the joints. Piccard, displaying a new sense of urgency, insisted in getting on to the Deep Dive as soon as possible.[11]

Nevertheless, Piccard departed for Christmas vacation in the United States immediately after the 18 December dive. Rechnitzer and Walsh decided that they would take the opportunity for some solo diving in the *Trieste* within Apra Harbor—the first independent dives each had ever made. On 29 December, Walsh and Philco technical representative Dennis C. Jensen (Dive #64), Shumaker and Chief Harlan DeGood (Dive #65), and Walsh and Rechnitzer (Dive #66) made dives to 100 feet.[12] The next day, Walsh and Chief Michel (Dive #67), and Shumaker and Rechnitzer (Dive #68) completed the pilot training dives in the harbor.[13]

In San Diego, early in the new year of 1959, Dietz accompanied Piccard to the office of Dr. Franz Kurie, technical director of NEL, to discuss the bathyscaph program and the progress of Project Nekton.[14] One of the concerns was the question of who would be the principals for the Deep Dive—to no resolution. On 29 December, six weeks after the *Trieste* had reached a world-record depth of 18,150 feet off Guam and ten days prior to its 23,000-foot dive into the Nero Deep, the *San Diego Union* ran an article:

> **Navy Bathyscaph to Attempt 7-Mile Plunge in Pacific**
> In a December San Diego newspaper article, Dr. Franz N.D. Kurie, NEL's Technical Director said the attempt to hit the bottom of the Challenger Deep, 35,000 feet down, will be made "as soon as we can." The plunge is scheduled for January or February "because those are usually the best weather months in that part of the Pacific," he said. "It is not in any way a stunt," declared Kurie. Neither Kurie nor Piccard would set a definite date for the deepest dive. Both emphasized that a decision to descend to the bottom would depend on analysis of preliminary findings.
>
> "The Navy said [that] Rechnitzer would make the bottom plunge in the bathyscaph's gondola with one other member of the team. The other 'hydronaut' may be Lt. Don Walsh, 27, a Navy submariner, Lt. Lawrence A. Shumaker, 27, another Navy submariner, or Piccard."[15]

This news violated the Navy Department's specific prohibition against announcing the planned Deep Dive, and appears to have been a device used by Kurie to get Rechnitzer approved as a participant in the dive.

When Piccard arrived back on Guam on 2 January 1960, Walsh records that "he was not happy with our self-qualification during his absence. His contract obligated him to instruct the Navy crew in operations and maintenance, yet it was becoming increasingly apparent that this would be a self-taught course."[16]

The Nero Dive

On 8 January 1960, Piccard and Walsh dived to 23,070 feet into the Nero Deep, about 60 miles southeast of Guam. This Dive #69 would be the final test dive prior to the Deep Dive, and it captured the second world record of Project Nekton. The Navy tug *Wandank* had departed Apra Harbor with the *Trieste* in tow on 6 January in company with the escort ship *Lewis*. During the long tows—two days to the dive site, two days returning—the *Lewis* provided berthing for the *Trieste* personnel as well as support at the dive site. The *Lewis* employed her fathometer and precision depth recorder to find an appropriate 24,000-foot location for the Nero Deep dive.

The tow to the dive site encountered heavy swells, and by the morning of 8 January the *Trieste* was facing a sea state of four, the highest yet during a bathyscaph tow. Although the weather and seas made conditions for the dive only "marginal," there was little to suggest a near-term improvement. Inspection of the bathyscaph prior to diving showed that the after flooding valve was inoperative; the deck railing and metal protective cover of the maneuvering tank had been torn away, and the telephone alert buzzers did not function—all damage caused by the tow through high seas. On the positive side, a new and much more capable underwater acoustic telephone had just been installed on the *Trieste*. This, it was hoped, would solve the chronic problem of poor communications between the surface ships and the *Trieste* while submerged.[17]

But the after flooding valve casualty was critical, and the only work-around would require someone to release the forward flooding valve while on the surface, then move to the aft valve and manually hold it open until the *Trieste* was submerging. That person would have to swim off the bathyscaph and be picked up by boat; a volunteer was required. The only man who could have been trusted for the task did volunteer—the indomitable Giuseppe Buono. The shark-infested water was no deterrent to a man as committed to the project as he was.

Descent commenced at 0954 on 8 January and proceeded without incident until 19,500 feet, when two implosions were felt. Piccard instantly dumped a ton of ballast to arrest descent so that the situation could be assessed. No critical instruments or controls had been affected, and the dive continued. At 23,100 feet a third implosion occurred. The descent continued, notwithstanding. (These implosions later were determined to be collapses of 0.75-inch welded tubing used as deck stanchions on the float, tubing that repair personnel had failed to drill for free-flooding during dives.)

The *Trieste*'s fathometer did not detect the ocean floor until it was 120 feet above the bottom; the fathometer's sensitivity had been degraded during the tow. Piccard immediately dumped ballast from both the fore and aft shot tubs and succeeded in arresting the descent 48 feet above the bottom. Neither Piccard nor Walsh visually sighted the bottom, and efforts to descend farther by releasing a bit of gasoline from the maneuvering tank failed. They returned to the surface without further incident, but the *Trieste*'s sluggishness during ascent indicated that gasoline had in fact been lost, and more ballast was dropped.

Subsequent inspection of the gasoline release valve revealed not only that it had been damaged during the tow to the dive site but also that it had been slowly bleeding gasoline throughout the dive.[18] A faster gasoline leak could have required emergency dumping of ballast or external equipment.

This dive demonstrated that Piccard no longer had a comprehensive and intimate knowledge of the rebuilt *Trieste* and also that long tows in the sea states to be expected near Guam during this season could critically damage the bathyscaph. Nevertheless, the primary, specific purpose for the Nero Dive had been to prove the bathyscaph ready for the Deep Dive, and Piccard declared it ready.[19]

Back on Guam, messages of congratulations flooded in from senior Navy officials.[20] It had become obvious that an assault on the Challenger Deep was in the offing, and newspaper and magazine reporters were making their way to Guam to get the story.[21] To be taken on board the diving support ships for the Deep Dive, the press corps agreed to hold sending in their copy until the Deep Dive was completed.[22]

Who Will Make the Deep Dive?

Project Nekton was operating under a curious organizational structure. Dr. Rechnitzer of the Navy Electronics Laboratory was the Nekton project director, under whom the officers-in-charge of the *Trieste*, Walsh and his assistant Shumaker were to manage, maintain, and operate the bathyscaph. Dr. Franz Kurie, technical director of NEL, was on scene for part of the period (January 1960), and the chief of the Bureau of Ships, Rear Admiral Ralph K. James, was on Guam just prior to the Deep Dive. Serving as a consultant with no legal or organizational responsibility (or authority) was Jacques Piccard, the previous owner and the only person with extensive experience as pilot of the bathyscaph.

Notwithstanding any claim by him or anyone else to the contrary, Piccard was making all significant operational decisions and piloting every ocean dive.

There were three men in contention for the two positions in the sphere for the Deep Dive. Walsh later related (referring to himself in the third person), "The original intent of BuShips and the Navy Electronics Laboratory since their first approval of Project Nekton for the *Trieste* was that an American scientist and an American Naval person should make the initial Deep Dive. The fact that Mr. Piccard would *not* be making the first dive was made quite clear to Lt. Walsh on his reporting to NEL in early 1959. In addition, Mr. Piccard himself was, of course, informed of this."[23]

The original Nekton plan provided for an incremental workup of several preparatory dives, then three dives into the Challenger Deep: the first with Rechnitzer and Walsh, the second with Piccard and Dietz, and the third with Shumaker and NEL acoustic scientist Mackenzie.[24]

Two Destinies

Jacques Piccard was not a man used to failing, surrendering, or withdrawing from the field in an unwinnable battle. He had seen his destiny opening before his eyes during his numerous conversations with Dietz, Lill, and Maxwell since 1956. Now, on the eve of the deep dive, he was informed that he would not be making the historic sortie into "inner space."

A man of extraordinary intellectual capacity, breadth of talents, and strength of will, Piccard would not be deflected. He recorded his emotions:

> I had many deep and personal reasons for wanting to make that dive to maximum depth off Guam. After all, my father had invented the bathyscaphe. As for myself, I had devoted a decade to make that dream a reality. *I felt destined for this culminating plunge.* There was also the compelling fact that I was technically responsible for the *Trieste* and I was still the only fully qualified pilot. Neither Walsh nor Rechnitzer had ever piloted the *Trieste* in open sea, let alone under the most difficult of all possible conditions. I felt my greater experience with the bathyscaphe increased the chances of a successful dive. By the time we were ready for the Big Dive, the chance of making more than one dive had become vanishingly small [emphasis added].[25]

Piccard realized that claiming some intrinsic priority as the son of the *Trieste*'s creator would carry no weight. He proceeded to fortify his position on two fronts: during the renegotiation for his second-year consulting contract with ONR in the autumn of 1959, Piccard had had a clause inserted stating that he had the right to participate in "any dive that presents special problems." This contract was with ONR and was not presented to either NEL or BuShips for review.

Piccard's backup plan was to ignore his contractual obligation to train Navy personnel to pilot the *Trieste*, or perhaps more nicely stated, to implement that training in a manner favorable to his destiny to make the Deep Dive. With the exception of five shallow harbor dives in Apra Harbor during the Christmas holidays, *Piccard had piloted every dive*—denying Walsh and Rechnitzer the opportunity to obtain a check-dive to qualify as bathyscaph pilots. Piccard could and eventually did claim that no one else but he was qualified to pilot the *Trieste*.

The penultimate Deep Dive on 8 January brought the matter to a head. Walsh had asked Piccard to allow him to pilot the Nero Deep Dive under instruction, and Piccard had agreed. However, as soon as the *Trieste* submerged, "Piccard's agreement vanished and Lt. Walsh did not pilot on this ocean dive."[26] It had become obvious that Piccard was not going to accede willingly to being bumped from the Challenger Deep Dive. Shortly thereafter, Rear Admiral James of BuShips arrived on Guam. Piccard recalled:

> I was working topside on the *Trieste* one morning when I was called aside [by Franz Kurie]. I was not scheduled for the Big Dive! I was incredulous, this could not be true! I pressed for more information. It was true, it was official, and there was no error. Walsh and Rechnitzer will make the dive. I argued to no avail.
>
> Finally, I invoked a clause from my consulting contract with the Navy. "My contract states that I reserve the right to make any dive that presents special problems. Wouldn't you say that a dive to the bottom of the Challenger Deep presents a special problem?" Did I have a copy of my contract? I did.[27]

The Final Decision

Kurie and Rechnitzer were stunned. They believed that the approval of the written Project Nekton plan by the Chief of Naval Operations ensured that

Rechnitzer and Walsh were "in solid" for the Deep Dive. On that assumption, Rechnitzer had made contact with *Life* magazine and offered his story for the princely sum of $100,000. Not only was Rechnitzer's place in history being threatened but so was a very lucrative writing contract.[28] Dietz would comment in a "Memorandum for Posterity" that it was curious that Kurie and Rechnitzer expressed ignorance of this contract clause, as he "had brought this fact to their attention many times over the previous six months." It seemed to be a purposeful loss of memory.[29]

Messages flew, and eventually the Office of Naval Research confirmed that the intent of this clause was to allow Piccard to make the Deep Dive. The provision had been written when it appeared unwise and premature to mention Project Nekton, thus it had been obliquely worded.[30] On 11 January Kurie received a message from the Chief of Naval Operations directing that Piccard and Walsh make the Deep Dive and informing him that ONR had concurred with this plan. Kurie temporized and did not pass this information on to Piccard or Dietz. Instead, he obtained Walsh's agreement to support a plan that would put Rechnitzer in uniform with his reserve rank of lieutenant commander and proposed a Piccard–*Lieutenant Commander* Rechnitzer dive.[31] This proposal was sent off to Washington on the 11th.

On 14 January, Piccard requested a meeting with Kurie, Rechnitzer, and Dietz. Both Kurie and Rechnitzer indicated that they had heard nothing from Washington and were still waiting for a final determination.[32] The meeting erupted into an intense argument, and it became obvious that bad blood had developed between Rechnitzer and Piccard. Later, at a reception hosted by Admiral James, Dietz pulled aside Walsh and Captain Norbert Frankenberger, the admiral's aide, to describe the impasse—only to learn from Walsh that the CNO's decision had been in hand since the 11th.[33]

The Navy wanted the officer-in-charge of the *Trieste* to make the dive, not a scientist temporarily in uniform. Walsh, a personable, highly intelligent, submarine-qualified officer, offered, from the public affairs point of view, a far better "look" for this dive than a reserve officer/scientist. Rechnitzer would have to stand aside.

Piccard had proven to be as formidable in gamesmanship as he was in so many other aspects of life. He had gone head-to-head against the Navy Electronics Laboratory, the Bureau of Ships, and the Chief of Naval Operations, and, with the aid of the Office of Naval Research, he had won. In an interview in San Diego, Piccard stated: "It [i.e., the Deep Dive] was something I had hoped to do since I was 24, when I first got interested in building the bathyscaph with my father."[34]

Approach and Localizing

A small task unit consisting of the *Lewis* and the *Wandank* towing the *Trieste*, cleared Apra Harbor at 1400 on 19 January 1960. The *Lewis* put on speed to arrive in the dive area early to pinpoint the deepest hole acoustically prior to the *Trieste*'s arrival. Walsh related: "Accurate position information was available from previous surveys; however, the reconciliation of the position information with the navigational accuracy of the *Lewis* made it necessary for the *Lewis* to survey the area once it arrived in the area of the known positions."[35]

With Rechnitzer personally supervising the survey, by the afternoon of 21 January the *Lewis* had expended her entire load of 300 explosive sounding charges and still had not determined the location of the deepest point. Lieutenant Commander Daniel L. Banks Jr., commanding the *Lewis*, radioed to Guam asking for additional sounding charges be sent out to complete the survey. At 1800 on 22 January, the large harbor tug *YTB 408* transferred the explosives to the *Lewis* by highline. By midnight of 22-23 January, the *Lewis* was back at work, banging away at the ocean depths.[36] On the morning of 23 January, after more than 800 explosions had kept the crews of both ships, *Lewis* and *Wandank*, awake throughout the night, Rechnitzer reported, "Deepest point located."[37] The *Lewis* led the *Wandank*, with the *Trieste* under tow, in a procession up the long axis of the depression and marked the exact center with a dye marker and floating flares.[38]

The *Wandank*'s tow had not been without incident: In late morning of 20 January, Lieutenant A. W. (Bill) Cooley, the tug's commanding officer, reported that the towline had broken. By 1300 he had been able to re-rig the tow and again was under way. Shumaker had been on the *Trieste* and reported that no topside damage was visible.

Dive or Scrub?

Sunrise on 23 January at the dive site found the sky overcast, the weather hot and humid, and the seas heavy with 25-foot swells, but only moderate winds.[39] The morning was filled with concerns. Piccard recorded his thoughts:

> The seas had not moderated perceptibly. If anything, the trade winds had freshened. Conditions didn't look auspicious. Should everything be risked under these circumstances? This was the decision I'd have to make soon. Lieutenant Shumaker and Giuseppe Buono joined me on the fantail [of the *Wandank*]. A try was made at launching the project boat. It was impossible. Twenty-five-foot waves forced us to abandon the attempt. A rubber raft was pressed into service. It was a touch-and-go operation, but finally we managed. We were seaborne. Lieutenant Shumaker, Buono and I. . . . The sight that met my eyes when we boarded the wallowing bathyscaph was discouraging. Broaching seas smothered her. The deck was a mess. It was apparent at once that the tow from Guam had taken a terrific toll. A hasty inspection revealed that the surface telephone had been carried away. The tachometer that measured our diving rate was badly damaged and inoperative. The vertical current meter was dangling by a few wires. I was worried.[40]

On board the *Lewis*, Rechnitzer told Walsh, "Son, we've really found you a hole—33,600 feet." (Rechnitzer was not quite seven years Walsh's senior; Walsh considered him his mentor.) Then Rechnitzer pleaded, "Just see one animal down there. That's all it takes, just one of anything." It was possible that this dive would settle one of the oldest and most rancorous questions of oceanography: Was there a depth below which complex life could not survive? The answer to that question might determine whether the deep-ocean trenches would be used for the long-term disposal of radioactive and other hazardous waste material.[41]

The *Lewis* was able to get a motor whaleboat into the water with a volunteer crew to take Walsh over to the *Trieste*. Transferring Walsh and Chief Michel from the *Lewis* to the motor whaleboat proved to be the most dangerous event of the entire operation. The chief recalled:

> Our . . . boat was having a hard time trying to come alongside to pick us up. At times it looked like they were going to go over the gunnel and come aboard, then they dropped down as far as the exposed bilge keel. We could not just straddle the rail or climb outboard, as the boat might crush our legs. It was going to be up and over in a carefully timed

> jump. The boat crew had spread out about 20 kapok life jackets so we could make a more or less soft landing. I jumped first. Timing was not too good. I fell the full vertical distance and hit my right shin on a seat edge. My shin was bleeding nicely as I arranged some kapoks on the motor cowling.
>
> Don [Walsh] was not looking very happy so I told him to jump when I cued him, and I would steer him to a soft spot. He jumped on cue and because his vertical position was off, I hit him with kapoks [life jackets] in hand and he landed on the motor cowl. Unhurt. I managed to get back on board, where I met Andy Rechnitzer who escorted me to sick bay to have my leg bandaged.[42]

When Walsh scrambled on board the *Trieste*, he too did not like what he saw: "What do you think, Jacques?" Piccard responded: "I'll know better after I've checked below." Climbing down into the sphere, Walsh checked out the electrical systems and the electromagnets. All circuits were in order.

Piccard mused, "It is all very well for a man seeking adventure to take chances. I wasn't looking for adventure. I wanted a successful and uneventful operation. I wanted to leave nothing to chance. I made the decision. We would dive."[43]

Buono followed Walsh and Piccard down the entry tube. Then all three men—Buono from outside, Piccard and Walsh from inside—lowered the 350-pound hatch into place and tightened the inside screws until grease squeezed out of the sphere's machined hatch socket. Once all was ready inside, Walsh signaled with a flashlight through the hatch porthole to Buono, crouched at the bottom of the entry tube where it turned to the horizontal to join the sphere. Two minutes later Buono was gone and the entry tube had been flooded.

Shortly after Walsh signaled Buono to flood the entry tube, a strange event occurred. The Piccards—father and son—had had a long relationship with Rolex SA, the watch company. As he climbed out of the entry tube, Buono secured a one-of-a-kind watch (the Rolex "Deep Sea Special" prototype) to the ladder. The unique watch had been especially made by Rolex for the dive, forwarded to Jacques Piccard, and kept hidden from Navy officials, who would have forbidden commercial use of a Navy activity. When the *Trieste* surfaced, Buono would surreptitiously retrieve the watch, and Jacques Piccard would return it to Rolex to be used as part of a special "Deep Diver" advertising campaign.[44] The watch is now in the possession of the Smithsonian Institution in Washington, D.C.

When the entry tube was flooded, Shumaker joined Buono atop the float to open the fore and aft topside ballast valves simultaneously, flooding the ballast tanks and providing the *Trieste* with negative buoyancy. They then scrambled onto a rubber raft as the bathyscaph slipped beneath the waves. The time was 0823 on Saturday, 23 January 1960.

"Cancel Diving; Come Home"

On board the *Lewis* a drama was in progress that might have ended the day's effort with a simple command by underwater telephone. Chief Michel was a witness:

> Andy [Rechnitzer] and I were now in search of a cup of coffee with not much to do but wait. After some time had passed, a ship's radioman found us and said he had a message for Dr. Rechnitzer and handed it to him. Andy read it, had no visible reaction, and then handed it to me:
>
> From: CO and Director US Navy Electronics Laboratory
> To: Director U.S. Navy Project Nekton
> "*Cancel Diving. Come Home.*"
>
> Andy said "Let's find some more coffee," and we made our way back to the mess. After a short time, we walked to the stern. We discussed the project and other things. Then he said, "Let's go to the radio room. I have to check in." We found the radioman and he [Andy] said "Send a message":
>
> From: Director Project Nekton
> To: CO and Director U.S. Navy Electronics Laboratory
> "Trieste *now passing 20,000 feet.*"[45]

It was a potentially career-destroying decision for a man who had just been edged out of a seat in history. Rechnitzer was a team player, a man of courage and rectitude. His was a bold decision.

7 The Deepest Dive

This dive was in the main a technical test. It has proved that a means has been created enabling the oceanographer to extend his research to the greatest depths in perfect safety.

—Jacques Piccard

On this historic dive Piccard had given Buono special instructions because of the loss of the topside telephone: "When I have closed the hatch, you may open the entrance-tube valves and proceed with normal operations. If at the last moment something doesn't go well, I will turn the propellers, and you will know that we must give up the dive." With this simple code all was prepared for the Deep Dive.[1]

The bobbing and weaving motion of the *Trieste* became less and less the deeper it dived, and at 80 feet nearly all of the surface wave action was left behind. The depth gauges and algae in the water were the only indications of motion, and that motion was directly down. As the bathyscaph exited the noisy surface layer, at about 250 feet, communication by underwater telephone became possible. Both the escort ship *Lewis* and the tug *Wandank* had underwater acoustic or UQC "telephones." Walsh was maintaining an audio log and dictating a running commentary throughout the dive, while Piccard was penning a log; both men were recording depth, amount and timing of ballast drops, temperature of the gasoline, and other data, including piloting decisions and times. Since the diving-rate tachometer was inoperative, Piccard was determining the speed of descent by the time interval between depth readings.

The first stop on this "elevator ride" was the thermocline, a depth just below that at which wave action mixes warm surface water with cool deeper water and, suddenly and sharply, the ocean temperature drops. The denser, cooler water reduced the *Trieste*'s displacement; thus the rate of descent slowed until the craft stopped and sat at the thermocline. To descend farther required that some of the gasoline in the maneuvering tank be released. This usually occurred once on each dive, somewhere between 250 and 500 feet. On this dive—unique in Piccard's experience of 65 dives—the rough sea conditions and internal waves had mixed the surface water to deeper levels, apparently forming four separate thermoclines, the main one at 340 feet, a second at 370 feet, a third at 420 feet, and the last barrier at 515 feet.

It took more than one-half hour to penetrate the thermoclines.[2] Each layer required a release of gasoline. When the last layer was breached the bathyscaph was "heavy" and gradually picked up speed to close to three feet per second (two-plus miles per hour). Even at that rate the bottom was still more than three hours straight down. The bathyscaph twice passed through layers of phosphorescent plankton, at 2,200 feet and at 20,000 feet, as it dropped farther into the depths.[3]

The First 23,000 Feet

It was starting to cool markedly within the sphere, and the two hydronauts changed into dry, warm clothing. Walsh explained, "It is quite an operation to see—two grown men changing clothes in a space 38 inches square and only five feet, eight inches high. Then we ate our first chocolate bars. On the last dive he had brought the [Swiss Nestlé] chocolate, so I told him that I would buy lunch for this one, and I had 15 bars of Hershey chocolate put aboard before we left Guam."[4]

Walsh's recorded commentary fails to mention whether the Swiss pilot appreciated the American chocolate. Piccard had secreted ten bars of Nestlé chocolate on board for his own consumption—just in case.[5]

At 5,750 feet they received a call from the *Wandank* on the newly installed underwater telephone, an amazingly loud and clear signal. Giuseppe reported, *"Tutto bene, signor."* All was well.[6]

At about 10,000 feet an "old friend" appeared—a small leak through one of the stuffing tubes by which electrical wires passed through the hull of the sphere. As in the past, the leak started at about 10,000 feet and self-sealed at about 15,000 feet. Almost simultaneously voice communications with the surface were lost. Until they were regained, Piccard and Walsh would, as planned, communicate by sending coded series of acoustic tones to the surface. The code was simple: even numbers of tones always indicated good news: two tones signaled "All's well," four meant "On the bottom," and six "On the way up." Odd numbers were for trouble: three tones signaled "Problem requiring a non-emergency surfacing," five tones "Emergency surfacing."

A second leak started at about 18,000 feet, and the slow seepage remained steady throughout the remainder of the dive. The leakage required them to move the Dräger cans—which chemically removed carbon dioxide from the atmosphere—out of the bilge.[7] They made note of the passing of 18,150 feet, a record set by Piccard and Rechnitzer on 15 November, and the 23,070-foot record set by themselves just 15 days before.

Once again they were entering depths never before visited by humans.

The two men were sensitive to the fact that they were diving into a "hole" only a mile in width, with no knowledge as to the character or the proximity of its walls. Given unknown and unknowable deep-water currents, there was a possibility of their being pushed into a wall. At 27,000 feet Piccard dumped enough ballast to slow the descent to about two feet per second. As they approached 32,400 feet they felt a strong shock and a muffled noise.[8] The time was 1206.

"The sphere rocked as though we were on land and going through a mild earthquake," Walsh recalled.[9] In the recording, immediate concern is apparent: "What happened?" from Piccard. "Have we touched bottom?" Walsh asked. Reading his depth gauges, Piccard could see no change in their rate of descent.[10] Walsh later related the next few minutes: "We waited anxiously for what might happen next. Nothing did. We turned off the instruments and the underwater telephone so that we could hear better. Still nothing."[11]

Although this event was definitely *not* an implosion, the three implosions just two weeks earlier made them all too familiar with unidentifiable mechanical failures. Piccard and Walsh locked eyes; both men shrugged, and without a formal discussion continued the dive.[12]

Rechnitzer had told Walsh that he had found a 33,600-foot hole. Thus, at 30,000 feet Piccard dumped ballast to slow descent to about one foot per second. At 33,000 feet the fathometer was blank—"no bottom." Slowly they descended, and at 36,000 feet Piccard joked, "Do you think we missed the floor?" Walsh replied in his characteristic dry manner, "Probably not."[13]

There was now anxiety that the fathometer might not be functioning properly. Piccard dumped more ballast to slow to one-half foot per second. They dropped through 36,000 feet, then through 37,000 feet. Still, there was no fathometer indication of the ocean floor.

Finally, at 1256, the fathometer started indicating bottom at 37,500 feet. They turned on the exterior lights. Piccard watched out the forward window while Walsh called out fathometer range to the bottom, in fathoms: "Thirty . . . twenty . . . ten. . . ." Piccard had an impression of a large rock face or ledge just out of visual range—it would seem that they were bottoming at a location very close to a wall of the trench.[14] At a "height" of eight fathoms (48 feet) Piccard could see the bottom. The guide rope lightly touched down at 1306 at 37,800 feet.[15]

The Hadal Depths

No historic words were spoken. This was an epochal event, and it was observed within the *Trieste* sphere with personal introspection.

As one of the first two human visitors to the "hadal" depths—the very deepest zones, below 20,000 feet—Piccard reported seeing a fleet of medusae, small jellyfish-like creatures about an inch in diameter, at 2,100 feet above the bottom.[16] Just prior to "landing" he had sighted a fish browsing in the silt, a flatfish—Piccard called it a "sole"—with both eyes on one side, about one foot long and six inches wide, white or silver in color. The creature that he observed seemed unaware of the lights, completely undisturbed by the fathometer's pinging, and calmly swam away. Later he saw a large red shrimp. This answered the great question; there was *no* depth of the ocean that was uncolonized by complex life forms. It was a finding of significance.[17] (Subsequent visits to hadal depths between 1960 and 2019 by both manned and remotely controlled vehicles have failed to corroborate Piccard's sighting of fish inhabiting Earth's greatest depths. Even if this sighting was in error, the two other sightings of hadal inhabitants—a shrimp and a fleet of madusae—proved that the ocean's deepest depths contain complex life forms.)

"After so many years of preparation, I was elated at being able to cast my eyes upon the deepest spot in the seven seas," Piccard wrote later.[18] The *Trieste* quietly hovered a few feet above the muddy bottom at 37,800 feet, as recorded by the *Trieste*'s on board manometer. (The device turned out to be incorrectly calibrated; corrected depth was about 35,800 feet, which has since been rendered more precisely as 35,797 feet.)

The descent had taken four hours, 38 minutes.[19] A plume of loose diatomaceous mud billowed up around the sphere, engulfing it in a thick white fog of silt caused not by the impact of their landing but by released ballast striking the bottom prior to and following their touchdown. Billowing bottom silt had been noted on earlier deep dives, but on this dive the silt was not swept clear by a deep-water current. In this hole there appeared to be no appreciable current, and the cloud would linger.

With no expectation of being heard, Walsh keyed his underwater telephone: "Pittsburg, Pittsburg, this is *Trieste*, we are on the bottom of the Challenger Deep at six-three hundred fathoms. Over." ("Pittsburg" was the radio call sign of the tug *Wandank*.)

To their astonishment, a reply came 30 seconds later: "*Trieste*, *Trieste*, this is Pittsburg. I hear you faint but clear. Will you repeat your present depth? Over."[20] Piccard and Walsh could hear the excitement in Shumaker's voice from the *Wandank*. The excitement was understandable: 37,800 feet was more than 2,000 feet deeper than any recent survey, and 4,000 feet deeper than Rechnitzer's calculations from two days of explosive soundings.

At 1320, while waiting for the cloud of silt to clear, Piccard sighted a third animal, an inch-long, dark red shrimp swimming about six feet above the bottom.[21] That three hadal animals were sighted under very unfavorable viewing conditions with relatively primitive equipment was astounding.

Walsh then took an opportunity to look through the hatch window aft, toward the curved three-by-four-foot plastic window on the after side of the flooded entry tube, which from that point led vertically to the "tower" on deck, and discovered the source of the unexplained vibrations at 32,400 feet. "I know what happened, that noise, that jolt," he said quietly. "It was the big viewing port of the entry tube that cracked."[22] The entry tube window had cracked by some shift of the metal structure, not pressure differential as the entry was free-flooded during a dive.

Here was a casualty with complex implications: while it would not affect the sphere or the float, or any of the controls or instruments of the bathyscaph, it could nevertheless have a severe impact on the operation. If the entry window failed and they were unable to blow the water out of the entry tube upon surfacing they would be trapped in the sphere, which would remain about ten feet below the surface, until the *Trieste* was towed to Guam (four days), de-fueled, and lifted from the water. Auguste Piccard had designed the *Trieste* with snorkels that would provide fresh air during such an entrapment. Possibly liquids could be sent down to a trapped crew. Still, it would mean a long and unpleasant return to port.

Piccard and Walsh decided that there was good reason to get to the surface as soon as possible to have maximum daylight in which to investigate the entry-tube window problem. Before departing the bottom the two hydronauts formally shook hands and pulled out flags, a Swiss flag for Piccard and an American flag for Walsh, and took pictures of their mini-ceremony with a fixed camera that *Life* magazine had installed in the sphere.[23] Piccard was concerned that the Navy would object to a photo of him with a Swiss flag at the bottom of the Challenger Deep. "Before we took the picture, I asked Don, 'Are you sure this is all right to take a picture of me with the Swiss flag?' I didn't ask to do it, Don suggested it. He said, 'It's good for you.' I said, 'Are you sure that there will be no trouble with the Navy?' He said, 'It's all right.' "[24]

Ascent and Surfacing

After 20 minutes on the ocean floor, Piccard dropped two tons of ballast to start the ascent while Walsh sounded six tones on the UQC indicating that they were on the way up. They were on an "express elevator" to the surface—three hours, 27 minutes at an average vertical speed nearly three feet per second (i.e., two miles per hour).

There was a general feeling of anti-climax. In the early minutes of the ascent Piccard once again sighted a fleet of what he identified as medusae, which now on closer inspection seemed to have phosphorescent spots inside their jelly-like bodies. He lacked the knowledge of a marine biologist to identify the animals properly.[25]

Once the excitement had passed the cold of the depths hit both of them, hard. They had been in a small, damp sphere for more than five hours with the inside temperature about 45-degrees Fahrenheit. They improvised a bit of relief by stuffing activated Dräger canisters under their sweaters, "which

Life at the Hadal Depths

Almost 60 years after Piccard's sighting of a "sole" at the bottom of the Challenger Deep, additional information emerged from Victor Vescovo's expeditions that systematically penetrated the deepest points of all of the world's oceans. Vescovo's deep submersible *Limiting Factor* made four manned dives into the Challenger Deep from 28 April to 5 May 2019 (see chapter 24). During that same period three automated "landers" were sent to the bottom with bait and color cameras. Review of their film showed that the bait drew many denizens into camera range—but no fish.

Don Walsh accompanied the Vescovo expedition and interviewed Chief Scientist Dr. Alan J. Jamieson of Newcastle University, a world expert on the biology of the deep trenches. Fish had been observed as deep as 26,700 feet—but could fish survive at 36,000 feet? Jamieson replied no, as "fish are biochemically constrained from reaching full ocean depths."

Ever since 1960, marine biologists have argued that Piccard's sighting of a fish at 36,000 feet was an error. Originally, their argument was based on the known depth limits of flatfish, which had been observed down to 10,000 feet, but no deeper. Fish species organize in depth zones, with almost 38,000 species at depths above 3,000 feet, declining to less than 20 species found at depths greater than 20,000. By 25,000 feet there are only a few species of fish to be found, including the cusk eel and the snailfish (the deepest live fish was found at 26,700 feet). Moreover, more recently, hadal biologists have found that the chemical trimethylamine oxide in the cells of fish increases with depth and mathematically would reach fatal/toxic levels at 27,500 feet.

Perhaps most tellingly, two remotely operated dive programs into the Mariana Trench after 1960—the Kaiko and the Nereus—obtained hundreds of hours of film of the Challenger Deep bottom with no fish. Fish appear at remotely laid bait sites down to 26,000 feet with no decline in relative populations but at greater depths are completely absent. The subsequent dives by manned submersibles into the Challenger Deep also have found no evidence of fish.

What then did Piccard see? The possibilities range from a large holothurian (sea cucumber), known to populate the Challenger Deep, to a disturbance of the bottom sediment by released ballast following as the bathyscaph descended.

helped a lot"; each would provide 30 to 45 minutes of warmth.[26] Like many who feel anxious, they reached for chocolate—both men consuming about one bar per hour. Walsh related that they both wanted more but that Piccard recommended saving their "food supply" in case they needed it while being towed to Guam.[27] They left the exterior lights on during much of the ascent in hopes of seeing more signs of marine life, but without success.[28]

The *Trieste* broke the surface at 1656.

Once on the surface the "trick" was to blow the water from the entry tube with the least possible stress to the acrylic window to prevent the crack from shattering and making exit from the sphere impossible. Walsh bled pressurized air into the entry tube very slowly, rather than the usual blast to clear the entry tube in a fountain of seawater up through the tower. They watched through the hatch's Plexiglas viewport: one full air bottle, no progress; two bottles, still no progress. Finally, as the third bottle bled into the entry the water level dropped below the viewport, and a stream of air bubbles broke through the airline indicating that the entry tube was clear.

The pair lost no time in exiting the sphere, pausing only to seal the hatch behind them. They climbed up on deck, first Piccard followed by Walsh, teeth chattering from hours in the chilled sphere, just as two Navy aircraft blasted by a few feet overhead, and into what seemed to be a carnival.[29]

On the Surface

When Rechnitzer had received the transmission from the *Trieste* on the bottom he had been able to estimate the time of surfacing. This information was transmitted to Guam so that search-and-rescue aircraft could be overhead to assist should the *Trieste* have drifted far from the surface ships.

Surfacing thus was expected at 1700. At 1658, Navy Electronics Laboratory photographer John Phflum was the first to sight the *Trieste* as it broke the surface. In the 8½-hour dive the *Trieste* had drifted approximately two miles from the diving position. Sea conditions upon surfacing were severe—25-foot swells with some whitecaps. Phflum spotted the orange-painted *Trieste* tower about 3,500 yards west of the escort ship *Lewis,* directly on her bow.

By this time the aircraft from Guam, an Air Force C-54 Skymaster transport, was overhead, and it directed the *Lewis* to the *Trieste*'s position. The *Wandank* was informed that the *Trieste* had been sighted and directed to follow the *Lewis.*

Two Navy F8U-1P Crusader photo-reconnaissance aircraft flying from Guam swooped in to record the event for history, passing just a few feet above the *Trieste*.

The *Lewis* launched a rubber raft in lieu of her motor whaleboat, for which the sea conditions were too severe. Two press photographers boarded it to take photos of Piccard and Walsh when they appeared. A second rubber raft was launched from the *Wandank* to pick up Piccard and Walsh and take them to the *Lewis*.

It was into this rush of activity that Piccard and Walsh emerged about 15 minutes after surfacing. Upon satisfying the photographers with memorable waves and gestures, they climbed on board the rubber raft to be transported to the *Lewis*. There they stood for photos and an interview by Lieutenant Commander Paul Trahan of the staff of the Commander-in-Chief, Pacific Fleet. Photos and newsreel motion-picture coverage recorded the interview.

In a brief ceremony, a large American flag encapsulated in hard, dense, clear plastic, was committed to the Challenger Deep in commemoration of the historic dive.[30] To celebrate Italy's contribution to the success, Giuseppe Buono joyfully tossed his hat, decorated in Italy's red, green, and white, and his ever-present police whistle, by which he had coordinated topside evolutions, into the deepest water of the oceans.[31]

To enable the reporters to file their stories as soon as possible the *Lewis* departed for Apra Harbor, arriving at 0800 on 24 January 1960. The *Wandank*, with the *Trieste* in tow, was delayed by the heavy seas and did not arrive until 1000 on the 28th. Two days later the *Trieste* was lifted out of the water and placed on its cradle, awaiting the start of Project Nekton, Phase II. On 1 February, Rechnitzer, Walsh, Piccard, and Shumaker departed on a special Navy flight from Guam for ceremonies in Washington, D.C.

One of the first things that Piccard had done after surfacing was to tell Walsh, Rechnitzer, and Shumaker that all three men were fully qualified as bathyscaph pilots.[32] Walsh had the distinction of being No. 1.

Piccard's Destiny

Photographs of Jacques Piccard taken shortly after the Deep Dive show a man emerging from intense and long-running psychological stress. During World War I soldiers called the expression on his face "the thousand-yard stare." How could this man, who claimed to be uninterested in risk-taking and adventure, have become so embroiled in a venture involving enormous risks to his life?

The sphere was the most critical component of the *Trieste.* With the new Krupp sphere Jacques had violated nearly every principle of safety trials and unmanned testing that his father had laid down with such emphasis in 1956. He had neither demanded nor conducted a single empty test of the Krupp sphere at any depth, despite the fact that its three-piece design was new and unique. The *Trieste* had then experienced an unexpected and potentially dangerous casualty to the sphere in November 1959, when the epoxy bonds broke. It had been a warning that the Krupp design may have been flawed.

Piccard had nevertheless allowed others to concoct and effect an unconventional fix, outside of his view and without his personal supervision. A man who had "inspected every nut and bolt" on the original *Trieste* was now risking his own life on each dive in this bathyscaph, which had been modified with an expanded float, designed by the Navy Electronics Laboratory and fabricated at the Naval Repair Facility in San Diego. What had brought about such a metamorphosis of a careful and conservative manager into a man willing to risk his life routinely and repeatedly in a diving and development program that violated his father's explicitly stated engineering-safety principles? It would appear that Piccard had seen his destiny and had resolved to pursue it—whatever the cost.

To protect his right to pilot the Deep Dive Piccard antagonized Navy officials of the Bureau of Ships and the Navy Electronics Laboratory, as well as the *Trieste* project director (Rechnitzer), and the officer-in-charge of the *Trieste* (Walsh). Piccard effectively burned his bridges to all of his American partners, with the exception of Dietz at NEL and two or three individuals at the Office of Naval Research. This left no reservoir of positive relationships for any future Piccard effort in the United States involving the Navy.

After the Deep Dive, Piccard sent out a flurry of telegrams to family, friends, supporters, and sponsors, including to Benvento Loser at the Monfalcone shipyard in Trieste: "To Engineer Benvento Loser, Monfalcone Shipyard—Depth Reached. World Record in Mariana Trench 11,800 [*sic*] Meters below Sea Level—Jacques Piccard."[33] Telegrams also were sent to Rolex in Switzerland, to Krupp in Germany, and to others. His telegram to Krupp stated, "Thanks to your excellent work, dived in perfect safety 11,000 meters—kindest regards—Jacques Piccard."[34] His telegram to Yolanda Versich was particularly revealing: "Congratulations to you, Yolanda. Without you, we would never have been able to build the bathyscaphe. You have always believed in me and in my destiny. Jacques Piccard."[35]

In June 1960, Piccard wrote to the Chief of Naval Operations that in his opinion the *Trieste* was not safe for further deep dives because of the condition of the sphere—the same sphere that he had taken to 35,797 feet in the same condition five months earlier. Subsequently, on 8 July 1960, Project Nekton II dives were restricted to depths no greater than 7,000 feet by the CNO and the Commander-in-Chief Pacific Fleet (see chapter 8).[36] There would be no further dives by the *Trieste* into the Challenger Deep.

The Krupp sphere configuration in three sections was evaluated by the Navy to be potentially hazardous and imprudently risky to employ. As Rear Admiral James stated upon being informed that the Krupp sphere was held together with glue, "Lieutenant Walsh, the Navy does not glue its ships together."[37] Nor would it in the future make an exception for the *Trieste*.

Honors and Awards

The Navy aircraft with Piccard, Walsh, Rechnitzer, and Shumaker on board proceeded from Guam to San Diego via Oahu. In Honolulu they received a hero's welcome, complete with flower leis provided by costumed wahines and a personal "Well Done" from Rear Admiral William E. Ferrall, Commander, Submarine Force Pacific. After one day with families in San Diego, the four men flew on to Washington, arriving on 3 February.

The morning of the 4th an official car picked them up at their hotel and drove them to the White House. There President Dwight D. Eisenhower presented awards to all four men.[38] This was a distinct pleasure for Eisenhower, who had bestowed the National Geographic Society's Hubbard Medal on Edmund Hillary in 1954, recognizing the conquest of the Earth's greatest height, Mount Everest, on 29 May 1953. Now, in his final year in office, Eisenhower was able to honor the conquerors of Earth's greatest depth, the Challenger Deep.

Walsh received the Legion of Merit, the Navy's second-highest noncombat award "for exceptionally meritorious conduct in the performance of outstanding services and achievements." (The Legion of Merit is typically reserved for senior officers—captains and admirals—in positions of major responsibility.) Jacques Piccard was awarded the Navy's Distinguished Public Service Award, given "for courageous or heroic acts or exceptionally outstanding service of substantial and long-term benefit to the Navy," a highly unusual and significant honor for a non-U.S. citizen. Dr. Rechnitzer, the Project Nekton director, was presented with the Navy's Distinguished Civilian Service Award, the highest honor the Secretary of the Navy can award, "for distinguished and extraordinary services to the Department of the Navy that far exceed the contributions and service of others with comparable responsibilities." Lieutenant Shumaker, the assistant officer-in-charge of the *Trieste*, was awarded the Navy Commendation Medal, "for meritorious achievement, outstanding and worthy of special recognition."[39]

The president commented that the dive into the Marianas Trench "impressively demonstrates that the United States is in the forefront of oceanographic research." Walsh presented to the president an American flag that had made the trip to the deepest pit in the ocean, a signed philatelic envelope, and a *Trieste* plaque.[40]

Admiral Arleigh Burke, the Chief of Naval Operations, was at the White House ceremony and in a much happier state of mind than when Walsh had stood before him in 1959 with the Project Nekton proposal in hand. Walsh presented him another of the American flags that had been to the bottom of the Challenger Deep. Another of the 100 flags that had been in the *Trieste* went to the National Geographic Society; most went to influential members of Congress and to senior Navy officers in the *Trieste*'s chain of command.[41]

Piccard was especially proud of a White House letter dated 9 February 1960 that stated in part, "As a citizen of Switzerland, a country admired by all the free world for its love of freedom and independence, you have the gratitude of all the people of the United States for helping to further open the doors of this important scientific field. Sincerely, Dwight D. Eisenhower."[42] On 12 February, back in San Diego, Piccard, Walsh, Rechnitzer, and Shumaker received the keys to the city from Mayor Charles Dail.[43]

The Project Nekton leaders were once again in Washington in April 1960, to testify before congressional committees. Walsh appeared before the House Science and Astronautics Committee on 28 April. Rechnitzer testified in the "Frontiers and Oceanographic Research" hearings along with Walsh, NEL's commanding officer Captain John Phelps, and NEL technical director Dr. Kurie.[44]

Piccard found a way to thank Dietz for his key role in bringing the *Trieste* to the United States and, perhaps no less, for his support in protecting his position as pilot on the Deep Dive. He contacted Bob Wilson, the congressional representative from San Diego, and expressed concern that Dietz had not been recognized for the pivotal role he played in the *Trieste* program: Dietz had "originally conceived the idea of the Navy acquiring the *Trieste* and helped immeasurably in the initial stages of the project." Wilson wrote to Admiral Burke requesting that Dietz's contributions to the program be considered. On 13 May 1960, Rear Admiral Rawson Bennett, the Chief of Naval Research, announced that Dietz would receive the Navy's Superior Civilian Service Award, the Navy's second-highest honor for civilian employees, for "meritorious service or contributions resulting in high value or benefits for the Navy."[45] Friendship with Dietz was one of the few benefits that Piccard retained from his years working with the U.S. Navy.[46]

Jacques Goes Home

Jacques Piccard never made another dive in the *Trieste*. His divorce from the bathyscaph was abrupt, decisive, final, and no doubt emotional. On 19 March 1960 Piccard, with his wife Marie Claude and two-year-old son Bertrand, departed from San Diego. They stopped in Minneapolis to visit Jean Felix Piccard, Auguste's twin brother, and his family. Leaving San Diego Piccard told the press, "My agreement with the Navy runs until November, so I probably will be back, but I don't know when." In New York on 27 March Piccard and family boarded the Italian liner *Christoforo Columbo* for the Atlantic crossing, after which they continued on to their home in Lausanne.[47]

In early November 1960, Jacques Piccard did indeed return to San Diego, to consult briefly with the Navy Electronics Laboratory concerning ongoing and planned modifications to the *Trieste*—a final obligation under his ONR contract.[48] This also was Piccard's last exposure to his beloved bathyscaph, as the projected changes were so fundamental and transformational that the *Trieste* would soon be almost unrecognizable as the Piccard submersible.

On 24 March 1962, Professor Auguste A. Piccard died of a heart attack at age 78 in his home in Lausanne. Jacques's future was to include the design, development, and piloting of unique new deep- and mid-depth submersibles, both scientific and commercial. That of the *Trieste* would see continued service as a highly significant U.S. Navy project.

The Navy Electronics Laboratory Years

Everything that is done in operating the Trieste *from an engineering, operational, or scientific point of view is breaking a trail in a virgin area, where man has never worked before.*

—Lieutenant Don Walsh, Officer-in-Charge, *Trieste*

What was the significance of the *Trieste*'s dive into the Challenger Deep?

In an echo of Dr. Robert Dietz's 1956 comparison of the then-impending Challenger Deep dive with the 1953 assault on Mount Everest, the German firm Friedrich Krupp AG described their Krupp sphere as the "Sputnik of the Deep."[1] Arthur E. Maxwell of the Office of Naval Research had a more prosaic view of Project Nekton: "It got the Navy started in the deep-submersion business."[2]

It took someone involved with deep submersibles his entire adult life to state succinctly the purport of Project Nekton: Jacques Piccard, who said, "This dive proved that a means has been created enabling the oceanographer to extend his researches to the greatest depths in perfect safety."[3]

Don Walsh expanded on that point: "The conquest of the Challenger Deep was the ultimate test of the bathyscaph as a research platform. The feat of the *Trieste* puts the U.S. on the threshold of a new era in oceanography and demonstrates the role of the Navy in its advancement. If the oceans' depths are to be the potential battlegrounds of the future, we must learn all we can about them. The bathyscaph's descent was the greatest advance to date in this effort."[4]

For the U.S. Navy, Nekton had shined a light on a blind spot in its strategic capabilities. The missions of the Navy included being able to operate in all dimensions of the ocean. Technology was opening the full range of the third dimension, and the need to protect the interests of the United States would involve exploiting deep-sea capabilities for a variety of activities.

When Rechnitzer, Walsh, and Shumaker returned to Guam from Washington on 19 February 1960, further *Trieste* operations were no longer possible. The impending loss of support ships to other Navy missions, continuing unfavorable weather, and the necessity to return the project's temporary personnel to their assignments at the Navy Electronics Laboratory combined to bring *Trieste* dives to a halt. It was decided to leave the bathyscaph and its support equipment at the repair facility at Guam while the project leaders returned to San Diego to reorganize and formulate plans for Project Nekton II.[5]

During March–May 1960, the NEL leadership not only planned the upcoming Nekton II but also began to study the *Trieste*'s subsequent future. Following the conclusion of Nekton II, the *Trieste* would be returned to San Diego, where a major overhaul and modification would commence.

Project Nekton II

Nekton II was designed to be a short program of workup dives leading to two additional dives into the Challenger Deep. The planners wanted to fulfill two goals, one acknowledged and the other unspoken. The acknowledged goal was to continue to employ the *Trieste* in deep waters for scientific research and, simultaneously, to explore the capabilities and limitations of the bathyscaph as a unique Navy resource. The unspoken objective was to train naval officers in the planning and execution of deep dives.

It had gone unacknowledged but not unobserved that the only experienced bathyscaph pilot, Jacques Piccard, had departed without a backward glance. Walsh and Shumaker would carefully approach their new and singular discipline without an experienced instructor to call upon. They were fortunate in being able to rely upon the vast experience and knowledge of Giuseppe Buono, who had agreed to take a permanent position with the *Trieste* team as a civilian maintenance and safety specialist.

It was the original plan of Nekton II to accomplish at least two more dives to 35,800 feet. Once NEL had the Nekton II plan approved by the Bureau of Ships and the Office of the Chief of Naval Operations, it was set in motion. On 18 May 1960, the Nekton II advance group arrived at Guam to re-activate the support office at the ship repair facility. The tug *Wandank* and other support ships were again made available, and the weather was now suitable for diving. By 1 June, the entire *Trieste* team had arrived on Guam (less Buono, who was visiting in Italy), and by working long days, seven days per week, the men had the bathyscaph ready for launch on 11 June. The launch was delayed until Buono's arrival the next morning.[6]

Also that day Lieutenant Commander James S. Kennedy and Lieutenant Commander C. Cornelius Winkler arrived on Guam for familiarization with bathyscaph operations. Kennedy was the *Trieste* liaison officer at the Office of Naval Research, while Winkler was the liaison for the commercial submersible *Aluminaut*.[7]

Three days later, on 15 June, the *Trieste* made two Apra Harbor dives for integrity checks and pilot training. On Dive #71, Walsh piloted to 100 feet with Winkler observing; on Dive #72, Shumaker piloted to 100 feet with Kennedy observing. Walsh arranged for the first deep-ocean dives for himself and Shumaker on 21 June, with both dives supported by the *Wandank*. On Dive #73, Walsh, with Rechnitzer observing, submerged to 1,070 feet for acoustic studies at midrange water depths; Dive #74 saw Shumaker take Kennedy down to 1,455 feet for familiarization. This was the first and only time that the *Trieste* made two open-ocean dives on the same day.[8]

The workup toward another Challenger Deep dive began with a third open-ocean dive. Dive #75 took Walsh and Rechnitzer to 8,525 feet. This dive, an acoustics measurement test, originally had been planned for Mackenzie, but a family illness required his presence in San Diego.[9]

Fire at 8,000 Feet!

While on the bottom, Walsh and Rechnitzer experienced an electrical fire in the sphere, in the high-voltage lighting circuit. They initiated an emergency surfacing without delay. A fire in an enclosed capsule one-half the size of a Volkswagen "beetle" some 8,000 feet below the surface of the ocean could have been catastrophic. The smoke had no way of dissipating, and the heat could affect the sphere's instruments and controls. Should both pilots succumb to the foul atmosphere in the sphere the *Trieste* might automatically return to the surface with a dead crew when the batteries eventually drained and all ballast automatically dumped.

Walsh and Rechnitzer immediately donned gas masks, whose limited oxygen supplies would protect them for well over the hour required to regain the surface. All of the emergency equipment and the pilot actions combined to bring the *Trieste* to the surface without injuries or further damage.[10]

Continued weeping of the sphere joints and concerns that the strength of the sphere might have been reduced by corrosion at the joints resulted in a decision that a deep dive into the Challenger Deep would not be attempted during this series. Rechnitzer decided that for safety reasons, the bathyscaph would dive only to "two-thirds" of its rated working depth.

The 18,920-foot Dive #76, nearly to that limit, marked the deepest dive of Nekton II. On 1 July Shumaker and Rechnitzer made that 7½-hour dive, with the *Wandank* providing the tow and the escort ship *Haverfield* (DER 393) assisting with security at the dive site and providing accommodations for the *Trieste* team. The dive was made onto the western slope of the Nero Deep, about 80 miles southeast of Guam.[11]

On 6 July, in Dive #77, Shumaker and Rechnitzer rode the *Trieste* down to 1,143 feet. Unexpectedly strong currents at 1,000 feet sent the *Trieste* so far from her dive point that the *Wandank* had difficulty in locating the bathyscaph upon its surfacing.

Piccard's letter to the Chief of Naval Operations in June 1960, stating his opinion that the *Trieste* sphere was unsafe for further deep dives resulted on 8 July in messages from both the Chief of Naval Operations and the Commander-in-Chief of the Pacific Fleet restricting the *Trieste*'s dives to 7,000 feet without specific CNO permission.[12] These messages were received just prior to the commencement of Dive #78.

The *Trieste* Dive #78 was to a depth of 7,200 feet with Walsh piloting and Rechnitzer as observer.[13] The *Wandank* provided the tow and the *Haverfield* sought to track the *Trieste* with her SQS-4 sonar. During this dive, about eight miles west of Guam, photographs of the bottom revealed an unexploded 5-inch Navy projectile, with a beer can leaning against it.[14] In addition to the planned acoustics tests, Rechnitzer had a portable gravimeter on board to ascertain if gravity tests were feasible in the *Trieste*. They were.[15] On this dive the European high-voltage external lighting system failed completely with the last spare bulb being blown.[16]

Project Nekton II was terminated the next day. While the reasons offered at the time were the loss of the external lighting and the shortage of the special European light bulbs it used, it is obvious in hindsight that effective denial of authorization to dive the deep trenches obviated any gain achieved by operating from a remote base.

On 13 July, the *Trieste* was lifted out of the water and onto its cradle at Agana, and the bathyscaph and its equipment were prepared for return to San Diego. The first increment of personnel left for San Diego on the 19th. By the 22nd all bathyscaph team members had departed except for Walsh, Shumaker, and Chief Michel. On 2 August, the bathyscaph and all support equipment were loaded on the Navy cargo ship *Sgt. Jack J. Pendleton* (T-AK 276) for shipment to San Diego. The last of the *Trieste* personnel departed Guam the following day.[17]

San Diego, 1960–1961

The *Pendleton* arrived in San Diego on 28 September 1960 and offloaded the *Trieste* with its equipment at the Navy Electronics Laboratory piers on Point Loma. The bathyscaph was scheduled for a comprehensive ten-month reconstruction involving replacement, repair, and modification of practically every system and piece of equipment. The objective of the rebuild was to apply the lessons of the past two years of Navy operation to equip the *Trieste* better for its scientific mission. Emphasis was placed on the design and use of locally available American materials rather than foreign. In addition, several contractors were invited to expand the bathyscaph's instrumentation suite to enable the maximum amount of information to be obtained on dives.[18]

The Krupp sphere was disassembled by Buono and Michel, who found the joints to be corroded, but only lightly; hand finishing would restore the joints to like-new condition. The sphere was stripped of its internal equipment and forwarded to the Mare Island Naval Shipyard in Vallejo to await decisions as to whether it would be rebuilt for future use and if so, how. (The Krupp sphere was retired in 1961, never to dive again. Attached to the *Trieste I*, it is on display at the U.S. Navy Museum, in the Navy Yard, Washington, D.C.)

The Terni sphere would be returned to service. It was stripped bare and rebuilt with a re-designed and re-equipped control and sensor suite. With the exchange of spheres, the weight-to-buoyancy equation needed to be revisited. The Terni sphere weighed several tons less than the Krupp sphere, weight that could be allocated to additional equipment.

The most pressing deficiency of the original Piccard-designed *Trieste* was battery power. Piccard's design was severely underpowered, marginally sufficient for the underwater lighting, fathometer, underwater telephone, and internal lighting but little else. Despite their high cost, Piccard had selected silver-zinc batteries because of their compact size and high power-to-weight ratio. All of them were inside the sphere, where they were a safety hazard and took up space needed for electronics, sensors, and crew support.

A prototype external battery array had been installed on the *Trieste* for Nekton II, and with it the Navy Electronics Laboratory had solved the problem that had beaten the Piccards: how to modify lead-acid batteries so that they were free-flooding and thus immune to the effects of water pressure at extreme depths. Automotive lead-acid, 12-volt batteries were installed along the top of the float, under removable deck plates that both provided more freeboard as well as greater safety for men working on deck at sea. The new battery array produced two and a one-half times the 15 kilowatt-hours available from the Piccard silver-zinc batteries. Also unlike the silver-zinc batteries, which had to be removed from the sphere after every deep dive for recharging, the lead-acid batteries could be recharged in place, allowing for a more rapid turnaround between dives. As a bonus, the lead-acid battery array cost $1,000 compared to $60,000 for the silver-zinc batteries.

The new lead-acid battery array provided sufficient power to drive the propellers and thus allow propulsion at the ocean bottom, potentially turning a scientific, deep-water "elevator" into an ocean-bottom work vehicle. (Later, upgraded 12-volt batteries were installed, increasing the *Trieste*'s electrical capacity by an additional 50 percent.)

The maintenance period also saw a major redesign of the controls to bring them up to Navy standards and labeling. Among the new components was a six-axis mechanical arm manufactured by General Mills.[19] The arm, which was controlled by a series of toggle switches inside of the sphere, had a four-foot reach and could lift 50 pounds.[20] It was mounted at the forward ballast tub; external baskets could be fitted within reach of the arm for storing objects taken from the ocean floor. Lieutenant George Martin, later a *Trieste* assistant officer in charge, related, "It worked great in the laboratory, but when we attached it to the forward shot tub and put the *Trieste* in salt water, it behaved like a battery [that is, galvanic corrosion resulted]. The many dissimilar metals of its construction were unsatisfactory for use in the ocean."[21]

The *Trieste*'s float also received major renovations. A re-fueling manifold system was installed so that all tanks could be fueled from a single connection; in Piccard's float each tank had to be fueled/de-fueled individually. This modification further reduced turnaround time between dives.

During this overhaul period the *Trieste* received its first sonar, an NEL-designed and built unit for the detection of obstacles as far as 150 feet in any direction.[22] Several months later an omni-directional, continuous-transmission, frequency-modulated sonar fabricated by Straza Industries was substituted for the NEL unit.[23] Better exterior lighting, fixed Edgerton cameras with strobes, and a closed-circuit TV completed the *Trieste*'s new sensor suite.[24]

The 1961 Dive Season

In November 1960, NEL received orders from BuShips to prepare the *Trieste* for a series of dives in the mid-Atlantic near Bermuda in early 1961. Project Mergo, as the program was named, was apparently in support of Navy seafloor acoustic arrays in the vicinity of Bermuda. Walsh flew out to Bermuda twice to determine what facilities were available and locations for the dives. Other trips to Washington, D.C., and to the Lemont Geological Observatory at Columbia University in New York were made for planning and coordination. NEL expended considerable effort to rush delivery of contractor-provided equipment and material for the *Trieste*'s modifications to meet the Project Mergo commitment. Planning and preparations continued throughout the winter of 1960–1961, but by late spring 1961 delays were making it increasingly unlikely that the *Trieste* could be made ready in time for the Bermuda dives. On 17 April 1961, the *Trieste*'s participation in Project Mergo was cancelled.[25]

When the bathyscaph emerged from overhaul that June, the *Trieste* was no longer a European craft, built to metric plans and with European equipment, but a U.S. Navy craft, with upgraded operational features, increased safety overdesign in buoyancy and ballast, more power, better instrumentation, and

American-standard fasteners, piping, controls, instruments, and components. There was space for three times as much equipment inside the sphere thanks to the removal of most of the batteries to the main deck. Five new propulsion motors made the bathyscaph capable of hovering and vertical, horizontal, and lateral movement. Despite all of these improvements, Walsh called the *Trieste* "the Model T of deep submersible vehicles"—meaning that it was first-generation technology but nonetheless a working vehicle with a future of profitable scientific work.[26]

Also in early June 1961, Rechnitzer, who had been NEL's *Trieste* project director and chief scientist became its Deep Submergence Research Programs Coordinator. Operational control of the *Trieste* shifted to the program research officer, Commander Albert C. Carson, and under him the Oceanographic Programs Officer, Dr. Gilbert H. Curl.[27]

On 26 June, the *Trieste* was launched and readied for a series of 60 planned dives in the waters off San Diego. The final modifications—the mounting, wiring, and testing of the five electrical propulsors—were delayed by their late delivery. Finally, in September, it was decided to start limited operations without them.

Other Missions for the *Trieste*

Classified operations for the *Trieste* were suggested in the summer of 1961. Walsh related:

> One year before I left [the *Trieste* program], an Air Force Lieutenant Colonel came to my office and asked, "Can you pick things up with that, that thing of yours, the bathy—whatever you call it?" I said, "Well it certainly can, Colonel, what did you have in mind?" He said, "We've got some stuff in pretty deep ocean that we'd like to have, that doesn't really belong to us, but we'd like to pick it up and have a look." I said, "Sure," and winged out some ideas and brainstormed with him. He asked, "Would you come to Wright Air Force Base in Dayton, Ohio and talk to our Science and Technology directorate"? They put me in an airplane and off I went. I didn't tell my commanding officer at the Lab that I was leaving—"the Lieutenant just wandered off somewhere," Walsh continued.

> "I ended up in Ohio and I briefed people and got up to a two-star general who asked, 'Will you go talk to our Intelligence directorate at Andrews Air Force Base in Washington, D.C.?' Walsh ended up talking to the head of Air Force Intelligence, a three-star general. Finally he said, 'General, you know everything that I know, but nobody knows where I am. I need to get back. You have propelled me across the United States from San Diego, here I am in Washington, D.C., and nobody in the Navy knows where I am. I think it's time for you guys [Air Force Intelligence and Navy Intelligence] to talk together."[28]

On 21 July, Astronaut Gus Grissom's space capsule *Liberty Bell* 7 splashed down some 300 miles east of Cape Canaveral and 90 miles northeast of Grand Bahama after a 15-minute suborbital flight. The mission was successful, but the capsule sank in "16,800 feet" of water when the main hatch's explosive bolts triggered, causing the capsule to flood. Grissom survived.

The first question asked at NEL was, Could the *Trieste* aid in the capsule's recovery? On 22 July Walsh advised the press that NEL had nothing available to reach that depth as the *Trieste* was then undergoing modifications and unavailable. Even when operational, the *Trieste* could "only make the recovery if it were lowered directly next to the capsule. It [the bathyscaph] has no mobile power underwater to search for the capsule."[29] (Almost 38 years to the day after it sank, the *Liberty Bell* 7 capsule was recovered from more than 15,000 feet by the salvage ship *Ocean Project*. The capsule had been located by sonar and a remotely operated vehicle. It is now on display at the Kansas Cosmosphere and Space Center.)

Fall 1961 Dives

On 7 September, Walsh piloted Dive #79 with a *Trieste* team member, Electrician's Mate 1st Class Curtis M. Adams, assisting in a 40-foot dip for a San Diego harbor test dive of the modified *Trieste*. Dive #80 on 12 September was a second harbor test dive, this time to 69 feet with Shumaker and *Trieste* teammate Chief Engineman Devoe. These first two dives checked the bathyscaph's new equipment.

Dive #81 on 14 September was the first ocean dive of the series, to 490 feet; Walsh was piloting, and Dr. George A. Shumway Jr., an NEL ocean geologist, was conducting observations. The tow was provided by the fleet tug *Tawasa* (ATF 92). Electrical difficulties with the *Trieste*'s ballast system were experienced during the dive, requiring a ten-day delay to the original diving schedule.[30] On 11 October, a harbor test was made to 48 feet, with Walsh as pilot assisted by Chief Electrician's Mate Chandler, to confirm that the electrical problems experienced on the previous dive had been corrected. On 13 October the fleet tug *Chickasaw* (ATF 83) towed the bathyscaph out for Dive #83, in which Shumaker took NEL scientist Robert F. Dill to 1,920 feet for geological research.

Dive #84, on 25 October, was for studying deep-ocean currents and biologics; Walsh piloted for Dr. Eugene C. LaFond, head of the marine environment division at NEL. At 480 feet the *Trieste* dropped through a clear, quiet zone, and Walsh paused the descent so that LaFond could record a variety of sea life at that depth. At 900 feet they found a deep scattering layer where LaFond used a specially designed external device to estimate the biologic density at 40 living creatures per cubic foot. When the external lights were extinguished, LaFond recorded, luminescent "brilliant circles and flashes of light suggested a submarine Broadway." At 300 feet above the ocean floor LaFond recorded another zone teeming with life.

The soup of small living, dying, and dead organisms at this depth elicited a comment from LaFond: "How filthy it is!" Just before touching bottom at 3,870 feet they entered a layer of clear water, the clarity apparently maintained by bottom dwellers scavenging the continually falling detritus. Another special instrument, this one he had designed himself, enabled LaFond to verify that the open ocean current at the bottom was in a northerly direction. That three-hour, 23-minute dive was supported by the fleet tug *Molala* (ATF 106).[31]

Dive #85 on 7 December was made in the harbor to certify correction of problems experienced on #84. Shumaker descended to 40 feet, providing John A. Beagles, NEL's lead diver, an opportunity to experience a bathyscaph descent. Also that December, Shumaker's relief, Lieutenant George W. Martin, visited NEL. He would not report on board until March 1962 but wanted to discuss his next assignment before deploying to the western Pacific on the diesel-electric submarine *Catfish* (SS 339).

1962 Scientific Dives

During 1962 the *Trieste* made six dives with NEL biologist Dr. Eric Barham, who was investigating the deep scattering layers near the San Diego Trough. The first, Dive #87 with Barham, was on 17 January, the second on 14 February, and a third on 28 March.[32] Barham was looking for the sea creatures responsible for the acoustic scattering layers at different depths. On the 14 February dive the *Trieste* broke into a new, unrecorded deep layer that Barham described "as more crowded with fish than any they had visited."[33] From 2,280 feet to the bottom at 4,107 feet the waters were sparsely populated. The *Trieste*'s ascent was made without external lighting, allowing Barham to see "a regular Fourth of July fireworks display" from luminescent creatures of the deep: red, orange, green, and blue. Barham stated that the bathyscaph had "proved itself a tool for observing deep ocean life without a peer."[34]

In March 1962, Martin returned from his deployment, detached from his submarine, and reported as prospective assistant officer-in-charge.

Scientific dives off San Diego in 1962 included a series of geophysical observations by NEL geologist Robert Dill. His led to a new understanding of a gigantic mechanism of nature—a process by which the ocean bottom is carved and changed. "There is no name for what we have seen. It is more like a glacier than anything else I can think of. This huge mass [of compacted sand] inches along with an intermittent movement and apparently grinds the sides and bottom of its bed as it moves."

On the 4 April dive, Shumaker and Dill took the *Trieste* to 1,920 feet some five miles northwest of La Jolla. As Shumaker moved down a slope into a canyon, he found the slope so steep that the stern kept bumping into the canyon wall. He developed a perfect solution, rotating the *Trieste* to lie parallel to the wall, and continued the descent. At the edge of the slope, the sediment appeared quite cohesive. When the bathyscaph set it into motion, it would "roll down the slopes in small balls, like snowballs." Dill could see an underlying body of sand and believed that there was a glacial-like movement down the canyon that was fed and pushed along by sand depositions near the canyon's head. Gravity was enough to propel the sand river. He opined that the bathyscaph "as it exists is not perfect, but it has proved to be the most powerful tool presented to marine geologists since the advent of echo sounders."[35]

Walsh published an article in *Naval Research Review* in May 1962 arguing that the *Trieste* was rapidly becoming an "operational" vehicle for scientific research, especially in the area of acoustical phenomena that affect submarine and anti-submarine warfare. He described the dive program to commence in the spring of 1962: "investigation of sound velocity in bottom sediments, operations as a deep target against all types of sonar equipment and vessels, the study of sound convergence zones, and the measurement of sound reflection and scattering from the ocean floor."[36]

A New Mission

Early in 1962, Walsh learned that the *Trieste* would be called upon to conduct its first classified dive, for an Air Force program known as Vela. The program was developing methods to monitor compliance with the partial nuclear test-ban treaty then being negotiated between the United States and the Soviet Union.

Hugh Bradner, a research geophysicist at the Institute of Geophysics and Planetary Physics at Scripps, was the lead scientist for emplacement of a test Vela underwater micro-seismograph off southern California.[37] Walsh related, "I was Hugh Bradner's pilot on that dive. Project Vela was set up by the U.S. Government to see which medium was best to monitor USSR nuclear weapons testing. One element of Vela was detecting from space, another using sensors plugged into rock structures on land and the third involved sensors on the seafloor. My dive with Hugh was for the last. We placed a very sensitive micro-seismograph on the seafloor to measure the background tectonic noise level there."[38] (Once the Vela detection system sensors were emplaced the United States conducted several nuclear tests, from October 1963 to July 1971, to build a library of nuclear events and calibrate the system.)

The article ran on page 1 of the *San Diego Union* on 28 May 1960, while the *Trieste* was operating off Guam during Nekton II: the *Army, Navy, Air Force Journal* had just reported a ruling by the General Accounting Office that submarine pay was only for duty on board an "armed combat ship." The ruling immediately denied "submarine pay" to Navy officers and enlisted personnel involved with *Trieste* operations.

The Navy's reclama pointed out that it assigned submariners to training and research submarines that were not "armed" and who were authorized for hazardous-duty pay; similarly, military aviators received flight pay whether they flew armed or unarmed aircraft. Don Walsh's situation became public and the Navy Department announced that it "still is convinced that [bathyscaph duty] is hazardous duty and intends to do everything it can to see that Lt. Walsh gets hazardous duty pay." Walsh's regular pay was about $480 per month, and hazardous duty pay added $190.

Bipartisan bills were sponsored in the U.S. House and Senate by the California delegation to grant hazardous duty pay to "operators" of undersea research vessels such as the bathyscaph *Trieste*. The Senate passed a bill on 29 June and sent it to the House, which passed it and sent it to the White House for the president's signature. There the bill sat.

The problem at the White House took more than two years to resolve. Finally, on 20 September 1962, President John F. Kennedy signed the bill, which included provisions specifically for the benefit of Don Walsh and Larry Shumaker—who were authorized hazardous duty pay retroactive to 1960. Walsh and Shumaker each receive a "windfall" of several thousand dollars.[39]

Building Foundations

The two and a half years between the Deep Dive on 23 January 1960 and the departure of Walsh as officer-in-charge of the *Trieste* in July 1962 was a pivotal period in the development of oceanography, the beginning of what would be called the "golden age" of manned submersibles. Not only was the science of high-pressure ocean engineering undergoing significant development, but the general public and government leaders were coming to realize that advancing undersea technology was vital to the security of the United States.

Dr. Rechnitzer was selling the bathyscaph as a unique tool for science and research, while Walsh was thinking deeper. As a Navy lieutenant in charge of a program to develop and operate a one-of-a-kind capability, Walsh had an acute grasp of the potential of deep submergence. He seemed intuitively to understand that manned deep submersibles could become a significant Navy resource, one with national security implications. Walsh presented the *Trieste* as a naval technology-development program complementary to the American space program. He painted a Navy future that captured the imagination of the general public and brought credit to Navy leaders for their foresight and acumen:

> In 1961, NEL began an investigation into the design of a second generation submersible. As there were then only six to eight manned submersibles in the world, this was very much a "clean sheet of paper" approach. In developing the proposal, the team got significant help from an unlikely source—Harold (Bud) Froehlich from General Mills, who was then at the Navy Electronics Laboratory discussing modifications to nuclear-related mechanical arms to meet deep submergence applications.
>
> During the time Bud was at NEL, he and the *Trieste* team brainstormed the idea of developing a small submersible. Eventually [we] worked out a conceptual design proposal.[40]

Called Sea Pup, the proposal was approved by the NEL leadership and sent to the Office of Naval Research's Submarine Warfare Branch, headed by Captain Charles (Swede) Momsen Jr. When briefed by Walsh on the Sea Pup concept, Momsen exclaimed, "We have got to get one of these!"

At that time, ONR was negotiating with the Reynolds Aluminum firm to lease their 80-ton deep submersible *Aluminaut* when it was ready in 1964. About $1 million in ONR funds had been committed to the Woods Hole Oceanographic Institution in anticipation of using the *Aluminaut*. Oceanographer Allyn C. Vine was one of the project leaders and a primary proponent of the initiative. When Momsen learned that Reynolds intended to raise significantly the lease rate, he initiated an internal program to build a craft based upon the Sea Pup concept. General Mills' Aeronautical Research Laboratory and Electronics Division was awarded that contract, but before the craft was completed that division was sold to Litton Industries, which delivered it in mid-1964. The craft was a larger, and improved, version of the Sea Pup. It would be owned by the Navy and operated by Woods Hole—and as an honor to Allyn Vine, it was named *Alvin*.

The *Alvin* remains in operation and as of the end of 2019 had made 5,050 dives.[41]

Command Changes and Challenges

Martin relieved Shumaker as "executive officer" on 22 May 1962, following his check-dive and qualification as bathyscaph pilot #3. Shumaker had orders to the San Diego–based diesel-electric submarine *Caiman* (SS 323) as executive officer.[42] He would return to the *Trieste* program in 1967.

Then, two weeks prior to his relief as officer-in-charge of the *Trieste,* Walsh was the honoree of a luncheon on 1 July 1962, given by the San Diego Chamber of Commerce, the City of San Diego, and the Kiwanis Club.[43] Walsh was relieved on 13 July by Lieutenant Commander Donald Keach, formerly commanding officer of the target/training submarine *Mackerel* (SST 1).[44] Walsh had orders to report as engineer officer on the San Diego–based diesel-electric submarine *Sea Fox* (SS 402).

Three days after Keach relieved Walsh on 16 July the press reported that Georges Houot and Pierre Henri Willm had taken the new French "super-bathyscaphe" *Archimède* to 31,350 feet, the bottom of the Kurile-Kamchatka Trench, for the second-deepest plunge ever made by man. They had remained on the bottom for three hours, as compared to Piccard and Walsh's 20-minute stay at the bottom of the Mariana Trench in 1960. On 12 August Houot and Henri Delauze plunged the *Archimède* to 30,511 feet in the Japan Deep.[45]

On 14 August, Keach announced that the next series of scientific dives for NEL and Scripps scientists would commence on 17 August, with a 4,000-foot dive into the Coronado Canyon, off southern California. Before that could occur, the first truly *operational* use for deep submergence vehicles would emerge, and the *Trieste* would be called upon to morph into a new role.

9 SubSunk

When Thresher *went down, our only hope was that we could find the hull and learn the cause of the disaster—and even that hope was slim.*

—James H. Wakelin, Assistant Secretary of the Navy for Research and Development

When the nuclear-propelled attack submarine *Thresher* (SSN 593) was lost 220 miles east of Boston on 10 April 1963, it triggered what was known as a "SubSunk" response. At the time the U.S. Navy had no search methods, no standard operating procedures to locate a ship on the deep-ocean floor, and no organization for attempting it. While 13 surface ships and submarines raced to the area of her last dive, early efforts devolved into a visual search for floating debris or an oil slick from the *Thresher*.[1]

The *Thresher* was the first of a new class of nuclear-propelled submarines designed to find and destroy enemy submarines. She was the Navy's deepest-diving combat submarine—with an approximately 1,300-foot test depth—and introduced new quieting measures and the advanced BQQ-2 sonar.

She was not carrying nuclear weapons when lost, but the *Thresher* was the world's first nuclear-powered ship to sink, and her nuclear reactor was on the ocean floor, potentially contaminating waters close to the U.S. eastern seaboard. And, the Navy had to examine the wreckage in an effort to determine the cause of her loss.

"We knew, however," wrote James Wakelin, the Assistant Secretary of the Navy for Research and Development, "that most of the search techniques and equipment that appeared most promising were still under development by scientists and engineers. The Navy's fighting ships could assist the search in only a limited manner. Research craft—oceanographic ships—were the most capable of investigating the area all the way to the ocean floor."[2]

During the search these oceanographic research ships operated under the command of Captain Frank A. Andrews, assigned as commander of Task Group 89.7. Andrews, a PhD physicist, was commander of Submarine Development Group 2, developing the tools and tactics for advanced submarine development and operations. Andrews was supported by an 11-man brain trust, the Chief of Naval Operations Technical Advisory Group (CNO TAG), co-chaired by Dr. Arthur E. Maxwell from the Office of Naval Research, and also by Captain Charles B. Bishop from the Office of the Chief of Naval Operations. There was a second technical advisory group, the TAG WHOI, chaired by Arthur E. Molloy of the Naval Oceanographic Office and staffed predominately from the Woods Hole Oceanographic Institution.[3] Ships from every East Coast laboratory and oceanographic office, civilian and military, were made available for the search effort.

Andrews developed a multi-phase search program to locate and examine the wreckage of the *Thresher*: (1) conduct a "fine-grained" bottom sonar charting of the area; (2) use towed sensors to investigate all possible wreckage positions (aborted because of sensor limitations); (3) employ towed still and television cameras to examine any wreckage identified by these efforts; and (4) use the *Trieste* for close examination and study of the wreckage. The bathyscaph would have to be transported from San Diego to Boston, and that transfer should take place as soon as practical.

San Diego, 1962–1963

With no foreknowledge or premonition of a deep-ocean emergency in their immediate future, the *Trieste* team, now led by Lieutenant Commander Donald Keach and Lieutenant George Martin, conducted a full schedule of scientific dives off San Diego from July to November 1962. The *Trieste* was then brought ashore for inspection and renovation of the Terni sphere.

In the second half of 1962 the *Trieste* team consisted of two officers and eight enlisted men plus one civilian (Giuseppe Buono)—an 11-man crew. Dr. Gilbert H. Curl remained as the Navy Electronics Laboratory official in operational control of the program, now the program research officer. After Rechnitzer left NEL in March 1962, Arthur L. Nelson acted as Deep

Submergence Research Program Coordinator for a few months until he was relieved by Arthur J. Schlosser, who departed in the fall; finally, in September 1962, Kenneth V. Mackenzie filled the position and remained so engaged through the remainder of the *Trieste*'s time with NEL.[4]

The planned *Trieste* dive schedule for the laboratory's scientists in 1962 was based on a highly optimistic presumption of one deep dive per week. Keach and Martin rotated responsibilities between pilot and safety officer. There also would be a few shallow harbor dips to check scientific instrumentation or confirm equipment repairs. The maintenance crew used the non-diving days to work on equipment and install the scientific instruments required for the next dive. These scientific research deep dives started on 6 July with Dive #101 and ended on 31 October with Dive #115.

The *Trieste* underwent maintenance at NEL from November 1962 to March 1963, during which the Terni sphere was the team's special focus. The sphere had never received an official Navy inspection, much less a formal certification. In the fall of 1962 the Navy sent Peter Palermo, an engineer from the David Taylor Model Basin in Potomac, Maryland, to examine the Terni sphere. He wrapped the sphere with strain gauges, and the *Trieste* made a dive to 4,000 feet. Subsequently, NEL sent the sphere to the Naval Repair Facility at San Diego for the intricate work of disassembly. Corrosion was found between flange faces, which was removed by grinding. The flange retaining rings were redesigned to allow easier removal in future overhauls (bolting instead of riveting). Palermo's report and the subsequent dismantling and rejoining of the sphere at San Diego resulted in certification to *dive to 4,000 feet!*

In a mission-oriented Navy, certification levels are not constraints when operations require action beyond approved limits. Operations in 1963 would require dives to more than twice the officially certified depth.

Meanwhile, the *Trieste* float was blasted and repainted, and the guide rope was converted into a heavier "trail ball." The mechanical arm was removed and returned to General Mills for redesign; it had failed, owing to faulty motors, saltwater leakage, inadequate pressure compensation, and the poor choice of materials that had led to bi-metallic electrolysis and excessive corrosion (see chapter 8). The on-deck battery array was modified to replace the aluminum battery boxes with non-corrosive fiberglass and was enlarged to accommodate 18 automotive/

truck 12-volt, lead-acid batteries. This change provided the *Trieste* with about 70 kilowatt-hours of electrical power for propulsion and external lighting.

Meanwhile, Keach, Martin, Mackenzie, Curl, and others at NEL were engaged in developing a "single-sheet" specification for a replacement float. Their requirements stressed increased safety, higher towing speed, increased submerged endurance, and faster turnaround between dives. These specification sheets were sent to the Mare Island Naval Shipyard, which was asked to develop the plans and drawings. This would be the first deep-diving craft to be designed and built in the United States, and there was required "a great deal of discussion to educate the designers [at Mare Island] in the fundamentals of bathyscaph design."[5] While the NEL team had decided that the Terni sphere would be adequate for near-term operations, the new float would be able to carry either the Terni or the Krupp sphere.[6] Plans were approved by the Bureau of Ships, funding was provided by the Office of Naval Research and BuShips, and construction of the float began at Mare Island in the spring of 1963.[7]

An Aborted Dive Season

The first dive of the 1963 season—Dive #116—was a 13-minute harbor "wetting" piloted by Keach to give Captain Harry C. Mason, the NEL commanding officer, exposure to the *Trieste*. The first scientific, open-ocean dive of 1963 was #117 on 4 April, with Keach piloting, NEL scientist Edwin C. Buffington observing, and *Trieste* crewmember Chief Electrician's Mate Forrest D. Barnett as a third man in the sphere. It was a three-hour dive to 336 feet. Three men made for a very tight fit. This was only the *Trieste*'s second three-man dive. A year earlier, on Dive #95 of 27 April 1962, Walsh had taken both Keach and Martin for an orientation dip to 64 feet off Ballast Point.

After the 4 April dive, the *Trieste*'s maintenance crew commenced preparations for the next scientific dive, scheduled for 11 April.

Almost the first action taken by the Office of the Chief of Naval Operations on 10 April 1963, when notified that the *Thresher* was possibly lost was a telephone call to Keach from Captain Isaac Kidd Jr., executive assistant to the CNO. Keach agreed that the *Trieste* was the only craft capable of diving on the wreck but expressed caution: "I wasn't concerned about the depth, but I did want him to make sure that they understood that it was basically an elevator and was very limited in maneuverability and very limited in speed and endurance."[8]

Hours after Kidd's telephone call, Admiral George W. Anderson, the Chief of Naval Operations, advised reporters that the bathyscaph *Trieste* was being put on alert for deployment to the East Coast to aid in search operations. In San Diego the bathyscaph team was evaluating the alternatives of air transport, truck, or rail to speed the *Trieste* to Boston. The *Trieste* crew knew only that the *Thresher* wreckage was about 270 miles east of Boston and that it was on the Atlantic continental slope at about 8,400 feet. Simultaneously, Captain Mason flew to Washington to brief authorities to the effect that "the *Trieste* was designed exclusively for oceanographic research" and, again, had very limited horizontal mobility underwater.[9]

By 12 April the decision was made to transport the *Trieste* by ship, a slower option but one that allowed the bathyscaph to travel in "one piece" and arrive in Boston ready to dive with minimum delay.[10] The dock landing ship *Point Defiance* (LSD 31) arrived in San Diego on 12 April and immediately began preparing to transport the *Trieste* to the East Coast.[11]

That same day, in Lausanne, Switzerland, Jacques Piccard was approached by journalists inquiring if he would be willing and ready to join the *Trieste* in a search for the wreckage of the *Thresher*—if asked. Piccard responded, "There is no question but that I would drop everything and go to help the *Trieste* for this search if the United States Navy asked me to. So far, however, I have received no such invitation." He added, "In any case, the Navy has several fully qualified pilots who are perfectly able to handle the *Trieste* without me."[12]

En Route to Boston

Keach and the *Trieste* team had only three days to pack up and make the bathyscaph ready for transport before they watched it lifted into the *Point Defiance*'s docking well. The most critical task in readying the bathyscaph for sea transport was removing more than 33,000 gallons of highly volatile aviation gasoline, most of it offloaded at North Island across the channel from NEL piers.[13] Then the *Trieste* fuel tanks were ventilated with compressed air for more than 12 hours and refilled with nitrogen to render them explosion proof.

Martin led the *Trieste* team of eight enlisted men that would accompany the bathyscaph on the voyage. Time was of the essence, and the *Point Defiance* was given expedited clearance through the Panama Canal, including an unusual high-speed transit permit through Gatun Lake.[14] (The *Point Defiance*'s top speed was 21 knots.)

With the bathyscaph and crew en route, Keach flew to Washington to brief officials on the bathyscaph's capabilities and limitations. He indicated that once the *Thresher* had been located, "the *Trieste* could move in close to the submarine's remains for visual observation." He further explained that the *Trieste* carried four external, 35-millimeter cameras that could take up to 2,000 photos and had sonar with a 400-yard range. He stressed to all that the *Trieste* had very limited horizontal mobility.[15]

Admiral Anderson explained the bathyscaph's limitations for this mission to the press: "The only thing the *Trieste* can do is probe the hulk in expectation of finding some clue to the cause of the disaster—the worst ever involving a submarine and the first involving a nuclear craft."[16] The 129 men lost with the *Thresher* represented the second-largest loss of life in any submarine disaster; in 1942 the French submarine *Surcouf* was sunk in a collision in the Caribbean with the loss of her entire crew of 130.

Keach was concerned about and frustrated at the fact that while the Terni sphere had just had a comprehensive, five-month inspection and rebuild, the outdated and much-used float had makeshift and temporary fixes in some of the gas tanks. Overall, the *Trieste*'s electrical and electronic systems were showing their age and could deteriorate markedly if operations off Boston were of high tempo or extended in duration. Giuseppe Buono was acutely aware of the mechanical deterioration of the bathyscaph and repeatedly advised Keach that it should not be employed to the 8,000-foot-plus depth at which the wreck was believed to lie. Keach, citing operational imperatives, overrode Buono's warnings.[17]

On the positive side, Keach and Captain Andrews bonded almost instantly, having similar histories in surface ships and submarines.[18] That close relationship would see them through the coming frustrations of technological limitations and mechanical casualties as they prepared for the search for the *Thresher*'s remains.

An Underwater Balloon Arrives

When the *Point Defiance* arrived in Boston on 26 April, the search for the *Thresher* was still involved in "fine-grained" bottom mapping to detect elongated "bumps" that might mark the wreck. The extreme difficulties of precisely mapping bottom features more than a mile and a half below the surface were multiplied by insufficient navigational accuracy to plot precision depth readings. In general, it did little good to discover and plot a rock outcropping (or potentially a wreck) if tides and currents, navigational inaccuracies, and coarse bathymetric fathometer readings made it impossible to return to a charted location and obtain the same soundings. Added to this situation was the fact that the initial search was based on the belief that the wreckage was in a single piece large enough to be detectable by echo ranging.

Soon everyone involved realized that the search for the *Thresher* wreckage would be long and difficult. Prior to leaving San Diego, Martin had told the enlisted crew, "Pack a full sea bag. Tell your wives we could be on this mission through September."[19]

Putting expert observers on the ocean floor using the *Trieste* to investigate the wreck was dependent upon first locating it, and then fixing it accurately enough that the bathyscaph could dive to a pre-selected and marked point, ideally homing on a pre-positioned acoustic beacon, and expect to find the wreck there when it arrived. The technical advisory groups were well aware of the *Trieste*'s inability to conduct area searches on the ocean bottom.

Thus the *Trieste* spent weeks in Boston awaiting the opportunity to begin diving on the wreckage. During that delay the Office of Naval Research, the Navy Electronics Laboratory, and several civilian contractors extensively modified and equipped the bathyscaph to help it perform as an ocean-floor search vehicle. Dr. Harold E. (Doc) Edgerton at the Massachusetts Institute of Technology, who had designed the deep-sea cameras and strobe lights for the *Trieste*, and the technology firm of which he was a founding partner (Edgerton, Germeshausen and Grier, later EG&G) of Boston were called on to improve the camera equipment and to develop new sonar and acoustic navigation equipment for the coming effort.

One of the major safety concerns was the possibility of radiation at the wreck site. The bathyscaph pilots, observers, and the float itself were fitted with radiation detectors that recorded instantaneous and cumulative exposure.

Captain Andrews' plan to exploit the capabilities of available oceanographic research ships met with the enthusiasm of the research community. Among the most important assets that responded were the oceanographic research ships *Robert D. Conrad* (T-AGOR 3), operated by the Lamont Geological Observatory for the Navy; the *Atlantis II*, operated by WHOI; the *James M. Gilliss* (T-AGOR 4), operated by the Naval Oceanographic Office; and the *Josiah Willard Gibbs* (T-AGOR 1), operated by the Hudson Laboratories of Columbia University. These were ships that Andrews called his "classifier ships," which were to employ deep-towed sleds with cameras, magnetometers, Geiger counters, and side-looking sonars to investigate each suspicious bump on the ocean floor.

Attempts to tow search equipment directly over suspected wreck sites were universally unsuccessful—the scientists and oceanographers had never developed a means of precisely tracking and controlling the paths of their deep-towed sensors. Andrews' plan thus died, and individual shipboard initiatives would have to drive the search.

The ships' efforts began to bear fruit on 14 May, when a camera towed by the *Atlantis II* captured images of debris. The *Atlantis* debris field was about 700 yards north of "Point D," a ship-size bump about two and a half miles east of the where the *Thresher's* surface escort, the submarine rescue ship *Skylark* (ASR 20), had been at the time of the submarine's sinking. The *Atlantis II's* finding was a stroke of good fortune that directed all attention to areas within two miles of Point D.[20]

On 20 May, oceanographers on the *Conrad* decided to experiment with a deep-ocean dredge. A dozen attempts came up empty before on the 27th the *Conrad* snared a packet of 19 O-rings. A check of the *Thresher's* spare parts records indicated that she was the only Navy ship with a supply-system allowance for those O-rings. The Navy released this information with a statement that there was "almost 100 per cent certainty" that the O-rings were from the *Thresher*.

Soon after, the scientists in the *Atlantis II*, a couple of miles south of the *Conrad*, decided to cobble together a deep-ocean dredge of their own and eventually brought up a section of battery plate six inches in length.[21] This plate later was identified being from a *Thresher*-class submarine. By 26 May, attention was focusing on the southeastern quadrant of a two-mile-by-two-mile box with Point D at the center.

The *Conrad* struck gold on 14 June: photos of a large compressed-air tank, a sonar hydrophone, and an eight-by-nine-foot piece of sheet metal, and several other pieces recognized as from a submarine. Soon thereafter the *Conrad* reported a large magnetometer reading in the same area.

Having established on the basis of this early information a "probability area" measuring 1,700 yards by 1,700 yards, Captain Andrews decided on 18 June to bring out the *Trieste.* Keach had indicated that the *Trieste* needed a search area no greater than two miles by two miles and now it had one.

Back in April, once the bathyscaph had been lifted from the well deck of the *Point Defiance* and lowered into the waters of Boston Harbor the *Trieste* team had been quick to make the bathyscaph ready for diving. On the 28th Keach took the *Trieste* on a 36-foot dockside dip; two days later Martin duplicated the dockside test, with Captain Frank C. Jones, commandant of the Boston Naval Shipyard, on board. These tests checked out the lights, cameras, sonars, and underwater telephone.[22]

Ten days after its arrival in Boston, the *Trieste* was towed by the salvage ship *Preserver* (ARS 8) to the Murray Basin area of the Gulf of Maine, 60 miles east of Boston, for Dive #118, a deep test dive on 5 May. Keach was piloting, and Lieutenant Charles Walker, commanding officer of the submarine rescue ship *Tringa* (ASR 16), was along for the ride.[23] The *Tringa* would support the *Trieste* dive operations in August. This dive, to 700 feet, was made to check sphere integrity, sensor readiness, and controls. Keach reported that the currents were so strong that it was all he could do to avoid seafloor rocks and holes. Keach was fearless but admitted he was hard-pressed to get back to the *Preserver.*[24]

The *Trieste* team was attempting to make the craft ready to accomplish a mission never before assigned it—ocean-bottom search. The bathyscaph was ready for operational dives but had only a rudimentary ability to conduct horizontal visual or sonar searches along the ocean floor. As an underwater search vehicle, the *Trieste* inherently suffered from four major deficiencies:

1. *Navigation.* Bottom navigation was a new problem. Electronic navigation systems such as Loran, Decca, and Omega did not penetrate water. Dead reckoning was difficult where deep currents were unknown or variable. Inertial navigation systems devised for submarines were too large for the bathyscaph. Something new was required, and it would have to be improvised on scene.
2. *Endurance.* The existing batteries produced about 70 kilowatt-hours of electricity; they could power three five-horsepower propellers for

about three hours at a speed through the water of about one knot. In Boston, after the first five dives were completed, the number of batteries was increased by more than one-quarter with the addition of two deck-mounted battery boxes, each containing ten 12-volt automotive batteries. Available power was increased to about 90 kilowatt-hours, with the addition specifically for propulsion.[25]

3. *Sensors.* The sensors mounted on the *Trieste* were inadequate for effective search. The total darkness at the sea floor permitted only three options: sonar, visual aided by high-intensity external lighting, and magnetometer. The *Trieste* was designed to carry magnetometers, but the tons of iron ballast forward and aft made the magnetometers useless, and they were removed.
4. *Turnaround time.* From 1959 to 1962 the *Trieste* team had proved that given direct access to shore-based support they could maintain a service rate of about one deep dive per week. After each dive the bathyscaph required charging of batteries, replacing of tons of ballast, topping-off of gasoline, and replenishing the sphere's life-support system.[26]

A number of solutions were forthcoming: For *navigation*, a system that involved "fortune cookies" was devised.[27] The cookies were colored plastic sheets to be laid in a grid "array" on the bottom, each with a different number large enough for the *Trieste* to read as it passed nearby and thereby orient itself if the cookie had unfolded on the bottom properly, which most did not.[28] The USS *Allegheny* (ATA 179), a Navy tug re-equipped in 1952 for research duty, was entrusted with seeding the cookie array. Lieutenant Edward E. (Buzz) Henifin, who would later have a major role in the Trieste program, was designated to handle the deployment of the cookies from the *Allegheny*. Henifin recalled that the idea had originated with Allyn Vine of Woods Hole but was organized and supported by the Hudson Laboratories. The *Allegheny* dropped each cookie into the water based on Decca navigation, each ideally to land within 300 yards of its intended bottom position.[29] However, because of unpredictable underwater currents and navigational inaccuracies, the cookies were scattered and formed only a rough approximation of a useful grid. Kenneth Mackenzie later stated that the cookies were "valuable for planning further search procedures and later when photo montages were prepared."[30]

A second aid to bottom navigation established for the second series of dives involved a floating "taut-wire" buoy, emplaced by the *Conrad* about 1,000 yards north of the magnetometer contacts she had reported in mid-June. This buoy included a radar reflector above water, and a sonar transponder at the buoy's anchor, allowing the salvage ship *Preserver* to position the *Trieste* for dives at specific ranges and bearings from the buoy.[31] Unfortunately, this system did nothing to account for the bathyscaph's horizontal drift from underwater currents as it dropped to the ocean floor. Nor did it account for the "watch circle," the diameter traced by the buoy around its anchor 8,400 feet below in response to winds, tides, and currents. If the wire was taut enough to restrict buoy drift to 100 yards, the buoy was vulnerable to loss in the first foul weather. Eventually the buoy was allowed almost 800 yards of drift, which imposed an identical error on programmed bottom searches. Strong currents later caused the relatively lightweight anchor to drag several miles, after which the buoy was moored on one side of the search area as a reference point.[32]

The *Gilliss* and the *Atlantis II* fielded two versions of a third, much more elaborate concept that was then on the cutting edge of bottom navigation. A ship transmitted an acoustic signal that was retransmitted by a transponder mounted on the bathyscaph; the response was received by hydrophones placed at the ship's bow, midsection, and stern. From time differences in receipt of the response at each hydrophone, the distances of each to the transponder (bathyscaphe) could be calculated; from them, the craft's bearing and range from the ship could be triangulated.[33] This "3-D" system, designed by the Applied Physics Laboratory at the University of Washington, later would be developed and called "short line acoustic navigation"[34]

But how could that information be of use to the *Trieste*? The *Trieste* and the Naval Research Laboratory towed-camera system streamed from the *Gilliss* each was fitted with a sonar transponder, the position of which could be plotted as a range and bearing from the *Gilliss*. Then, by underwater telephone, the *Gilliss* could vector the *Trieste* toward a target or area of interest. Andrews later wrote, "Unfortunately, this system was just beginning to work when we all had to go home."[35]

With respect to *sensors*, a Straza continuous-transmission, frequency-modulated sonar was fitted to the *Trieste* that scanned from dead ahead to ten degrees abaft of either beam and could resolve targets out to a maximum effective operational range of 400 yards.[36] This sonar represented cutting-edge technology and often broke down after a few minutes of use.[37] The developmental sonar disappointed as a deep ocean-search tool.

Captain Andrews later wrote, "The major mystery of the entire *Thresher* search operation became the complete lack of detection of any of the *Thresher* wreckage by sonar. Not once was a sonar echo obtained and identified from any portion of the *Thresher*'s hulk."[38] His comment applied not only to the *Trieste*'s sonars but also to those of the various towed systems deployed by the other ships.

Visual search "by eyeball," by television, and/or fixed camera systems was possible, but only within the limitations of the external lighting installed below the *Trieste*'s float. Water within 20 feet of the deep-ocean bottom was quite clear, and unless the silt became disturbed, particulate matter in suspension did not inhibit visual examination of the ocean floor within 20 to 30 feet of the sphere. The four Edgerton 35-millimeter external cameras with synchronized strobes each had film for about 500 exposures. There was a closed-circuit TV system, manufactured by Ocean Engineering Corporation of San Diego, a single TV camera with pan-and-tilt capability mounted outside the sphere and a single five-inch monitor inside.[39]

The bottom could be observed from the Terni sphere only at the pilot's window at the lower center of the front of the sphere. Sightings were possible perhaps 30 feet ahead and 20 feet to either side of the *Trieste*'s track along the bottom. One man positioned himself at the front observation window; a second man watched the sonar and the TV monitor, navigated, communicated, and cogitated. When a third man crowded into the sphere, the work became a bit less of a one-handed juggling act.

The external radiation monitoring devices were fragile and had to be mounted by divers before each descent and removed after each surfacing. Over the program of *Thresher* dives these devices were constantly modified, repositioned, replaced, upgraded, and generally "fiddled with" by scientists and technical representatives. In addition, the pilot and crew in the bathyscaph sphere wore dosimeters and film badges, and a gamma ray–sensing Radiac instrument was carried inside the sphere.[40]

Reducing *turnaround time* was vital. The *Trieste* team had never attempted multiple dives in rapid succession, with all support and replenishment conducted at sea. Boston being a three-day tow from the *Thresher* dive site, major modifications to turnaround procedures had to be devised. To that end the *Trieste* team built a shot-tub loading machine that poured shot into the tub;[41] also, it made several refinements to its replenishing procedures.

Supporting the *Trieste*

The Office of Naval Research, the Navy Electronics Laboratory, and the Boston Naval Shipyard provided the *Trieste* almost unlimited access to necessary resources. Direct support of the bathyscaph was the province of the salvage ship *Preserver*—replenishment and maintenance between dives, as well as towing. If the weather and sea state allowed, the *Trieste* would be brought alongside the *Preserver,* where replenishment of the bathyscaph would be relatively speedy, battery recharging requiring the most time.

When seas demanded that the *Trieste*'s fragile float be kept at a more respectful distance, the *Preserver* would tow the bathyscaph close astern and float electrical, fueling, and ballasting lines to it. This evolution could take all night. With Keach and Martin being the only officers attached, Captain Andrews soon detailed additional officers to oversee replenishments to allow the diving officers some sleep. Among those so tasked was future *Trieste* officer-in-charge Buzz Henifin.

At the third level of support was the dock landing ship *Fort Snelling* (LSD 30), which could provide repairs, medical support, living accommodations, communications, and space for staffs and for civilian technical representatives. If necessary, the *Fort Snelling* could shelter the *Trieste*'s small dive tender, an LCM-8 landing craft, in her docking well. The *Fort Snelling* also provided dive security, interposing herself protectively between the dive site and any transiting civilian ship (or curious Soviet ship).[42]

The replenishment procedure had been tested off Boston from 24 June through 1 July 1963 in conjunction with five successful dives in seven days.[43] Captain Andrews would observe, "The wisdom of using a large ship—the *Fort Snelling*—as a seagoing base of operations, and the salvage ship *Preserver* as an immediate mother and tow ship became increasingly apparent throughout the operation. The whole operation would have been hopeless without the housekeeping, communications, and repair facilities provided by these two ships."[44]

10 The *Thresher* Search, 1963

The bottom was a mass of debris everywhere we looked. There could be no question—this was Thresher.

—Lieutenant Commander Donald Keach,
Officer-in-Charge, *Trieste*

After a three-day tow from Boston to the dive site by the salvage ship *Preserver*, the *Trieste* conducted its first *Thresher* dive on 24 June 1963, Dive #119. Don Keach and NEL's Kenneth V. Mackenzie conducted a visual and TV search for the debris field that had been reported by oceanographic research ships at approximately 41°-45'N / 64°-58'W, some 270 miles east of Boston Harbor.

The officer in tactical command of the search, Captain Frank Andrews, was on board the oceanographic research ship *Conrad*, while both the salvage ship *Preserver* and the dock landing ship *Fort Snelling* provided support. The *Trieste*'s goal was to produce conclusive photographs—possibly, "the clincher would be a view of the number '593' painted in white on the *Thresher*'s hull."[1]

The *Preserver* navigated to the selected dive site with a combination of Loran C, Decca, and Omega lines of position taken by the research ships every three to five minutes in advance of the dive time. Once submerged, the *Trieste* would navigate largely by dead reckoning. The dive commenced at 1035; a 37-kilocycle acoustic marker on the *Trieste* sent signals to the LCM-8 dive tender. In this manner the landing craft followed the bathyscaph's movements and tried to remain "on top."

The *Trieste* was diving to Point D, a ridge at a depth of 8,370 feet, where the *Atlantis II* had surveyed the fourth fathometer bottom trace that suggested the presence of a sunken ship. The area between the June magnetometer readings of the *Conrad* and the debris photographed by the *Atlantis II* north of Point D offered the highest probability of finding the debris field.[2]

On the bottom, Keach and Mackenzie discovered an undulating silted surface, hills and valleys, ravines and crags, with depth differences of as much as 100 feet. There were pits and holes large enough to hide wrecks, hindering visual search even with high-intensity external lighting. The *Trieste*'s sonar search—during those few minutes that the sonar was working—was of limited value; "Huge boulders and steep ridges frequently obscure the picture on our sonar," Keach reported.[3]

The debris field of a submarine imploding several thousand feet above the bottom would be spread across thousands of square yards of ocean bottom. Keach wrote, "What happens when a submarine goes down is it implodes, and then explodes: the lightweight material is carried by the current further than the heavy material, and the pressure hull and the reactor and all that stuff dropped straight down. But the rest of it goes on an angle, so you find aluminum, then you find paper and all that stuff downstream. So you've got a debris pattern that is elongated by as much as a mile or more."[4]

If the *Thresher*'s debris field could be found and mapped, the search for major hull components could be narrowed down to a reasonably sized area. The visual search, which had to contend with a cloud of reflective particles about 25 feet off the bottom continued until 1430, when low-battery alarms required the bathyscaph to surface. It appeared on the surface at 1555. This first deep dive of the operation uncovered no debris and lasted only three hours.

Dive #120—the second dive at the *Thresher* site—was delayed by both material and personnel casualties. One of the *Trieste*'s electrical inverters that changed the batteries' direct-current output into alternating current failed. The inverter could be repaired within 24 hours but required parts to be flown in. The personnel injury was to Lieutenant George Martin, the assistant officer-in-charge, who sprained his right ankle climbing from the *Trieste*'s LCM-8 to the *Preserver*. Initially it was recommended that Martin not work in the cramped sphere for an extended period, but he was able to talk himself back onto the dive schedule.

Landing Next to the Reactor

On 25 June the *Trieste* submerged at 0830, with Martin piloting and Mackenzie observing. The fathometer failed at about 2,000 feet, more than 6,000 feet above the ocean floor. With only a water-pressure depth gauge and the speed of the "snow" falling upward past his window to guide him, Martin started dumping ballast well before he saw the bottom. Some finesse was required: if too much ballast was dropped the bathyscaph would become positively buoyant before reaching the ocean floor.

Feeling his way down into the dark, Martin was unable to reduce his speed quickly enough when he suddenly saw light reflect off the bottom. The sphere hit the ocean floor too hard and buried itself so deeply that silt covered the forward window. The rear window remained clear, but the *Trieste* was well and truly stuck in the mud. Attempts to de-ballast and vertically pull out of the ooze were unsuccessful. The released ballast pellets just piled up on the ocean floor. Martin thought, "I'm not going to spend my whole life down here, dammit."

In times of stress memories come quickly, and Martin recalled learning in high school science that the resistance between two objects in contact sliding against each other—that is, sliding friction—is less than static friction. He could try to twist the craft so it would slide out.[5] But it would have to break free from the ocean floor without launching into an unstoppable ascent to the surface. Plus, the voice of Don Walsh, who had trained him in bathyscaph diving less than a year earlier, now reverberated in Martin's brain: "If you ever get stuck, you can always dump the tubs," followed by sardonic laughter. Martin remembered and understood the laughter very clearly. He definitely did not want to dump the shot tubs, which would mean a return to Boston to fit new ones.[6] Finally twisting free after 45 minutes of jostling, the *Trieste* shot up 500 feet above the bottom, but there Martin regained depth control.

He and Mackenzie had detected relatively high levels of radiation while stuck on the bottom and now speculated that the *Trieste* must have been *very close* to the submarine's reactor.[7] However, after lifting out and then returning to the ocean floor, Martin could not determine how far they had drifted. After two hours of further searching, the most that they could say was that they had been close but had sighted no debris.

The total time submerged was six hours, five minutes. An inspection of the bathyscaph float and sphere by divers back at the surface revealed only minor scrapes and paint damage from the rough bottoming.

The First Debris

The third dive of the search, on 27 June, Dive #121, demonstrated how electrical power management could extend endurance, adding almost 45 minutes of bottom search compared to the first dive. Once again the *Trieste* was positioned to dive near Point D, but this time would probe westward to seek the eastern edge of the debris field.

Again searching from about 20 feet off the bottom, Keach and Mackenzie saw nothing of interest for four hours. Then Keach spotted a flash of yellow in the mud. It was the first confirmed item of the *Thresher*'s debris field: a yellow plastic shoe cover of the kind worn in the submarine's reactor compartment. Visible on the plastic "bootie" was a stenciled marking: "SSN 5." The *Thresher*'s hull number, of course, was SSN 593; the plastic had been folded just where the last two numerals would have been. Adjacent to the bootie were paint chips and fragments of metal. As Keach maneuvered in the area, more scraps of paper and rags appeared—more than 100 individual items.[8]

Keach thus confirmed the location of the *Thresher*'s debris field reported by the *Atlantis II* and the *Conrad*. Significantly, this sighting was of light debris, which would be downstream of the major hull sections. If the general drift from currents could be estimated, the team could guesstimate a direction from this light debris to the much heavier hull sections. It later was determined that this field of light debris had fallen several hundred yards north of the heavy wreckage.

The *Trieste* surfaced at 1745 after six hours submerged. Keach later recalled, "I could not help but feel that it was only a matter of time before we would find the final resting place of *Thresher*. We had seen enough firm sediment to make it appear unlikely that *Thresher* could be buried to any extent. We had pieced together a picture of deep currents in the area. And last, but not least, we had shown beyond question that *Trieste* could do more than just dive; she could search effectively."[9]

An Incident at Sea

While Keach and Mackenzie were searching the bottom during Dive #121, a confrontation was taking place on the surface between the *Fort Snelling* and the Soviet fishing trawler *Kuprin*. The United States had issued a notice to aviators and mariners that established a navigation exclusion area within a 12-nautical-mile radius of the site. This notice appeared to attract Soviet ships. The most aggressive of these snoopers was the *Kuprin*, which on 27 June aggressively sought to get a close look at the search activities.[10]

When the *Trieste* was underwater, the greatest danger to her crew's safety was the possibility of surfacing into the bottom of a passing ship—with catastrophic results for the bathyscaph and its crew. The commanding officer of the *Fort Snelling* had orders to ram if necessary any ship threatening to put the bathyscaph at risk in that way, and on the 27th found himself in a "tango" with the *Kuprin* that might have resulted in a collision. What appears finally to have convinced the master of the *Kuprin* that he was pushing his luck too far may have been the actions of a gunner's mate at one of the *Fort Snelling*'s forward 3-inch/50-caliber twin gun mounts.

Conducting regular maintenance on the mount, the senior gunner's mate noticed the aggressiveness of the *Kuprin* and decided to conduct an unauthorized "test" of the mount's accuracy. Fixing the trawler's bridge in his gun sights may have suggested to the Soviet ship's master the one risk he was unwilling to ignore, as he had the *Fort Snelling*'s repeated signals by light, flags, and horn. Perhaps it now, finally, penetrated the mind of a determined master that the possible intelligence gain was not worth the risk of being shot. The *Kuprin* opened to more than 7,000 yards from the expected surfacing position of the *Trieste,* and thereafter the *Fort Snelling* successfully kept her there—and the reporters on board the LSD had a great "sea story." There was a general rush to the radio room to get their copy out to their editors.

But the *Fort Snelling*'s commanding officer declared a "casualty" to all of the ship's radio transmitters. That evening the reporters, who were gathered into the wardroom, were informed that the gunner's mate's actions had not been authorized and that inflammatory articles could cause an international incident. The mass transmitter "casualty" was soon corrected, and the reporters' stories about the day's events contained nothing about an overzealous gunner and his guns.[11]

Dive #121 on 21 June had convinced Keach and Andrews that the bathyscaph could make one more dive despite the continuing deterioration of electrical systems. Reconstruction of the *Trieste*'s track along the bottom had given Andrews' operations officer, Lieutenant Commander Arthur H. Gilmore, reason to believe that the next dive would be productive. Keach and Andrews decided to undertake it.

In a news conference in Washington on 26 June, a Navy spokesman indicated that while the Navy was reluctant to give up the search, almost

$1.9 million had already been spent out of the Navy's research and development budget for the search, a figure that did not include significant expenditures from the Navy's operating budget. The headline the next day in *The San Diego Union* read, "Deadline Nearing in *Thresher* Search."[12]

The fourth *Thresher* dive, #122 on 29 June 1963, took Martin and two observers down to the debris field. With Martin were Mackenzie and Lieutenant Commander Eugene J. Cash from the staff of the Commander, Submarine Force Atlantic. Cash had volunteered to help with the navigation and TV search, releasing Mackenzie for visual search.[13] This marked the first time three men had undertaken a deep dive in the *Trieste* and would pave the way for routine dives with three in the sphere. (Earlier Walsh and Keach each had piloted a three-man dive to relatively shallow depths.)

This three-man dive to 8,400 feet required the specific approval of the Chief of Naval Operations and Commander, Submarine Force Atlantic. The initial replies from both admirals were negative, because of the lack of emergency breathing gear for a third person. A civilian diver in the *Trieste* crew offered to put together an emergency air container that Cash could carry. This proposal was radioed to the CNO and ComSubLant, and both consented.

The jury-rig was the work of John R. Houchen, with the Navy Electronics Laboratory, who had joined the *Trieste* team specifically for the *Thresher* search. He had come up with a rather primitive solution—essentially a scuba air tank, a scuba mouthpiece, and a clip to pinch Cash's nose closed.[14] Just before Cash boarded the *Trieste*, Houchen pulled him aside to warn that there was only 30 minutes of air in the cylinder. As an emergency surfacing from 8,400 feet would take more than an hour, Houchen advised Cash in that case to breathe for a minute and then hold his breath for a minute!

Martin, Mackenzie, and Cash's bottom search on the 29th extended the known limits of the debris field, entering it from the northeast and discovering larger pieces and heavier parts, the largest a five- or six-inch gate valve probably weighing about 200 pounds. Heavier debris, as Keach explained, was less affected by ocean currents during its plunge and its presence indicated the *Trieste* was probing toward the area where major hull sections would be found.

Martin felt that the search was highly productive and decided to continue it until the main batteries ran down to zero. Once again, they were close to the main wreckage—but no cigar.

Each dive was adding to the mosaic of data that was being put together by Gilmore, Cash, and Andrews and that eventually would lead to the major hull wreckage of the *Thresher*. However, navigational inaccuracies were making the pattern less of a mosaic than a patchwork, whose pieces could not be precisely correlated. Dive #122 had been stretched to six hours, 16 minutes.

Meanwhile, the *Trieste* was becoming a maintenance nightmare. Electrical systems were failing, and electronics systems, such as the fathometer, were increasingly unreliable. This three-man dive was accordingly planned to be the last for this series. But although in later years Martin would express disappointment in the dive, its findings had caused a surge in optimism on the *Fort Snelling*—the results had been too good to halt the search. Andrews leaned on Keach: "I want one more dive. I'm going down personally." Keach relented.

The Last June Dive

Keach piloted what would really be the last dive of the series, Dive #123 on 30 June, with Mackenzie observing and Captain Andrews as the third man in the sphere. Andrews was there in the hope that a major hull section would be sighted. Just before Andrews boarded a small boat for the trip over to the bathyscaph, Cash pulled him aside to inform him of the 30-minute limitation on his emergency air supply. Andrews' reply included salty language: "You #@&*%, I've got nine kids at home!" Nevertheless, he went.[15]

Previous dives and reports from the research ships had consistently located the debris field in a one-mile square, and the bathyscaph's planned track on Dive #123 was designed to cross its area of highest probability. The *Trieste* submerged at 1315, and after only 90 minutes on the ocean floor the bathyscaph suffered a casualty to the starboard propulsion motor.

"We believe we were within 1,000 yards of the *Thresher* when the starboard motor failed," reported Mackenzie. The bathyscaph had no adjustable rudder, and imbalanced thrust would drive them in circles. The dive ended early, having failed to relocate the debris field.[16] In addition to the propulsion casualty, the bathyscaph also lost the use of its compass. The *Trieste*'s casualties were mission-ending and required shore-based repair. The *Preserver* towed the bathyscaph to Boston on 2–4 July.

Andrews later disclosed the effect of the navigational problems experienced during this series of dives: "At best the calculated position of a towed sensor was probably [accurate to only] 700 yards. It was pure chance, therefore, that enabled the towed cameras and magnetometers to be brought near enough [to a given point] to obtain readings. A revisit for further study was virtually impossible."[17] The log of the *Trieste*'s courses and speeds, with estimated bottom currents, was adjusted by a Decca radar fix taken upon the bathyscaph's surfacing. Total elapsed time between fixes had been six to eight hours. A reconstructed location of the *Thresher*'s debris sighted by the *Trieste* "was probably in error by at least 750 yards," according to Andrews.[18]

Modifications and Back to Sea

After almost two weeks at sea, including time under tow, the *Trieste*'s crew had plenty of work to do in Boston. The Boston shipyard was directed to extend a blank check to the *Trieste*; the shipyard workers were proud to be part of the *Thresher* search effort. Assisted by technical representatives and the yard workers, the *Trieste* crew undertook the repair or replacement of electronic and electrical equipment. Also, they repaired the compass and replaced the starboard thruster motor with a spare.

All personnel felt the pressure, none more so than Keach and Buono. One morning Keach told Martin through gritted teeth that Buono had reverted to his Italian shipyard habits of the old Jacques Piccard days by "temporarily" fixing several minor leaks in the gasoline floatation tanks with wooden plugs.[19] That definitely was not the "Navy way." Both Keach and Martin remained silent, however, and entrusted their lives over the next five dives to Buono's experience and expertise. After all, he had helped "write the book"—such as it was.

The refit extended into early August. To allow more time on the ocean floor upon return to the search site, the yard added two ten-battery boxes to the after portion of the weather deck, increasing propulsion power more than 25 percent. The redesigned and rebuilt manipulating arm was installed at the forward ballast tub for recovery of debris.

Upon the completion of the yard work, on 14 August, the *Preserver* towed the *Trieste* to a rendezvous with the *Fort Snelling* in the *Thresher* dive area.

Arriving in the dive area on the 18th, the *Trieste* found marginal weather conditions, heavy seas and increasing winds. Keach decided to prepare for the first dive anyway, only to have a freak wave swamp the access tube and drench the sphere's interior. It took five days with mops and sponges to dry out the sphere and make it ready to dive again.

An area 2,000 yards south and east of Point D had been seeded with 1,441 fortune-cookie markers during the *Trieste*'s July–August repair, to allow, it was hoped, more methodical navigation. However, both Dives #124 and #125 were to find them of no help. On this second series of five dives the *Preserver* positioned the *Trieste* by reference to the *Conrad*'s bottom-anchored, taut-wire, open-ocean buoy, which they christened the *Trieste*'s "lighthouse."[20]

On Dive #124, 23 August, Keach and Mackenzie took the *Trieste* to the bottom. They could spend only two hours, 30 minutes there, because of new casualties to navigation equipment. After that dive the weather continued to worsen, and the *Trieste* occasionally encountered 15- to 20-foot waves. They were being caused by the approach of Hurricane Beulah. Andrews maneuvered his Task Group 89.7, with the *Trieste* in tow, out of the dive area and to the north, toward Nova Scotia, to avoid it, then returned to the dive area.

Dive #125 occurred on 27 August, with Martin piloting and Mackenzie and Lieutenant Commander Gilmore of Andrews' staff as observers. The dive was to the southeast of Point D, where the debris field had been encountered. This dive generated no positive results; navigational inaccuracies continued to affect the search. After this dive Mackenzie returned to San Diego, and Commander James W. Davies from NEL arrived to take his place.

The rough sea state made the steel-hulled LCM-8 unusable as the dive tender, and for the remainder of this series the safety officer and divers employed a fiberglass Boston Whaler instead.

Success at Last!

On Dive #126, on 28 August, Keach took Davies and Gilmore down to a position north of a large air flask photographed by the towed camera of the *Conrad* on 14 June. On a southeasterly track, Keach searched for three hours before noticing small paper and paint flakes of the type sighted by Dive #121 on 27 June. Keach advised the others, "We must be close."

With an hour of battery endurance remaining, Keach slowed: "Within minutes we were in the midst of enormous masses of torn, twisted wreckage of all sizes and shapes. Battery plates, lead ballast, shredded cables, a section of superstructure—the bottom was a mass of debris everywhere we looked. There could be no question—this was *Thresher*. . . . I pointed out one large plate. We could read the numbers '0,' '3,' and '4' on it." Gilmore declared, "It's a bow section. Those are her 20-, 23-, and 24-foot draft marks." Keach continued:

> Again, our batteries were running low; we would have to surface soon. While we still could maneuver, I decided to retrieve some part of *Thresher* for surface analysis. I landed *Trieste* in a clear area alongside a section of twisted pipe about five feet long. Our newly installed manipulator was designed to perform the functions of a human arm with a claw, wrist, elbow, and shoulder joint, each controlled by a separate electric motor.
>
> Lowering, raising, twisting, grasping—for fifteen minutes I attempted the pickup. Each time the claw gripped the pipe, it failed to hold. It seemed to have arthritis. Finally, I managed to slide the elbow of the device under a crook in the pipe. I raised my catch gently until the arm held it against the forward ballast silo.
>
> Scuba divers retrieved the twisted pipe as soon as we surfaced. Close examination showed a scratched shipyard code marking—"593 Boat." That provided final identification. After months of backbreaking effort, we had found her.[21]

Lieutenant Buzz Henifin of Andrews' staff, who was waiting on the surface when the bathyscaph surfaced recalled retrieving pieces of a *Thresher* battery plate embedded in the nylon rope of the trail ball. Henifin retained a four-inch "T-shaped" piece and later had it framed. One of the officers lost on the *Thresher* had been a roommate of Henifin at the Naval Academy.[22]

There were two more dives on the *Thresher* remains in 1963. Both were anti-climactic and added nothing to the story. On Dive #127 on 29 August, Martin as pilot took Davies and Lieutenant Commander A. Dalton James, a medical officer on Captain Andrews' staff, on a four-hour bottom search. On the final effort, Dive #128 on 1 September, Keach took Dr. James and Henifin for a brief dive of one hour, 40 minutes on the ocean floor. A partial battery failure cut it short. Neither dive re-entered the debris field.[23]

These final dives highlighted two deficiencies, one new and specific, the other systemic. Dive #126 had discovered a major segment of the *Thresher*'s debris field, but the *Trieste* had nothing with which to mark the location for revisits, resulting in the failure of the next two dives to relocate the wreck. The *Trieste* would in the future be fitted to launch sonar transponders and acoustic beacons to mark items or locations requiring revisits.

The systemic problem—that in fact had affected the entire ten-dive dive program east of Boston in 1963—was the lack of any means for accurate and repeatable bottom navigation. Captain Andrews listed as his major conclusion of the 1963 dive series "the requirement to develop improved search techniques with the capability to locate the search sensors precisely in a geographic frame of reference." Andrews argued that the *Trieste*, with its navigational accuracy of probably no better than 750 yards, had been trying to locate and relocate a wreck site no larger than 400 yards in diameter.[24] Mackenzie assessed that the *Trieste* "only traversed a total of about 30 miles during the ten dives in 1963 and visually covered an area less than 0.3 square miles."[25] Mackenzie also reported that "attempts to utilize bottom pingers and transponders proved unsuccessful during 1963."

Correcting this fundamental inability to navigate accurately at the deep ocean floor would become the highest priority at several oceanographic laboratories and institutions over the coming years. The problem would be solved by the Naval Research Laboratory, and the methodology would be available within a year.

More than a month before the *Thresher* search, Captain Andrews had indicated that the goal of the *Trieste*'s dives was to produce one conclusive photograph.[26] Keach had done better. In addition to several photos of the wreckage of the sonar dome and related structural elements of the bow, he had brought back physical proof that the *Thresher* wreck had been located.

Following Dive #128, the *Trieste*'s battery boxes were found to be damaged, making it was necessary to halt. When the *Fort Snelling*, *Preserver*, and the *Trieste* arrived in Boston on 4 September, Andrews and Keach immediately boarded a flight to Washington. There they assisted the Secretary of the Navy's staff in drafting a news release, issued on 5 September:

> The latest series of five dives by the *Trieste* have been tremendously successful. In her third dive, on 24 August 1963, the bathyscaph took a number of extremely valuable photographs and made a unique recovery from the ocean floor. The item recovered was a length of copper piping and fitting with markings that definitely established that it came from the *Thresher*. The piping was picked up by a mechanical arm operated from inside the *Trieste*'s gondola in the first successful test of this device.[27]
>
> Tremendously valuable information has been developed. The results have been outstanding, [given] that the *Trieste* was not designed for tasks such as those that she has recently completed. We are planning to continue with exploration work in the *Thresher* area. As a part of this effort, periodic surveys will be made in the area using oceanographic research ships and new deep submergence vehicles and systems as they are developed.[28]

Secretary of the Navy Fred Korth concluded, "I have today directed that the associated operational aspects of the search for the nuclear submarine *Thresher* be terminated."[29]

On 6 September 1963, Secretary Korth pinned Navy Commendation Medals on Captain Andrews and Lieutenant Commander Keach and awarded the Navy Superior Civilian Service Award to Arthur Maxwell of ONR.[30] On 8 October 1963, Korth awarded the *Trieste*—the first time to a "research" craft—the Navy Unit Commendation for the 1963 *Thresher* search operation.

In recognition of a research craft's unusual status, the Navy specifically included civilian members of the crew in the award, authorizing them to wear the commendation ribbon.[31] Among the civilians so honored were Kenneth Mackenzie, John Houchen, Giuseppe Buono, and Manuel M. Medina of NEL. Separately, Andrews sent letters of commendation to the *Trieste*'s officers and enlisted crew. In February 1964, the other, non-NEL officers who had gone down in the *Trieste* during the *Thresher* search were added to the recipients of the Navy Unit Commendation.

Captain Andrews summarized the summer's accomplishments:

> The *Thresher* Search Operation shows that, with patience, the deep search and study problem can be solved. There are major problems yet to be solved, however, if the Navy is to become proficient at this type of an operation, whether it be search for a bottomed submarine, satellite, missile, or any other object. Most important, we must develop improved search techniques with the capability to locate the search sensors precisely in a geographic frame of reference.[32]

When the *Trieste* was examined at the Boston Naval Shipyard, deterioration of the topside electrical systems and the battery array was found to be so extensive that the *Trieste* was sent back to San Diego to await the new Mare Island–built float.

However, Keach, Martin, and Mackenzie were heading east. The Office of Naval Research had arranged an invitation from the French Navy for discussions with officials in their own bathyscaph program, the *Archimède*. Departing in mid-September, the trio, accompanied by Allyn Vine from Woods Hole, spent a week in Marseilles with Lieutenant Commander Georges Houot, his assistant, Lieutenant Gerard de Froberville, and the *Archimède* crew. The *Trieste* officers were interested in how the French had solved the problems of batteries and hull penetrations, both of which continued to plague the *Trieste*. They returned to San Diego in the last week of September.

On 23 September, Keach received orders to command the new diesel-electric submarine *Darter* (SS 576). He would be relieved of duty on the *Trieste* and depart in February 1964.[33]

The *Trieste* left Boston for San Diego on 16 October on board the Navy cargo ship *Francis X. McGraw* (T-AK 241), while the bathyscaph's eight enlisted men and three civilians returned there independently.[34] The *McGraw* arrived in San Diego on 29 October. There the Terni sphere was removed, overhauled, and prepared for attachment to the new Mare Island–built float.[35] The original float was then retired, having made 129 dives, 81 of them for the U.S. Navy and logged, and one more that was classified and not logged.[36]

Upon their return to the Navy Electronics Laboratory, Keach, Martin, Davies, Mackenzie, and Curl recorded their lessons learned and recommended program modifications to be forwarded to the Mare Island yard to be applied to the new float.[37]

Lieutenant Commander Gilmore from Captain Andrews' staff recorded the 1963 *Thresher* episode most succinctly: "The fact that *Thresher* was located *at all* using the crude equipment that was available in 1963 had important long term National Security implications. The 1963 effort to find *Thresher* brought many concepts and ideas to the fore and provided the seed for future underwater search and recovery efforts."[38]

Trieste Legacies

The Navy lent the Krupp sphere to the German pavilion for display during Expo '67 in Montreal, Canada. In September 1967 the Philadelphia Maritime Museum initiated a campaign to find a permanent home for the "old" *Trieste*. When they examined the old float, rusting away at NEL, the museum staff realized that they lacked the funds to restore the whole bathyscaph to display condition but offered to take the Krupp sphere.[39]

Instead, on 30 March 1967 Secretary of the Navy Paul H. Nitze proposed to Dr. S. Dillon Ripley of the Smithsonian Institution that the Navy loan the *Trieste* to the Smithsonian for display, contributing $10,000 worth of refurbishment toward an estimated $50,000 of required work. On 1 May the Smithsonian accepted the offer, and on 20 December 1967, the original *Trieste* float was placed on board a freighter for transport to Washington, to be reunited there with the Krupp sphere upon its return from Montreal.[40]

The Smithsonian withdrew the *Trieste* from public display in 1973 and in February 1978 returned the bathyscaph to the Navy, transferring her to the museum in the Washington Navy Yard, in Washington, D.C. The *Trieste* is now on permanent display at the museum, with the Krupp sphere mounted and the Terni sphere resting alongside.

11 The *Trieste II*

The basic design philosophy, which is heartily endorsed by the operators, is that the number of surfacings must be exactly equal to the number of dives.

—Herbert L. Graybeal, Chief Design Engineer,
Mare Island Naval Shipyard

The Mare Island Naval Shipyard had begun studies of an improved float for the bathyscaph *Trieste* in November 1962, under the supervision of chief design engineer Herbert L. Graybeal.[1] The rebuild and recertification of the Terni sphere during the winter 1962–1963 had made obvious the need for a more advanced float. The naval architects at Mare Island undertook the design of a *scientific* deep diver with better towing ability, greater underwater endurance, improved safety, and less vulnerability to damage from waves and weather. The Mare Island float would replace the aging Piccard float that had made 129 dives—including the Walsh-Piccard dive to the deepest location on Earth.

In the spring of 1963, upon completion of specifications and drawings, and with Bureau of Ships approval, the design shop at the Mare Island set about producing the construction plans and drawings needed to cut steel.[2] Soon there were "shipalts"—alterations to the original plans—reflecting lessons learned in the *Thresher* search. Graybeal wrote that the *Trieste II* "was designed as an oceanographic research vehicle with the capacity to dive to any depth with limited maneuverability and endurance. The design characteristics were determined before the *Thresher* accident, so it was not really suited for that mission."[3] In light of the *Thresher* loss, the basic premise that the *Trieste II* would be only a scientific research tool was re-examined, and a deep submergence-search-and-recovery capability was provided for.

Tows up to ten knots would require a more boat-like hull form, a reinforced bow structure, and a raised prow to cut through waves. New safety measures ensured the bathyscaph would float if two of its twelve gasoline floatation tanks were damaged and their contents lost. For this purpose the *Trieste II* carried a reserve of 15,600 pounds of releasable ballast. Safety also was the reason for greater freeboard. The sphere was placed farther forward in the hull and partially recessed into the float to provide protection during towing.

These changes resulted in a much larger and heavier float: 78 feet long, 95 tons dry and 336 tons fully loaded with the sphere, ballast, and gasoline. The new float was designed to be married to either the Terni or the Krupp sphere, but the Terni sphere was the default, with a designed 20,000-foot operating depth—which covered 98 percent of the ocean floor. If married to the Krupp sphere the operating depth would be unlimited. The assembled float plus sphere had a 19-foot draft when ready to dive. Model tests of the new configuration were conducted at both the University of California and the Navy's David Taylor Ship Research and Development Center in Bethesda, Maryland.[4]

The new float was under construction and testing from August to October 1963. In the first week of November it was lifted into the floating dry dock *ARD 11,* which was towed by the fleet tug *Apache* (ATF 67) to San Diego. The float arrived at the Navy Electronics Laboratory in mid-November.[5]

On 17 January 1964, the Navy formally dedicated and "launched" the new bathyscaph at the NEL piers on Point Loma. The Mare Island float with the refurbished Terni sphere were jointly christened as the *Trieste II.* There had been a move afoot to call this new craft the *Point Loma*, but the Navy preferred to build on the exceptional public recognition of the name *Trieste*.[6]

More than 400 persons attended the ceremony, including the crew of the *Trieste II*, many with their wives, officials from the Mare Island shipyard, and senior personnel from NEL. Rear Admiral Denys W. Knoll, the Oceanographer of the Navy, delivered the dedication: "The conquest of inner space, our oceans, is yet to be glamorized; nevertheless, achievements in ocean exploration have been as awe-inspiring as those in outer space." Knoll continued, "Our oceanographic program requires many vehicles like *Trieste* to conduct scientific research, to identify phenomena for investigation and to collect valuable data to improve man's knowledge of the ocean's depths. Deep research vehicles, capable of transporting men to the depths of the oceans, are needed to successfully implement our national oceanographic program."[7]

At the conclusion of the dedication, Elizabeth Keach, wife of the officer-in-charge of the *Trieste II*, pulled a line to pour a bottle of deep-sea water over the new bathyscaph, symbolizing its rededication to the sea. A few minutes later a floating crane lifted the bathyscaph from its cradle and lowered it gently into San Diego Bay.[8] This was the only time any incarnation of the *Trieste* bathyscaph was officially dedicated and launched with public attention and fanfare.

Two Officers-in-Charge?

In January 1964, two officers arrived at NEL, both clutching orders to relieve Lieutenant Commander Keach as officer-in-charge of the *Trieste II*: Lieutenant Commander John B. (Brad) Mooney Jr. and Lieutenant Shumaker, who had been Don Walsh's assistant officer-in-charge in 1960–1962. Both officers were eminently qualified, but the kerfuffle was decided in Mooney's favor. Washington officials wanted Mooney at least in part because he already held high-level security clearances that could take months to provide to Shumaker.[9] Instead, Shumaker was given the option of another set of orders or staying on; he chose to stay on board as "chief pilot."

Mooney was the second Naval Academy graduate (Class of 1953) to snare command of the high-visibility *Trieste* program. He was a qualified submariner and came from a staff position that had involved high-level security accesses. He was already known as a "comer," and he would become the only *Trieste* alumnus to reach flag rank.

After a few shallow dives in February 1964, Mooney relieved Keach as *Trieste* officer-in-charge.[10] The next assistant officer-in-charge, Lieutenant John H. Howland, reported in April but would not relieve Lieutenant Martin until June 1964.

The decision of whether to redeploy the *Trieste II* to Boston for the 1964 summer diving season was awaiting the upshot of developments already in motion.[11] There had been consideration at the highest Navy level of the potential value of a second season of dives on the *Thresher* wreckage. Some officials voiced concern that a second search would simply expose the Navy to addition criticism about the loss of the *Thresher* and Navy's inability to locate the remains more promptly, the year before.

Admiral Claude V. Ricketts, the Vice Chief of Naval Operations, on 5 April 1964, publicly expressed the Navy's impatience: "One great lesson is the urgent requirement for a deeper probe into the ocean. For example, isn't it incongruous that we can track man-made satellites orbiting the Earth at hundreds of miles altitude, but we have not been able to locate [the main components of] *Thresher,* 8,400 feet under the surface of the ocean?"[12] The counter-argument, expressed most strongly by Assistant Secretary of the Navy James Wakelin, was that the embarrassment would be best resolved by solving the problems exposed by the 1963 search.[13]

It was not until mid-March, when the Naval Research Laboratory's research ship *Mizar* (T-AGOR 11) was about to be modified with advanced oceanographic and deep-search equipment, that the final decision was made to send the *Trieste II* to Boston. The *Mizar* and the *Trieste II* would apply new equipment and procedures to address the difficulties experienced in the earlier *Thresher* search.

Return to Boston

On 14 April 1964, the *Trieste II* left San Diego on board the Navy cargo ship *Francis X. McGraw.*[14] Lieutenant Howland and four enlisted men embarked with the bathyscaph. The *McGraw* arrived at the Boston Naval Shipyard on 3 May; there the bathyscaph was offloaded and prepared for the installation of additional equipment and sensors.[15] On 25 May the *Trieste II* was lowered into the water.[16] The crew loaded the shot tubs the same day and filled the float with avgas on the following day. On 5 June, Howland took the bathyscaph on a dip pierside in Boston Harbor, primarily to test the new sonar against a target of known size and shape.[17] Two days later Howland relieved Martin as assistant officer-in-charge, and Martin departed for the diesel-electric submarine *Trutta* (SS 421) in Key West, Florida, as the executive officer.[18]

On 8 June, the salvage ship *Hoist* (ARS 40) towed the bathyscaph out to a position about 70 miles off the coast, in the Gulf of Maine. Three dives to between 600 and 1,000 feet were planned to familiarize the *Hoist*'s crew with the *Trieste*'s pre- and post-dive procedures, test sonar systems, and acquaint new bathyscaph pilots with Atlantic waters.[19] Mooney reported that on the first dive, on 9 June, "one of the shot chambers froze up when we were down 600 feet. It was spooky, but we managed to make a normal ascent by releasing the shot from the other chamber."[20]

Dive #12 on 9 June, with Mooney, Shumaker, and Mackenzie on board, encountered a problem with rusty shot that clogged the magnetic ballast-release valves, making it impossible to release ballast.[21] Mooney would have to dump the shot tubs themselves, with their contents stuck inside, but he made the conservative decision to dump only the after tub, not both. That action made for a wild, corkscrewing ascent to the surface in an uncomfortably out-of-trim bathyscaph. A review of the loading determined that even when the shot tubs were loaded to capacity, enough water and air remained in the tubs to rust the shot, seal shut the magnetic release valve, and force the jettisoning of an entire shot tub.[22] Thereafter, the crew flooded the shot tubs to expel all of the air before loading shot. Problem solved. That was only time that a Navy bathyscaph pilot had to eject a ballast tub to return to the surface. (Jacques Piccard had ejected both tubs in an emergency surfacing on Dive #7, on 2 October 1953—before the U.S. Navy acquired the *Trieste*.)

After jettisoning of the shot tub the *Hoist* had to return the *Trieste II* to Boston, on 11 June, after only one of the three planned practice dives.[23] The *Trieste* was under way again on 15 June with a replacement shot tub and made practice dives on the 18th and 19th. The highlight of the former, T-II Dive #13, was acquiring by sonar and homing on a Straza transponder at 1,700 yards. This was repeated the next day on T-II Dive #14. The *Hoist* and *Trieste* arrived back in Boston on 20 June, having successfully completed all the "shallow" dive checks.[24]

Preparations to Return

Captain Frank Andrews was again officer in tactical command of the *Thresher* search force, this year designated Task Group 168.1, consisting of the oceanographic research ship *Mizar* operated by NRL, the salvage ship *Hoist*, and the *Trieste II*. Using the *Hoist* as the mother ship brought a critical disadvantage, in that the salvage ship had a limited avgas capacity and the *Trieste* had to top off before every dive. This problem was solved by filling large rubber fenders with avgas and lashing them to the sides of the *Hoist*.[25]

The Navy took pains to avoid the media circus of the previous year. The 1964 dives were announced as oceanographic research to further Navy development of deep-ocean technological systems and equipment. Reporters were *not* invited to the site: "The primary purpose will be to develop the difficult techniques for locating any future submarine that may be lost in the ocean depths."[26]

Again, colored and numbered bottom markers were seeded in the target area. Nine north–south lanes, each with 85 markers, were dropped sequentially from the *Hoist* between 41°-44'N / 64°-56'W and 41°-45'N / 64°-57'W, for a total of 935 markers. Lieutenants Ray Stoetzer, Paul Peterson, and Millard Firebaugh, all active-duty Navy officers and students at the Massachusetts Institute of Technology, designed this new system of bottom markers. They had set up a mini-factory to produce them and organized a labor force of volunteers from the First Naval District's Boston brig.

Stoetzer's new design—for which Peterson coined the term "stoetzeroonies," likening them to a form of pasta—used solid Plexiglas fins.[27] The stoetzeroonies were useful only if within visual range (20 to 40 feet) of the bathyscaph's track over the bottom; as they were emplaced an average of 60 feet apart in rows 180 feet apart, they once again proved of limited value for visual search.[28] The system showed potential in salvage or archaeological situations but for present purposes proved inferior to the acoustic system employed by the *Mizar*.[29]

The new Mare Island float promised to make the new bathyscaph a far more capable bottom-search vehicle than the earlier craft had been. The *Trieste II* could average six and one-half hours per dive, only slightly exceeding the average experienced in 1963 but extending bottom search time from four hours to five hours per dive. And, the *Trieste II* could search at speeds in excess of three knots, although Mooney advised that such speeds were "too fast" for effective visual search.[30] While both Martin and Shumaker had extensive experience in the original *Trieste*, in the operations now coming up the new float would present unexpected challenges and expose several hidden flaws. For example, engineers festooned the float with radiological sensors and devices to capture water and bottom samples. The theory was that the *Trieste II* would stir up mud and sediment from the bottom whose radioactivity could be measured and of which samples could be captured by deck-mounted equipment. The concept was flawed, and it was not until the *Thresher* site was revisited in later years that a full set of baseline radiological measurements could be obtained.

The Columbia Broadcasting System (CBS) installed a movie camera with a 160-degree lens in the sphere with which the hydronauts could film a typical ascent, descent, and transit of the bottom. When time was available, they also took photos with the CBS camera through the forward window. Also introduced were throat microphones connected to an audio tape recorder that eliminated the written logs kept during previous dives.[31]

The real star of the upcoming show was not the *Trieste II*, but the oceanographic research ship *Mizar*, under the operational control of Chester (Buck) Buchanan of the Naval Research Laboratory. He was the remarkably practical scientist who had found the small Navy cargo ship *Mizar,* built as an Arctic supply ship and placed in Navy service in 1958 as T-AK 212. Buchanan had got NRL to acquire the ship, sponsor her 1963–1964 conversion, provide special search and oceanographic systems, train her crew and technical team, and then sail her on a triumphant demonstration of his vision and acumen, all within a period of nine months.

In 1963, Buchanan had led the NRL search team on the oceanographic research ship *James M. Gilliss*, one of the four "classifier ships" engaged in investigating "possibles" of the *Thresher* wreckage detected by precision-mapping ships. Buchanan's team had deployed an early-version towed camera sled from the *Gilliss*, eventually taking more than 30,000 exposures at depth. One of those photographs showed unidentifiable debris, which was (in similar 1964 photography) only later recognized as the *Thresher*'s sail. Also in 1963, the *Gilliss* fielded an embryonic "short-baseline" acoustic navigation system (see chapter 9).

With this exposure to the situation, when the 1963 search terminated Buchanan was convinced that he knew the problem and could field the solution. The problem was that the cameras, magnetometers, and sonars towed by surface ships could not be controlled to pass over precise points, nor even be monitored as to their exact locations above the bottom.

Dr. Stanley Murphy from the University of Washington's Applied Physics Laboratory arrived to assist in installing the 3-D short-baseline navigation system on the ship. Buchanan was also able to elicit a "donation" by the HRB Singer Company of a new-model underwater magnetometer for their towed "fish."[32] After all this equipment had been fitted, the *Mizar* sailed to Boston for a shakedown of the equipment, including an underwater tracking system to track the location and movements of the towed fish and of the *Trieste II*. This system consisted of the Applied Physics Laboratory's three hydrophones mounted on the ship, a transponder on the bottom, another on the towed fish, and a third on the *Trieste II*. Within a week, all hands were confident that they knew their craft well enough to begin bottom photography.[33]

On 25 June the *Mizar* sailed from Boston to Point D. Theoretically, Buchanan's underwater tracking equipment could plot the towed fish with an accuracy of about 250 feet (a ten-fold improvement over the 750-yard accuracy of 1963). The plotted locations of photographs were good to within 100 feet, and the new magnetometer could sweep a path about 300 feet in width.[34] Buchanan's technique was to run random sweeps through the probability area, using only the magnetometer to detect and plot large *Thresher* debris. Then the camera would be activated, and sweeps through the area would continue until the fish was directly over the magnetic source, the target. Later in the *Thresher* search a camera was rigged that was activated by the magnetometer.

On the *Mizar*, Captain Andrews' staff included the three summer volunteers on temporary additional duty from the Navy's MIT program and Lieutenant Dennis E. Curtis. Curtis served with Andrews during the 1963 *Thresher* search, then had resigned from active service to finish his first year of law school. In May 1964, Curtis received a telephone call from Andrews inviting him to return to active duty for the new search. He leapt at the offer.

Curtiss became expert at using the *Mizar*'s acoustic navigation system: interrogating a transponder on the bottom, another on the towed fish, and one on the *Trieste II*. He was designated chief of the at-sea analysis group, responsible for plotting the location of the photographic information and developing a large photo mosaic of the debris field.[35]

The staff used an empty hold in the *Mizar* to lay out the photos in a gigantic mosaic.[36] The photographic montage was then taped together and glued onto butcher paper.[37]

Astonishingly, on 27 June 1964, within the first eight hours of search, Buchanan's team on the *Mizar* located and photographed the first major segment of the *Thresher* hulk.[38] By 21 July almost the entire wreck, which had broken into five major sections, had been photographed. On the 23rd the ships returned to Boston with a broom affixed to the *Mizar*'s signal halyard, the traditional naval symbol of a "clean sweep."[39] The *Mizar* ultimately took 119,000 photos of the bottom during the 1964 search.[40] In less than nine months Buchanan and his team had erased the key deficiency that had plagued the 1963 *Thresher* search, the kind of achievement that Buchanan and the *Mizar* would be repeating in subsequent deep-ocean search operations.

Trieste II on Fire

While the *Mizar*'s survey was only partially complete by 27 June, her photography of that day was sufficient to sail the *Hoist* towing the *Trieste II* into the area. The bathyscaph's fourth 1964 dive off Boston and the first dive on the *Thresher* site was at 41°-44.73'N / 64°-55.05'W on 3 July—that is, T-II Dive #15 with Mooney, Shumaker, and Mackenzie. Early in the dive the manipulating-arm control box shorted out when the crew was attempting to operate the arm. For the remainder of the dive the arm was dragged through the bottom silt along with the trail ball, leaving a uniquely identifiable trace on the bottom.

Captain Andrews described the termination of the 3 July dive:

> While on the bottom, and near the end of the six-hour dive, the *Trieste II*'s main [propulsion] motors developed a severe short circuit. Unfortunately, the construction and setting of the overload relays in the 120-volt battery circuit was such that they failed to open, but instead arced over to each other. The arcing apparently caused a carbonization of the compensation oil used inside the relay boxes. There were no fuses in the battery box, and hence excessive current continued to flow until the battery cables melted. The battery continued to discharge to ground so that no electrical energy was left in the main power batteries when the *Trieste II* surfaced.
>
> The fortunate aspect of the situation was that the fire occurred submerged at 8,300 feet and that the battery was completely discharged

> before surfacing. Had there been a source of oxygen, it is entirely possible that the arcing could have touched off the 46,000 gallons of high-octane gasoline in the tanks, roughly four feet from the main motors and battery tank.[41]

Later, Mooney would quip, "We had an electrical short at 8,400 feet, it almost spoiled our afternoon." While such sangfroid could be summoned much later, at the time the *Trieste*'s officers and crew were greatly concerned over the safety and reliability of the new Mare Island float.

The *Hoist* towed the bathyscaph back to Boston, arriving on 8 July. Correcting the problems delayed the search operation for more than a month. With only 15 dives in the *Trieste II*, two of which had put the lives of the crew at risk, a complete reexamination of the bathyscaph and its systems was required.

Mooney recalled:

> During the period immediately following [the electrical fire], I realized more and more how lucky we were to come back from Dive #15. I also recalled Dr. [James] Wakelin telling me not to dive *Trieste* if I was unsure of its safely because the loss of any vehicle would set back the entire deep submergence program.
>
> I [over-]reacted a bit to get *Trieste* more support and get it out of the "do-it-yourself" category. I was called to Washington and sat before Dr. [Robert] Morse, Admiral [John] Leydon [the new Chief of Naval Research], Admiral [William] Brockett [the Chief of the Bureau of Ships] and about 20 other assorted people to state my case.[42]

Mooney's screams for help were potentially career destroying. His summons to appear before a panel of admirals and senior civilians in Washington was a case of massive organizations defending themselves from the claims of a lieutenant commander. Such confrontations rarely end well for the officer being examined. However, Mooney was no average officer—he was comfortable dealing with admirals on a professional level and had the facts on his side.

Upon conclusion of the Washington meeting, Mooney reported, "All agreed *Trieste* had not received the support it deserved. All agreed it was time to get engineers and experts to assist us. All my doubts were answered satisfactorily or fixed prior to our going to sea."[43]

Reassessment and Repairs

The Bureau of Ships became more directly involved in the *Trieste* program. BuShips engineers reviewed—perhaps with more rigor than before—Mare Island's design and engineering program for the new float for other potentially life-threatening flaws.

They found one. Mare Island had erred in the stability calculations. BuShips engineers calculated that under certain circumstances, if the *Trieste II* dropped both shot tubs the center of buoyancy would suddenly be above the center of gravity, and the craft could flip upside-down on the way to the surface.[44] Further, it was an open question whether such a flip would disconnect the sphere from the float—no one wanted to conduct that experiment. BuShips arranged to load 7,000 pounds of lead into the bottom of the ballast tank, which helped to solve the problem, but that was a temporary fix. Until something final could be completed the *Trieste II* pilots would have to refrain from dropping both shot tubs near the surface before the avgas in the tanks had time to cool.[45] They were only "one-point" safe, as any interruption of power to the magnets holding the tubs in place would instantly drop both without human intervention.

Personnel from the Portsmouth Naval Shipyard in Kittery, Maine, were brought in to remove both of the Hoover five-horsepower propulsion motors and replace them with one-horsepower General Electric motors, which would reduce amperage loads. This too was a stopgap measure, accepted in order to proceed with the dives at the *Thresher* site. Portsmouth electricians also rewired the propulsion power circuit and installed the necessary fusing. The result was an under-powered, sluggish, and nearly unmaneuverable vehicle, capable of less than one-half knot on the ocean floor. As bottom currents could average almost that much, navigating and maneuvering the bathyscaph at the ocean floor could be a difficult and frustrating task.[46]

The poor performance of both sonars—the Straza continuous-transmission/frequency-modulated 35–45 kilocycle sonar and the newly installed Edgerton 12-kilocycle side-scanning sonar—brought in technical representatives from both firms to help with them. Similarly, the mechanical arm that had shorted out was completely rebuilt. On 4 August, EG&G's Marty Klein assisted Buchanan on the *Mizar* with the Straza transponder problem. A pinger supplied by Ocean Research Corporation also was tested and worked well.[47] With the *Trieste II*'s problems largely solved, the *Mizar* and the *Hoist* sailed from Boston on 10 August with the bathyscaph in tow.

Back into the Depths

The 14 August dive, #16, piloted by Mooney with Howland and Fitzgerald in the sphere, experienced gyrocompass problems.[48] Mooney related, "We sighted 'fortune cookies' almost immediately. We wandered generally westward and were in light debris all of the five hours on the bottom. Our photographs tied Keach's 'junkyard' and the *Atlantis* area together. They are only 150 yards apart. Apparently, navigation inaccuracies separated them by 8,000 yards in 1963."[49]

The repetition of the same gyrocompass casualty disrupted the next dive, #17 on 16 August, with Shumaker piloting with Howland and Mackenzie on board. They spent seven hours at depth, but their only sightings were a few of the stoetzeroonies. The *Mizar*'s 3-D acoustic bottom navigation system was working and provided vectors, but the *Trieste II* was so under-powered that it could not make headway into the current. The start point for the next dive would have to be up-current from the target area.

Finally, on 18 August, everything came together. Mooney piloted with Howland and Andrews on Dive #18. Andrews and Buchanan selected a site for the dive; Curtis and Firebaugh "vectored" the *Trieste II* into the appropriate location on the surface to commence descent. After two dives spoiled by gyrocompass problems, there was some doubt on the *Mizar* about the bathyscaph's readiness for the task in its current condition. Buchanan recalled: "This had never been attempted before, and we really had no way to determine the accuracy of the system. Finally, the little pencil dots on our plotting board reached the point where we had previously judged the *Thresher* hulk to lie."

Mooney continued the story of the 18 August dive:

> Upon landing on the bottom, we moved in a westerly direction. The [*Mizar*'s] 3-D system confirmed that this was a desirable course. After

> 15 minutes, we sighted the unique double track [the dragging trail ball and mechanical arm] from Dive #15. We followed it for maybe 100 yards then sighted the scar [a track of a towed sled that had earlier impacted the bottom]. Photographs from a towed sled indicated that our best course of action was to turn left and follow the scar. This we did and after traveling about 30 yards, we saw violently disturbed bottom sediment. It looked like a sandpit. Sediment was piled up in it in irregularly placed mounds, maybe 30 feet high. We took *Trieste* into the sandpit and looked around.[50]

Communications between Captain Andrews and Lieutenant Curtis by the UQC underwater telephone became tense:

> *Andrews:* I need steering to the target.
> *Curtis:* You are as close as we can get you.
> *Andrews:* We don't see anything. Give me something.
> *Curtis:* I can't give you anything. Our data shows you in the right area.
> *Andrews* [frustrated and irritated]: Give me something—*anything.*
> *Curtis:* There is no data to give you. Just look around. Give it a 360-degree sweep.[51]

Mooney had set the bathyscaph on the bottom at Andrews' direction and was looking through the forward window while Howell watched the TV monitor and Andrews handled communications. Howland suddenly yelled, "Hold it! Holy cow, there's metal, it's all metal, it's the hulk!" Mooney realized that "we had been resting 30 feet above the bottom on a section of the *Thresher*'s hull."[52] The operations center on the *Mizar* got the word at once as Andrews had left the UQC microphone keyed open. They could hear Howland's excited yells: "There!" "Over there, too!" "It's the hulk!" Mooney concluded his dive summary:

> One side of that hull was butted up against the side of the sand pit, and the other side of the hull had no sand around it at all. You could see down to where it was sitting on the bottom.

> It looked like it was a swirling impact. As it swirled, it sort of dug its own hole there.
>
> So, we moved around on top of the hull and could see the safety track that's on top of the submarine where you have safety lines. We could see a large piece of metal that stuck up, the point of it sort of stuck back toward us like a beer can opener, like a church key. We, of course, with a three-sixteenth-inch hull, wanted to stay fairly far away from that, so we gave it a lot of berth.
>
> We did attempt to see if there were any additional high-level [radiation] readings down on the side of the hull, so we moved off the hull, eventually went down between it and the side of the sandpit. It was a tight squeeze. It turned out that we scraped paint off both sides of the *Trieste* as we descended down into that hole, and down there, there were still no elevated readings.[53]

Later, at the dive debriefing, apparently still piqued, Andrews asked Curtis why he had not provided better vectoring information. Curtis replied, "We dropped you right on top of it. How much better could we have done?" Andrews smiled. All was forgiven.[54]

There was one further dive, on 19 August, the fifth in that series, piloted by Shumaker, with Howland and Lieutenant Edward Kingston from the Submarine Development Group 2 staff. Once again they found themselves fighting the deep current with a nearly unmanageable submersible. The *Mizar* tracked them for most of the dive, to about 200 yards west of the main debris field, which checked against sighted fortune cookies. They were unable to close the distance. Dive #19 did not add to the existing data on the wreckage location or condition.

Captain Andrews summarized the Navy's improvements in navigating and deep-ocean bottom searches in this way:

> The 1964 search operation indicated that the *Mizar* . . . underwater navigation scheme was highly successful, although random errors of 250 feet were still considered larger than desirable. The significant problem yet to be solved, however, was that of navigating the bathyscaph independently

> of a surface ship. The 1964 operation ended with significant progress in navigation at 8,400 feet, but with more engineering sophistication and new concepts still required.[55]

In June 1963, Captain Andrews had told the press that the "clincher would be a view of the number '593' painted in white on the *Thresher*'s hull."[56] The *Mizar* captured that photo in July 1964, and then employed her new underwater tracking system to vector the *Trieste II* into physical contact with one of the major hull sections. On 1 October, a press release from Secretary of the Navy Paul Nitze announced that the *Trieste II* had spent 37 hours submerged on five dives in the area.

Captain Andrews spoke well of the *Trieste II* and her crew, despite the problems experienced during the summer of 1964:

> In summary, the 1963 and 1964 operations demonstrated rather conclusively that the manned vehicle is not merely a piece of hardware which can be sent indiscriminately below the surface of the water. It is rather a vehicle with human beings inside, and hence all precautions must be taken to ensure that the safety of operator's lives is maximized. . . . It also is important to state that the entire *Trieste* operation required considerable skill, courage, and devotion to duty on the part of its officers, civilian and enlisted crew. None of these effective and pioneering operations could have been conducted had it not been for the superb leadership of Lieutenant Commander Keach, the officer-in-charge of *Trieste I* in 1963, and Lieutenant Commander Mooney, the officer-in-charge of *Trieste II* in 1964.[57]

Captain Andrews retired from active service shortly after the 1964 *Thresher* search. However, because of his unique experience in underwater search, he was called on by the Navy to consult during both the searches for the H-bomb off of Palomares, Spain, in 1966 and for the sunken submarine *Scorpion* (SSN 589) in 1968.

In 1965, the Navy honored Buchanan's NRL sonar branch with the Navy Merit Award for the 1964 *Thresher* search, naming 24 awardees. In 1966, the

Marine Technology Society recognized and commended "Mr. C. L. Buchanan and his team of scientists and engineers from the U.S. Naval Research Laboratory, who by their ingenuity and persistence under the most adverse conditions, made it possible to locate and identify the submarine *Thresher* over eight thousand feet below the surface of the North Atlantic Ocean."[58]

After the *Thresher* search, Buchanan and his *Mizar* team went from success to success. From 1963 to 1974, the *Mizar* participated in several important searches: for the H-bomb off Palomares, for the *Scorpion* wreckage, for the sunken submersible *Alvin*, and for the sunken French submarines *Eurydice* and *Minerve*, among many other deep-sea search operations.

The *Trieste II* was loaded on board the commercial freighter *President Buchanan*, which left Boston on 7 October 1964, and arrived in San Diego on the 29th.[59] Mooney told onlookers, "We plan some minor adjustments; then hope to begin scientific dives off the coast here in January."[60]

Mooney was being less than candid.

12 Between Disasters

What's impressed me most is the improved undersea operational capability that's been developed in less than three years since the efforts to find the submarine Thresher.

—Paul M. Fry, Director, Woods Hole Oceanographic Institution

Two weeks after the loss of the *Thresher,* on 24 April 1961, Secretary of the Navy Fred Korth established the Deep Submergence Systems Review Group (DSSRG).[1] Under the leadership of a former Oceanographer of the Navy, Rear Adm. Edward C. Stephan, the group was charged with developing a five-year program to establish a capability to save personnel from sunken submarines and recover objects from the deep ocean floor.[2] Stephan specifically was charged to recommend means and organization for implementing the DSSRG's proposals.[3]

The DSSRG brought together leaders of government, academia, and industry. Among the 58 individuals invited to participate in the study were Allyn Vine from Woods Hole, Fred Spiess from Scripps, Ed Link from industry, submariner Captain John H. Dolan, oceanography specialist Captain Charles N. G. Hendrix, and Captain William R. Anderson, former commanding officer of the first nuclear-propelled submarine, the *Nautilus* (SSN 571). Another submariner, Captain Robert E. Dornin of the Joint Chiefs of Staff, joined the effort as Admiral Stephan's deputy.[4] Each of these men would head study panels. And on 14 May 1963, less than a month into the DSSRG effort, Stephan's charter was expanded to include U.S. Air Force requirements to recover aerospace objects from the ocean floor. On the "white side" was recovery of U.S. missile components; the "black side" reflected the Air Force's interest in recovering Soviet missile components from open-ocean test impact areas.

On 1 March 1964, the Stephan committee published an unclassified, 14-page summary of its 1,134-page, Top Secret report. The summary listed four major program recommendations:

(1) *Rescue:* A program to recover survivors from ocean depths down to the collapse depth of current submarines (less than 2,000 feet).
(2) *Man in the Sea:* A program to allow men to perform useful work on the ocean floor down to 600 feet.
(3) *Location and recovery:* Location and recovery of small objects—that is, as small as "the size of a basketball" or as large as ten tons—from depths to 20,000 feet; specifically recommended were two Deep Submergence Search Vehicles (DSSV) that could operate to 20,000 feet and thus reach 98 percent of the ocean's floor.
(4) *Large-object recovery:* Recovering large objects (read, intact submarines) of up to 1,000 tons from depths to 2,000 feet; within five years develop the capability to lift 2,000 to 10,000 tons from depths of 20,000 feet and explore possible means to lift 10,000 tons using only submerged systems.[5]

As for the issue of organization, Admiral Stephan "consider[ed] this to be a matter of paramount importance."[6] Fearing that starting a program from scratch would slow implementation, he recommended against establishing a new semi-independent empire within the Navy.[7] Instead, Stephan "strongly recommend[ed] that the Director, Special Projects, under the Chief of Naval Material, be assigned the responsibilities for the management and technical control of the DSSRG program within the Navy."[8] Special Projects was the agency responsible for the development of the Navy's submarine-launched ballistic missile systems—Polaris and Poseidon.

On 28 May 1964, Secretary of the Navy Paul H. Nitze accordingly assigned the Deep Submergence Program to the Special Projects Office.[9] For the first time in U.S. naval history the responsibility for the design and development of ships and submarines was not to be in the hands of the Bureau of Ships or its predecessor organizations.[10]

In the fall of 1964, however, Rear Admiral William F. Raborn Jr., Director of Special Projects, found that his organization was overloaded already with the Poseidon missile system. He decided to calve off what he renamed the

"Deep Submergence Systems Project" (DSSP) as a separate agency, temporarily under the direction of his chief scientist, Dr. John P. Craven. Thus, DSSP became only the second—after Polaris/Poseidon missile systems—Navy project independent of the "bureau system" that had controlled naval weapons and ship development since 1842.

The Deep Submergence Group

On 1 March 1965, Dr. Craven engineered the shift of operational control of the *Trieste II* from the Navy Electronics Laboratory to DSSP. The title of officer-in-charge of a new, informal "Deep Submergence Group" would reside with the officer-in-charge of the *Trieste* until the creation of Submarine Development Group 1 in August 1967. The Deep Submergence Group acted as the San Diego arm of the newly established DSSP.

The operational Navy—as represented by the Deep Submergence Group—now had control of the *Trieste II.* The research and development community—personified by the Navy Electronics Laboratory—was less than pleased. According to Dr. Craven, "This did not placate the researchers at the Navy Electronics Laboratory. They believed that they had lost their deep submergence laboratory capability to a nonmilitary mission."[11] After March 1965, DSSP was exercising operational control of the *Trieste II*, through Submarine Squadron 3 at Point Loma in San Diego. An unintended consequence was that new officers and crew for the *Trieste* were ordered in as "staff" for SubRon 3. Thus, the special act of Congress awarding submarine pay to the *Trieste*'s officers and enlisted crew was unintentionally subverted by Craven's bureaucratic maneuverings. Once again it took the personal intervention of Representative Bob Wilson to intercede with the Navy to correct that administrative quirk. Some of the personnel ordered to *Trieste* lost as much as a year's submarine pay prior to reinstatement of hazardous-duty pay for the *Trieste.*[12]

The *Trieste*'s Future

Upon completion of the 1963 *Thresher* dives, Lieutenant Commander Brad Mooney had been briefed on the changes in the wind by Dr. Craven at Special Projects. In a letter dated 15 October 1964, two weeks before the *Trieste II* arrived back in San Diego from Boston, Mooney summarized the central points of that briefing as it applied to the bathyscaph: "In the future, we have new problems to conquer, as well as smooth out those existing: *Trieste III*, the Air Force Project, a mid-depth vehicle for NEL."

The "*Trieste III*" referred to by Mooney would be called the *DSV-1* in its design and construction phases and for its first few years as a covert deep submersible, then given the name *Trieste II* (with the hull number—not name—DSV 1). The "Air Force Project" was Winterwind, a Navy/Air Force special-access project that would shape the *Trieste* program's development and activity for the next four years (see chapter 13).

In early 1965, Mooney and Lieutenant John H. Howland made several trips to the Bureau of Naval Personnel in Washington to hand-select a dozen officers to be sent to the Deep Submergence Group as the foundation of the future body of submersible pilots. The selected officers trickled in over the next nine months, intended for the fleet of deep submersibles planned by DSSP. It was intended that all would qualify as "hydronauts" (i.e., deep submersible pilots) during their 24-month assignment, but the limited availability of the bathyscaph would make that goal impossible.

After being cradled at the Navy Electronics Laboratory piers since November 1964, for cleaning and light maintenance, the *Trieste II* was returned to the water on 19 April 1965, under the auspices of the Deep Submergence Group. Five dives for post-overhaul testing and acceptance trials proceeded through late May 1965. Most of the remainder of the year was devoted to training pilots and crews in the bathyscaph's new mission of bottom search and recovery. On 12 May, Mooney arranged for Lieutenant Commander Harris E. Steinke, the inventor of the Steinke Hood for submarine escape, to dive with him to 2,500 feet. Soon thereafter Mooney had one of the sphere's equipment bays removed to allow a seat for a fourth man, thus speeding the training cycles. Of the 26 dives into San Diego areas in 1965, most to depths of about 3,500 feet, one-half of them carried four men.

At the Mare Island Naval Shipyard a third bathyscaph, intended for project Winterwind, already was being designed. On 21 May 1965, the *Mare Island Grapevine* published an artist's concept of the "*Bathyscaph III*" and explained:

> H. L. (Herb) Graybeal unveiled the latest design produced at Mare Island for a third-generation deep-diving bathyscaph. Graybeal said that preliminary plans are nearing completion and construction will

> follow final approval by the Bureau of Ships, of designs prepared here. What makes Mare Island's continuing engagement with the Navy's deep-submergence program real news is that the sphere for the third bathyscaph (which will be built at M.I.) is the first such to be built in this country. Both deep-diver gondolas on the other bathyscaphs were built in Europe. The new one will be seven feet in diameter.[13]

On 2 August, Howard R. Talkington of the Naval Oceanographic Systems Command in San Diego—the project manager for the Cable-controlled Underwater Research Vehicle (CURV), one of the Navy's first and most successful remotely operated vehicles—was given an orientation dive on the *Trieste II*, with Larry Shumaker piloting. Apparently Dr. Craven had tasked Talkington with designing some of the mission support devices for the upcoming Winterwind project.[14]

During this period three "special" individuals were given "guest" rides to about 3,400 feet. The first was on 23 June 1965, a "CDR [Commander] Boling"; then on 3 August an "ENS [Ensign] Asbury"; and finally on 6 August an "ENS Sudakadis." These noms de guerre probably hid the identities of Air Force intelligence officers involved in the Winterwind project. The *Trieste*'s 1965 diving season was entirely operational, with no dives for the research scientists.

Chief pilot Larry Shumaker departed without relief prior to year's end. (Shumaker left active duty in 1966, remaining in the Naval Reserve. He joined the Lockheed Missile and Space Company in San Diego, where he was chief pilot of the new deep submersible *Deep Quest*. When Lockheed was awarded the contract for the Navy's deep submergence rescue vehicles he became test pilot for that program.)

A Changing Program

The submarine force was now supporting the *Trieste II* with a much-enlarged maintenance crew of more than a score of enlisted men, including four chief petty officers, plus six officers. To assist Mooney and Howland, Senior Chief John Michel was recalled to provide more corporate memory and beef up the engineering lineup after the loss of Buono. Michel's participation in special projects at Pearl Harbor during the period 1962–1965 further enhanced his value by helping to manage the design of the bathyscaph at Mare Island and reactivating a floating dry dock at Long Beach for conversion to a dedicated bathyscaph mother ship.[15]

Three new officers arrived as 1965 pilot trainees: Lieutenant Donald E. Saner, who qualified as pilot #7, and became the *Trieste II*'s engineer officer; Lieutenant Richard D. Waer, pilot #8, would later become the officer-in-charge of the *Turtle* (DSV 3); and David F. Helms, pilot #9, would be assigned as liaison officer to the submersible *Aluminaut* during the Palomares bomb-recovery operations in 1966. Later, Waer would become the commanding officer of the bathyscaph support ship *Point Loma* (AGDS 2). (In 1968, Saner retired from the Navy and joined Shumaker at Lockheed as a pilot for the *Deep Quest*.)[16]

In September 1965, with the completion of the year's diving season, the *Trieste II* was sent back to Mare Island for what was expected to be two and one-half months of inspection and modification. It stretched into more than eight months. Those modifications included a major revamping of instrumentation and control equipment in the sphere; DSSP was using the overhaul to prototype new systems for the proposed Deep Submergence Rescue Vehicles (DSRV) and for the nuclear-propelled submersible *NR-1*. A summary of this work included the statement that the *Trieste II* had been "permanently assigned to [the DSSP] as a subsystem test and DSRV pilot training vehicle."[17]

The *Trieste* sphere was provided with an integrated precision navigation, piloting, guidance, autopilot, and control package, including the very first "microelectronic digital computer," a Sperry Mark 15. This device had originally been designed for the Boeing 707 aircraft; the bathyscaph got the first production unit, demonstrating the high priority of DSSP programs. Doppler close-in navigation sonar, a gyrocompass, an analog X-Y plotter, and a sonar interrogator for use with bottom-fixed transponders were among the new systems.

Also during the overhaul the *Trieste* was provided with a new Straza continuous-transmitting, frequency-modulated sonar, useful out to 800 yards; five television cameras (up from three), with three monitors (previously there was one); and a removable periscope that allowed one operator binocular vision, or two operators simultaneous monocular vision, without their having to kneel on the bottom of the sphere as previously had been necessary.[18]

Modifications to the float and propulsion included skegs, or "feet," added to either side of the sphere and filled with additional ballast to prevent the *Trieste II* from flipping upside down if both tubs were jettisoned. The float was enlarged to add avgas and thus buoyancy to compensate for the weight of the skegs. Automatic computer controls were added to the propulsors, which were brought back up to five horsepower each. A third five-horsepower propulsor was added. [19]

Palomares Interlude

On 17 January 1966, a U.S. Air Force B-52G strategic bomber with four hydrogen bombs collided with a KC-135 refueling tanker aircraft off the Mediterranean coast of Spain. Both aircraft were lost, only four of the 11 crewmen in the two aircraft surviving. The four weapons, which were unarmed, fell near the Spanish coastal town of Palomares: one, supported by a parachute, was found intact on the ground almost undamaged; two fell without parachutes and were smashed on impact with the ground, releasing radiation contamination at the sites; and the fourth fell into Mediterranean Sea.[20]

Dr. Craven of DSSP was named to head the recovery effort for the hydrogen bomb that had fallen offshore. DSSP "had not had time to develop such search doctrine or equipment"; nevertheless, Dr. Craven worked with the Navy's former Supervisor of Salvage, Captain Willard F. (Bill) Searle Jr., to organize a search and recovery effort.[21] Craven would provide the search equipment and specialists, while Searle would supervise the recovery resources.

The Navy could not send the *Trieste II,* still in overhaul at Mare Island. Craven did, however, dispatch the *Trieste*'s officer-in-charge, Lieutenant Commander Mooney, his assistant, Lieutenant Howland, and two other *Trieste* officers to help coordinate the deep submersibles being used in the search for the errant bomb.

The *Mizar* arrived on 19 February, contributing her capability for relatively precise navigation of towed search sleds.[22] By day the *Mizar* would track and vector the submersible *Alvin* in an assigned search area; by night the *Mizar* would launch her fish for bottom photography. Unlike in the 1964 *Thresher* search, the *Mizar* was attempting to locate a single bomb and parachute, not thousands of tons of scattered submarine debris. Several Navy and commercial submersibles participated in the lengthy search for the errant hydrogen bomb.

During her six weeks on the site the *Mizar*'s technicians exposed more than 65,000 photographs of the ocean floor, but it was the *Alvin* that finally located the bomb, at a depth of some 2,500 feet.

In mid-February, the prospective officer-in-charge of the *Trieste*, Lieutenant Commander "Buzz" Henifin, arrived and made one search dive in the commercial submersible *Aluminaut.* Also on scene for brief periods were Lieutenant Commander George Martin, former assistant officer-in-charge of the *Trieste,* and four lieutenants of the *Trieste* team: Richard F. Huebner, David F. Helms, James T. Worthington, and Duane Tollaksen.

When on 15 March the *Alvin* finally ran down the bomb she had no means of marking it. Accordingly, Mooney managed the first-ever underwater rendezvous of two submersibles, a potentially hazardous evolution that required the *Alvin* to sit on the bottom near the bomb until the *Aluminaut* could arrive. The *Alvin* "babysat" the bomb for more than eight hours.

After almost 24 hours the *Alvin* returned to rendezvous with the *Aluminaut* and the bomb, now ready to emplace a transponder and a pinger. The *Mizar* could thus establish the precise geographic position of the bomb. On her next dive the *Alvin* succeeded in snaring the weapon's parachute with grapples lowered from the *Mizar,* positioned directly above. The lift from the surface commenced. Some 30 minutes into the lift, a line parted and the bomb fell back to the ocean floor.

It took nine days to relocate the bomb. On 2 April the *Alvin*, with George Martin as an observer, found the weapon and watched it until the *Aluminaut* arrived. The larger submersible than stood by until the *Alvin*, as before, returned to mark the target with a transponder and pingers.

An unmanned system finally snagged and retrieved the bomb: the Cable-controlled Underwater Research Vehicle (CURV), remotely operated from the submarine rescue ship *Petrel* (ASR 14). The CURV already had a history of more than 50 successful torpedo recoveries off San Clemente Island. It previously had been limited to a 2,000-foot depth; a crash program had increased the limit to 3,000 feet for the bomb recovery.

Trieste II, 1966

Although the *Trieste* did not participate in the Palomares bomb recovery, the operation provided several of the bathyscaph's officers with invaluable experience in deep-sea search and recovery It was during the Palomares drama, on 9 February 1966, that the Deep Submergence Systems Project, was moved out of the Special Projects Office, becoming a separate program office under the Chief of Naval Material. At the same time the nuclear-propelled deep submersible *NR-1* was assigned to DSSP.[23] Dr. Craven, "seconded" from the Special Projects Office, while retaining his responsibilities there as chief scientist, became director of DSSP, pending assignment of a naval officer to that position.

Meanwhile, on 14 April 1966, in San Diego, Vice Adm. Lawson P. Ramage, Commander, First Fleet, pinned the Navy Commendation Medal on Lieutenant Commander Mooney for his service with the *Trieste II* during the 1964 *Thresher* dives off Boston.[24] Also that April "Buzz" Henifin reported to Submarine Squadron 3 following his visit to the Palomares search operation. His orders identified him as the future officer-in-charge of the "*Trieste III*." On 20 May, Henifin relieved Mooney of command of the *Trieste II* and the Deep Submergence Group. Henifin would become the Navy's tenth deep submergence pilot. Mooney had orders to command the diesel-electric submarine *Menhaden* (SS 377).[25]

The *Trieste II* returned to San Diego on 25 May 1966, from the extended overhaul at Mare Island, transported home in the amphibious ship *Monticello* (LSD 35). The five-month-plus extension to the overhaul had been the result of the new systems installed and—significantly—the new Bureau of Ships requirement that each bathyscaph component be individually tested and certified.[26] BuShips was keeping Mare Island under a tight leash.

The *Trieste II* re-entered the water on 22 June.[27] The 1966 diving season, originally scheduled to commence on 13 July, was delayed by electrical problems.[28] The first deep dive for the year thus occurred on 21 August, off San Clemente. Howland was piloting, with Henifin and Lieutenant (Junior Grade) Mike Staehle on board. Soon after submerging and beginning the descent, the float started making loud "bangs" and "pops." Howland had never experienced anything like those sounds in his many *Trieste* dives. He immediately aborted the dive, emptied both shot tubs, and sped to the surface. The craft had reached about 600 feet. Henifin was later to observe, "We got to the surface, thanks to Howland, but the 32 pounds of mercury expended into the ocean was a great way to show that Papa Piccard was right about the safety of the ship and people against the sea." The mercury was part of an emergency system designed by Auguste Piccard for the first *Trieste* and included in subsequent versions of the craft.[29] Yet again, *Trieste* officers had reason to suspect the competence of design and quality control at Mare Island.

The problem took weeks to isolate and correct, with the result that only five dives would be made for the year, rather than the 40 that had been originally planned. The fault was in a valve that automatically compensated for the compression of the gasoline in the tanks by decreased temperature and increased water pressure at depth. The Mare Island staff had redesigned the valve and had machined a lifter shaft an inch and a half too long, owing to an error in converting Piccard's metric measurements to English equivalents. When the valve tried to actuate, its valve stem struck the cover dome and prevented it from fully opening. As the bathyscaph descended the increasing water pressure was pushing in the sides of each tank as the volume of the avgas in it decreased. Eventually—perhaps at as little as 2,000-foot depth—the tanks would have ruptured. That event was as close as the Navy would come to losing the *Trieste*.

All of the 1966 dives were for evaluation of newly installed systems and for pilot training. And all of the dives made off San Clemente were to 1,300 feet or less.

A May 1966 document states, "Once test and evaluation is complete, the craft will be permanently assigned to the DSSP's Deep Submergence Group at ComSubRon 3, Ballast Point, San Diego, where 40 DSRV pilots will use it for pilot training—at the same time cooperating in further system development and debugging."[30]

New Faces and New Tasks

The deep submergence "pilot trainees" reported for duty in San Diego between September 1965 and May 1966. The first of the planned DSRV rescue submersibles was scheduled for delivery in 1968.[31] These first trainees were Huebner, J. Worthington, Tollaksen, Kenneth B. Killen, Malcolm Bartels, Robert R. Denis, Joseph B. Berkley Jr., and John B. Field, all lieutenants; Lieutenant (Junior Grade) Staehle; plus Thomas A. Hedgecoth and John H. Carroll, warrant officers. Of these men, only two would qualify as *Trieste* pilots—Staehle (#12) and Bartels (#24).

The major problem in training this large group was, as noted above, the limited availability of the *Trieste II.* With scheduled and unscheduled maintenance, and the lengthy overhaul at Mare Island, the bathyscaph logged only 34 dives between May 1965 and March 1967. Even with two trainees in the sphere on each dive it was impossible to qualify more than a handful of pilots in a year. All the trainees spent time in a bathyscaph trainer, or simulator, at the Sperry Rand facility in Charlottesville, Virginia, but certification required about five "real" dives under instruction plus a deep qualification dive.[32] Thus, Saner, Waer, Helms, and Henifin were all that could be certified.[33] The Commander, Submarine Squadron 3 had a letter placed in the service records of the other student pilots explaining the situation:

> During the period of your assignment to Submarine Squadron THREE as a student, operator/pilot of the bathyscaph *Trieste,* operational and material contingencies developed which prevented you from fulfilling the requirements to be designated as a "qualified" operator. These circumstances were through no fault of your own. During your tour of duty, the *Trieste* was subjected to extensive modification and overhaul which minimized your opportunity to apply the classroom skills and knowledge you had achieved. In spite of the reduced operational time available to *Trieste,* you successfully completed [number] dives as an observer. In addition, you amassed a total of [number] man-hours in the *Trieste* Simulator, wherein, you displayed a proficient and competent skill in simulated problems of bottom navigation and deep submersible ship control. . . .

In mid-August 1966, Vice President Hubert H. Humphrey indicated that he "wouldn't mind" going down in the *Trieste II* while he was on a tour of major oceanographic installations. Apparently, neither the Navy nor the Secret Service reacted positively, and nothing came of it.[34]

The *Trieste II* continued to receive an influx of new officers: the craft's assigned complement grew to 14 officers and 35 enlisted men. Lieutenant Commander Howland remained assistant officer-in-charge until January 1967, when Lieutenant Frederick Forst relieved him. Howland then went to the Navy's most recently commissioned diesel-electric attack submarine (and the penultimate to be built), the *Bonefish* (SS 582), as executive officer, and then went on to command the *Harder* (SS 568).

Lieutenant Commander Henifin was able to squeeze in four dives in the fall of 1966, of which one, as has been seen, was aborted after five minutes and another was cut short at one hour, 32 minutes submerged. Still, the two other *Trieste* dives qualified Lieutenant Waer on 27 September, and, finally, Henifin on 23 November.

Trieste II, 1967

An abbreviated 1967 diving season of nine dives for pilot training was terminated on 31 March 1967, after the *Trieste II* Dive #54. These nine dives included qualification dives for Saner and Helms, as well as a familiarization dive for the Commander, Submarine Squadron 3, Captain Allen E. May. The very last dive of the *Trieste II*, on 31 March 1967, allowed Forst and Saner to escort to the bottom, at 2,000 feet, everyone's favorite technical representative—John B. Van Voorhis—the senior Sperry techrep, who had been involved in the *Trieste* program for several years and would remain with it another decade. He represented the program's deepest corporate memory and was a most inventive and responsive field engineer.

The *Trieste II* was then lifted back to Mare Island, ostensibly for overhaul. During the summer months of 1967 the fleet tug *Apache* (ATF 67) was engaged with the new floating dry dock/mother ship—the *ARD 20*—with a bathyscaph stand-in, called the "pig," off the coast of San Clemente Island. By mid-summer, most of the *Trieste* officers and crew had moved with their families to temporary quarters in Vallejo, to be present for the "overhaul." Yet there was little or no work for most of them. The bathyscaph was under the care of the shipyard, and the men had little to do except check for morning roll call and be dismissed for the remainder of the day.

A New Mother Ship

A dedicated mother ship for the *Trieste* had first been requested in the ten-year program recommended by the National Academy of Sciences in 1959: "Engineering needs for ocean exploration: (a) An improved bathyscaph should be designed and built immediately; (b) it is recommended that funds be made available for a mother ship."[35] However, not until seven years later did a dedicated mother ship become a reality.

In the fall of 1965, Senior Chief Michel received temporary orders from the Deep Submergence Group in San Diego to the Long Beach Naval Shipyard to select from the inactivated vessels there, the "mothball fleet," the floating dry dock most suitable for modification to support the Winterwind project. He was provided a single three-by-five-inch index card with a phone number. Card in hand, he bulled his way past two levels of secretarial screening and into the presence of the shipyard commander, Captain Morton H. Lytle.[36]

> *Lytle:* What can I do for you, Senior Chief?
>
> *Michel:* Sir, I have been ordered to pick out an ARD for special modifications.
>
> *Lytle:* Chief, look out that window and tell me what you see.
>
> *Michel:* Well, sir, I see a couple carriers, a cruiser, and a whole bunch of destroyers.
>
> *Lytle:* That is correct, and they are all preparing for Vietnam. Don't you think I have better things to do than talk to you about the mothball fleet?
>
> *Michel:* I don't know, sir. But, perhaps if you would call this number, we might find out.

Captain Lytle took the card with the phone number and handed it to a secretary to make the call. Michel continues the story: "The secretary made the call, the captain picked up the phone and identified himself, then there was two minutes of complete silence as the captain listened. He then hung up without a word and instructed his secretary to get the security officer in charge of the mothball fleet into his office immediately. We waited a few minutes without conversation, and when the chief warrant in charge of the reserve fleet arrived, Captain Lytle pointed to me and said 'Give Chief Michel whatever he wants.'"[37]

Such was the power of the Deep Submergence System Project's priority for Winterwind—Brickbat/Priority-1.[38]

The *ARD 20* was one of the scores of floating dry docks built by the U.S. Navy during World War II. Completed in 1944, the dock was 491⅔ feet long and 81 feet wide, and had a lift capacity of 3,500 tons—sufficient for docking destroyers and submarines. Like most other floating dry docks, the *ARD 20* was non-self-propelled, intended to be towed to forward bases to help maintain the fighting fleet. Having selected the *ARD 20*, Chief Michel arranged for the stripping of unwanted equipment and ordered new material. The dock's conversion added berthing and messing facilities for the dock's crew and the *Trieste*'s crew, and modernized the dock's repair shops. The modifications began in October 1965.

Undersecretary of the Navy Robert H. B. Baldwin revealed on 12 January 1966 that work was "under way on a classified advanced base repair dock . . . the Navy's first mother ship for a bathyscaph or deep-sea vehicle."[39] Later in the month a newspaper article identified the dry dock: "It is called *ARD 20*. The project comes under the Navy's many-faceted Deep Sea Submergence Program [*sic*] and carries a multi-million dollar price tag."[40]

Dr. Craven later indicated that "the dry dock modifications proved to be the most difficult of all the ship modifications undertaken by the DSSP. The major design problem revolved around the docking and undocking of the *Trieste* [in the open ocean]. It had a float that consisted of eighty tons [*sic*] of high-octane aviation gas contained in a light eggshell structure. There was always the possibility of a rupture during the docking process and a spill. Everything on board had to be made spark-proof."[41] Actually, the 66,000 gallons of avgas in the *DSV-1* float weighed 197 tons. The forward half of the *ARD 20*'s dock was converted into a fuel tank for 185,000 gallons of aviation gasoline (surrounded by tanks charged with carbon dioxide); a firefighting foam suppressant system similar to those found on aircraft carriers was installed. The *ARD 20* also was fitted with three thrusters, one installed transversely in the bow and two at the stern. Thus the dock would have the ability to "hover" in a specific position against wind and current.[42]

The unclassified, "public" reason for the reactivation of the World War II–built floating dry dock was to support experimental submersibles such as the *Trieste* and commercial craft like Electric Boat's *Star I* and *Star II*, Westinghouse's *Deep Star 4000*, and Lockheed's *Deep Quest*.

On 14 September 1966, the dry dock was placed "in service" as the *ARD(BS) 20*, the modifying designation indicating "bathyscaph support."[43]

On 9 March 1968, she was named *White Sands* and the hull number—no longer her name—shortened back to ARD 20. (The dock lived with her old designation for five years until on 1 August 1973, the *White Sands* was redesignated AGDS 1 for miscellaneous auxiliary, deep submergence support.)

In September 1967, the tug *Apache* was assigned semi-permanently to Submarine Development Group 1 to support the *Trieste* program and the new mother ship. That month the *Apache* towed the new mother ship to Mare Island, where the shipyard modified the ARD's docking well. In late September the *Trieste II*, with her "boat float" shape newly painted white with vertical black strips, was loaded into the docking well of her new mother ship.

The *Apache* then towed the whole "assembly" through San Pablo and San Francisco bays to the Hunters Point Naval Shipyard, for a few days of crew liberty in San Francisco over the long Labor Day weekend. This was followed by post-overhaul trials and training off San Clemente. Something "big" was in the wind, and some of the *Trieste* crew felt the vaguest feelings of apprehension. What was happening? Only a few of the officers and Senior Chief John Michel knew the real situation.

13 A "Trieste III"

Pay no attention to that man behind the curtain.

—The Wizard of Oz

The director of the Deep Submergence Systems Project, Dr. John P. Craven, "received a call from the Pentagon from a man who said he was a naval intelligence officer. He asked me to meet with him at the Pentagon and not to tell anyone on my staff, including my secretary. When I arrived, a man whom I knew well as a submarine officer involved with the Polaris program greeted me."[1]

Dr. Craven soon found himself in an inconspicuous office in the Pentagon being briefed into a Navy "compartment" of highly classified "Special Intelligence." The program in question was to generate a capability to work secretly on the ocean floor, primarily to recover objects that "don't necessarily belong to the United States." The program was then called Low Flyer.

The appointment of Dr. Craven to head DSSP in 1964 had been considered temporary, but Craven retained that position for a full year until replaced by Captain William M. Nicholson. Craven remained with DSSP as chief scientist while simultaneously serving in that position with the Navy's Special Project Office.

Low Flyer / Sand Dollar

The Low Flyer program—changed to Sand Dollar in mid-1964—was focused on Soviet hardware believed to be on the ocean floor, a comprehensive cornucopia of lost sensitive military objects. Here was a list of opportunities to obtain samples of advanced Soviet military weapons and aerospace technology, objects that could provide critical intelligence. Sand Dollar was the overarching program for the effort to recover some of that "tantalizing trash" and to develop special systems to effect those recoveries.

The U.S. Intelligence Community posed the question: Could Dr. Craven organize and manage a program to collect those objects littering the deep ocean floor? Craven already was creating systems that within five years could rescue trapped crewmen in sunken submarines down to crush depth, allow divers to work at 600 feet, and locate and recover small objects down to 20,000 feet. Naval intelligence officials were asking for these systems *now,* to be used in highly classified projects.

Dr. Craven's initial response was that DSSP was in fact developing some of the technology to conduct the missions visualized in the Sand Dollar program but not complete systems. In any case, he added, "We don't have anything that could do your operation, because that requires things to be clandestine. So it's really not worth doing Sand Dollar unless you do it from a submarine."[2] Brainstorming out loud in front of a very critical audience, Craven had succinctly defined a program that the Navy would pursue until the end of the Cold War and beyond.[3]

In March–April 1965 Craven submitted plans to create specialized undersea vehicles and systems for deep-ocean search, recovery of small objects, and work at a depth of 600 feet. Approval was expeditious, and Craven received a major budget addition for DSSP. Craven had the will and the means to create the systems that would satisfy the almost impossible dreams of the Intelligence Community.

Project Winterwind

Earlier, in 1963, Dr. Harold Brown, the Deputy Secretary of Defense for Research and Engineering, had tasked the Navy and Air Force to collaborate in a comprehensive study of, and develop a plan for, the recovery of nose cones from test-fired Soviet missiles that came down on the ocean floor. Such testing of ballistic missiles into the northern Pacific had begun in 1960. Subsequently, the Chief of Naval Operations assigned the Bureau of Ships to study deep-sea search and recovery. Both those study efforts eventually were folded into the Low Flyer/Sand Dollar program and constituted as a new project within it—Winterwind. Of several covert projects developed to meet Sand Dollar objectives, two were very closely related: Underdog and Winterwind.

Meanwhile, Project Underdog was designed to locate and recover Soviet *cruise* missile debris and practice torpedoes in the Sea of Japan. Its search component was originally to be a submarine-mounted digital camera fitted for laser illumination of the bottom, for searches no deeper that the submarine's maximum operating depth, approximately 1,200 feet.

Winterwind was a much more extensive and ambitious project, seeking *ballistic* missile nose cones on the ocean floor. It would require more than four years of planning and technology development before its first planned recovery in the summer of 1969. One of the proposals was to employ the special-mission submarine *Halibut* (SSN 587) for the covert deep-ocean search, using a deep-towed sled to locate Soviet ICBM re-entry vehicles. The *Halibut* would require extensive modifications for that mission.[4]

Moreover, how could a target once located be recovered from depths down to 20,000 feet? BuShips evaluated the *Trieste*'s potential for the recovery of Winterwind targets. However, the modifications to the *Trieste II* that would be needed would exceed the cost of designing and constructing a purpose-built submersible. Thus, the Winterwind program would cover the search efforts both to locate a target and (1964–1968) to recover it down to 20,000 feet by a specially designed and covertly constructed deep submersible. This craft would be a third version of the *Trieste*, designated the Deep Submergence Vehicle No. 1—*DSV-1*, which, on the "white side," would occasionally be referred to as the *Trieste III*, both formally and informally.[5]

The original, 1965 Winterwind concept of operations was straightforward:

- Precision bottom mapping and surveys of Soviet ballistic missile open-ocean impact areas by the Navy's navigation research ship *Compass Island* (AG 153).
- Clandestine submarine search operations conducted by the extensively modified *Halibut* using a deep-towed sensor sled.
- Recovery operations by the covertly constructed, manned deep submersible *DSV-1*.
- Surface support provided to the *DSV-1* by a floating dry dock, eventually the *White Sands* (ARD 20).

Winterwind would require major modifications to the *White Sands* as well as to the *Halibut,* and the new *DSV-1* bathyscaph would be covertly designed and constructed at the Mare Island Shipyard. Related and support systems, such as the submarine deep-towed search sled (later called an unmanned instrumented platform) and the submarine-deployed, deep-ocean transponders, also would have to be developed.

In mid-1968, both Underdog and Winterwind were approaching operational status and were shifted to the sponsorship of the Director of Naval Intelligence. Through bureaucratic maneuvering, however, the two projects eventually came under the control of the undersea warfare desk in the Office of the Chief of Naval Operations in the Pentagon. Later, speaking of Winterwind in an oral history interview, Captain John E. Bennett said:

> This program was to recover Russian nose cones that were being fired over the Pacific and dropping in the Central Pacific where they sank to twelve thousand or sixteen thousand feet. Our object would be to get the nose cones. All the telemetry had been copied. The Air Force had airplanes up in the Aleutians and they kept one airborne all the time just to tape all the telemetry. Of course, nobody could decipher it. They had a full warehouse full of tapes up there, "Project Daisy Mae." If we could recover a nose cone, we could decipher the tapes.[6]

The ability to record and decipher Soviet missile telemetry in real time would represent a significant advance in U.S. understanding of Soviet strategic missile developments. Before 1962, Air Force intelligence had recognized that these Soviet missile tests presented a rare opportunity and made contact with naval intelligence that year. The Soviet nose cones from the re-entry vehicles were lying on the ocean floor in international waters—apparently free for the taking. The key question: Was it possible to recover them?

The Winterwind plan required a new bathyscaph, a support ship, training off the California coast, the dry dock to be towed with the new bathyscaph on board to a point several hundred miles north of Midway Island, and dives to locate and retrieve the re-entry vehicle. The entire operation would be sub-rosa, or "black." Finally, the new bathyscaph would be a one-mission vehicle—perhaps conducting multiple recoveries over several months or years *but then destroyed.*

A necessary pre-condition for success was the covert location and marking of the target. The Navy's planned modifications to the *Halibut* would enable her to employ a deep-towed fish fitted with magnetometer, sonar, television, and special cameras. The work on the *Halibut* for her clandestine search role was undertaken at the Pearl Harbor Naval Shipyard.

Developing a New Bathyscaph

The Mare Island shipyard, with its extensive submarine-building experience—including construction of the *Halibut*—would design and fabricate the Winterwind bathyscaph. Design work on the "*Trieste III* " at Mare Island occupied the second half of 1964, and construction took place from September 1965 through August 1967.[7] The work was done in the covered docking well of the *ARD 10*, a World War II–built floating dry dock tied up at Mare Island, the covering to defeat Soviet spy satellites. Overall security was tight, but that was not unusual for a shipyard involved with nuclear-propelled submarines.[8]

Meanwhile, Dr. Craven, in his DSSP capacity, "requisitioned" the existing *Trieste II* in March 1965, outwardly as a test platform for the electronics and controls for the project's planned 20,000-foot Deep Submergence Search Vehicle (DSSV).[9] The Navy Electronics Laboratory in San Diego was greatly aggrieved by the loss of the *Trieste II*'s services for scientific and oceanographic investigations.[10] According to then-Lieutenant Michael Staehle at NEL, "When I got to the Deep Submergence Group we were told that we would get a new bathyscaph to be built at Mare Island, the T-III-1; and almost in parallel the T-III-2 would be built for the Navy Electronics Laboratory. I don't know how the T-III-2 died; possibly during the subsequent budget crises."

To quiet this unrest at the NEL over the loss of the *Trieste*, the Navy released a cover story that would inadvertently be the first disclosure in, as Craven would call it, the "premature exposure of the Navy's role in Special Intelligence."[11] An Associated Press story of 28 April 1964, headlined "Warhead Recovery System Proposed," may be the "premature exposure" to which he referred:

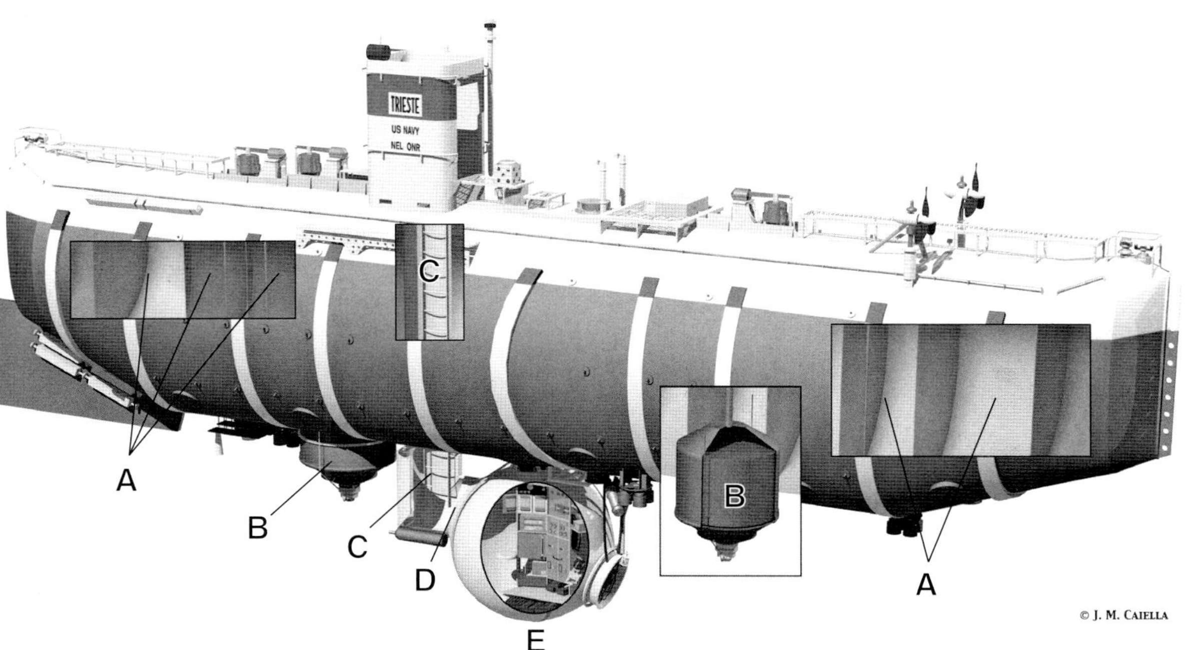

The *Trieste* in 1960. *J. M. Caiella*

A Fuel tank (12 sections)

B Ballast (shot) tubs

C Access trunk

D Entrance to sphere

E Pressure sphere

Auguste Piccard at a celebration that marked his setting of one of his atmospheric balloon records. His ballooning experiences led to his design of the *Trieste*—an "underwater balloon." *Piccard Family Archive*

The bathyscaphes *Archimède* (*left*) and *FNRS 3* at Toulon. The French Navy showed a keen interest in deep-water exploration. *French Navy*

Auguste Piccard and his son Jacques—the inventors of the bathyscaph. Their record-achieving craft was named *Trieste* for the city where it was constructed. *U.S. Navy*

The bathyscaph *Trieste* being lowered into the water in 1959. Note the L-shaped entry tunnel to the left of the sphere. Jacques Piccard and his trusted mechanic Giuseppe Buono accompanied the *Trieste* to the United States. *U.S. Navy*

The *Trieste* high and dry. The large ballast "shot tubs" forward and aft of the sphere are evident in this view. In an emergency the tubs could be entirely jettisoned, among the several "fail-safe" features built into the *Trieste*. *U.S. Navy*

The interior of the bathyscaph *Trieste*'s sphere—a space 38 inches square and five feet, eight inches high. Don Walsh and Jacques Piccard occupied this space for more than seven hours during the "deep dive." *U.S. Navy*

Three key participants on the deepest dive. From left are Dr. Robert Dietz and Dr. Andreas Rechnitzer of the Navy Electronics Laboratory in San Diego, and Lieutenant Don Walsh—the U.S. Navy's first "hydronaut." *U.S. Navy*

Jacques Piccard, Giuseppe Buono, and sailors inspecting the *Trieste* after the craft's first test dive off the coast of Guam. Note the bathyscaph's minimal freeboard. *U.S. Navy*

Jacques Piccard (*right*) and Ernest Virgil load iron pellets into the *Trieste*'s "shot tubs" prior to a deep descent. The *Trieste* left numerous "piles" of shot on the ocean floor. *U.S. Navy*

The *Trieste* at sea moments before the dive into the Mariana Trench, the deepest point in the world's oceans. The escort ship *Lewis* (DE 535) stands by. The *Trieste* was towed to and from the dive location from the naval base at Apra Harbor, Guam. *U.S. Navy*

Georges Houot and Pierre Henri Willm in 1962 took the French bathyscaphe *Archimède* six miles down to 31,350 feet, the bottom of the Kurile-Kamchatka Trench, for the second-deepest plunge made by man. This is a 1964 photo taken when the craft dived to 21,000 feet in the Puerto Rican Trench. *U.S. Navy*

The *Trieste* resting in the well deck of the landing ship *Point Defiance* (LSD 31) as the ship entered Boston Harbor on 26 April 1963, while the bathyscaph was being prepared to dive on the wreckage of the nuclear-propelled submarine *Thresher* (SSN 593). *U.S. Navy*

The *Trieste* rests on her cradle in the docking well of the *Point Defiance.* The need for surface transport and support greatly limited the ability of the *Trieste* to conduct clandestine search-and-recovery operations. *U.S. Navy*

Divers prepare the *Trieste* for dives on the wreckage of the submarine *Thresher*, lost with all on board during post-overhaul trials, resting at a depth of some 8,400 feet. *U.S. Navy*

The *Trieste* in Boston Harbor after the first series of dives on the *Thresher* wreckage, showing the marine growth on the float. The craft was cleaned and refurbished in preparation for subsequent dives on the wreckage. *U.S. Navy*

Lieutenant Commander Donald Keach was the second officer-in-charge of the *Trieste*. *U.S. Navy*

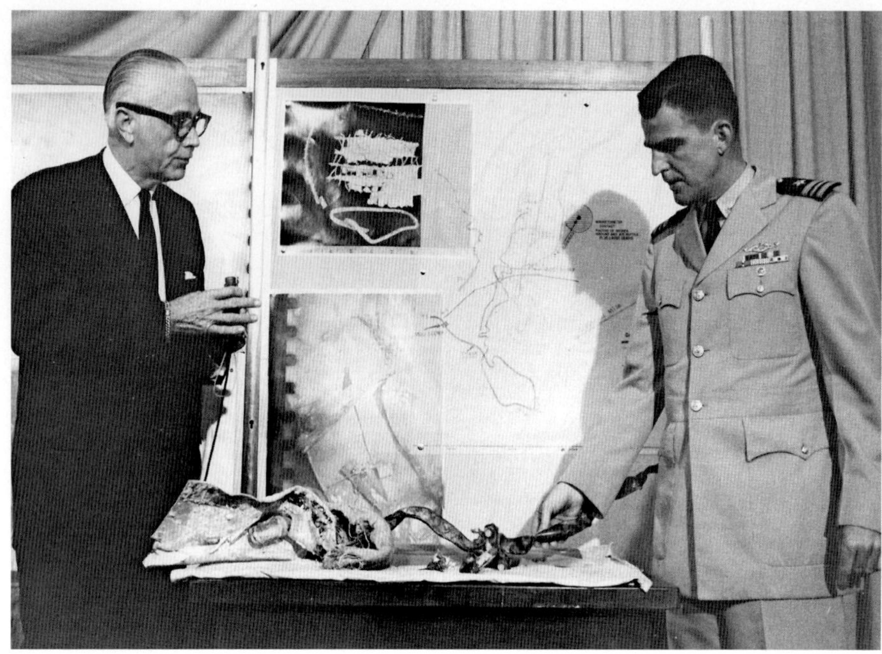

Secretary of the Navy Fred Korth and Lieutenant Commander Donald Keach display a length of piping recovered by the *Trieste*'s mechanical arm at a Pentagon press conference. The piping had markings establishing that it came from the submarine *Thresher. U.S. Navy*

The *Trieste II* (DSV 1) was an extensively modified bathyscaph provided with a "boat hull" for improved towing and seakeeping. *U.S. Navy*

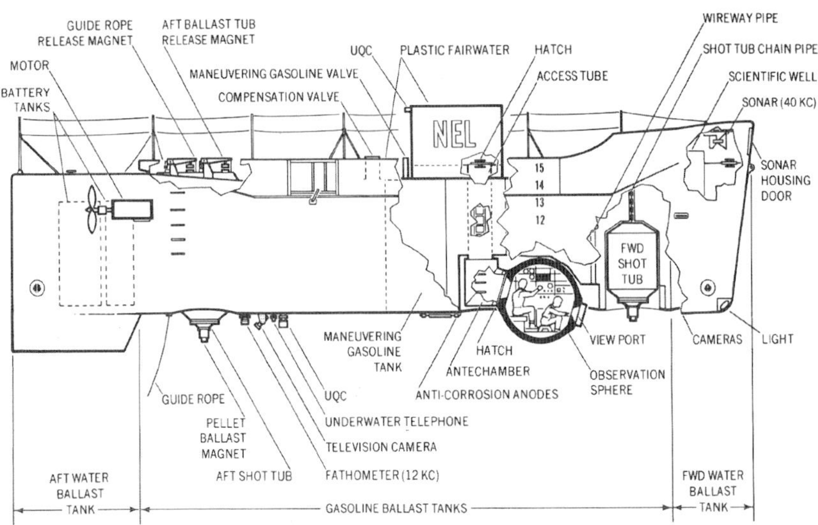

The *Trieste II* configuration. *U.S. Navy*

The oceanographic research ship *Mizar* (T-AGOR 11), directed by Buck Buchanan's Naval Research Laboratory team and operated by a civilian crew for the Navy's Military Sealift Command, was a key component of the *Trieste* "team." She located targets for the bathyscaph's bottom searches. *U.S. Navy*

The commercial deep submergence vehicles *Aluminaut* and the Navy-developed *Alvin* had key roles in the recovery of the nuclear weapon lost in the Mediterranean off the coast of Palomares, Spain. This is the *Aluminaut*. The *Trieste* was unavailable for that effort. *U.S. Navy*

On board the submarine rescue ship *Petrel* (ASR 14), Major General Delmar E. Wilson, U.S. Air Force, and Rear Admiral William S. Guest, commander of the bomb search effort, look at the recovered nuclear weapon. Behind them is the Cable-controlled Research Vehicle (CURV) that brought the bomb to the surface. *U.S. Navy*

The "launching" of the "new" *Trieste II*—also known unofficially as the *Trieste III*—with "skids" fitted to the sides of the sphere, the X-tail configuration, additional propulsors, and more sophisticated recovery devices. *U.S. Navy*

Stern aspect of the "new" bathyscaph *Trieste II. U.S. Navy*

Lieutenant Commander John (Brad) Mooney Jr. was the only *Trieste* officer-in-charge to attain flag rank. Other *Trieste* OINCs, however, did later hold senior Navy positions, primarily in the research and development fields and in submarine-related intelligence. *U.S. Navy*

The *Trieste* maneuvering close aboard the mother ship *White Sands* (ARD 20, later designated AGDS 1). With the fleet tug *Apache* (ATF 67) they formed the Integral Operating Unit (IOU). *U.S. Navy*

The *Point Loma* (AGDS 2) was converted from the specialized cargo ship *Point Barrow* (T-AKD 1) specifically to support the *Trieste* program. Here she is ballasted down off Santa Catalina, California, in 1977. *U.S. Navy*

Partially replacing the *Trieste*'s capabilities were the DSVs *Sea Cliff* (*left*) and *Turtle*. Here they are "launched" at the General Dynamics / Electric Boat submarine-building yard in Groton, Connecticut. Later they were extensively modified to increase their depth capabilities. *U.S. Navy*

The original *Trieste* with the Krupp sphere is preserved at the Washington Navy Yard, with the Terni sphere on display alongside. There, on 23 January 2020, standing beneath the *Trieste*, Dr. Don Walsh, Navy officials, and Norman Polmar, one of the authors of this book, celebrated the 60th anniversary of the bathyscaph's "deep dive." *U.S. Navy*

> The United States could acquire the means for snagging both Soviet and American test missile warheads from the ocean floor under a broad new plan for improving U.S. ability to work deep undersea.
>
> One goal spelled out in the report calls for "efficient search, investigation, and recovery of aerospace hardware down to 20,000 feet." That is the maximum depth in over 98 percent of the world's oceans.
>
> Navy Capt. John H. Dolan, who heads the study group, said the Air Force is greatly interested in this program because it might permit more accurate and swifter analysis of what went wrong with a missile or space rocket in an unsuccessful test.
>
> *Officials refused to talk about the possible use of deep submergence "working submarines" and other equipment to salvage Soviet warheads fired into the Pacific or other ocean areas in missile tests* [emphasis added].
>
> Such tests, even if unannounced, are watched on U.S. radar. American patrol planes and ships have noted the splash of Soviet test warheads in past tests into the Pacific, according to reliable reports.
>
> *There would be obvious value in being able to recover such Soviet warheads. Study of these devices could tell much about the state of the Soviet art in missiles, including possible advances in the vital field of decoys.*[12]

In February 1965, the DSSP budget, previously planned for $333 million over five years but slashed to $200 million in December 1964, was further reduced to $185 million for fiscal years 1966 through 1970.[13] These cuts and other factors would lead to the DSRV rescue program being reduced from six to two vehicles and the DSSV search program being dropped, along with other program impacts. Simultaneously, however, DSSP's "black" funding was increasing by a somewhat offsetting amount.

Entirely independent of about $70 million for the modification and outfitting of the *Halibut,* DSSP would receive about $90 million for fabrication of the bathyscaph *DSV-1* at Mare Island; it was this money that had funded the ARD 20 conversion into the *Trieste* mother ship at Long Beach. The deep-towed, unmanned sled was contracted to Westinghouse Oceanics Division, in Annapolis: six fish with magnetometers, cameras, television, and sonar.[14] Additional, classified contracts to Westinghouse, Lockheed, and other firms would produce devices to grab, net, and capture missile re-entry vehicles, torpedoes, or other targets and develop submarine-launched deep-sea acoustic transponders.

In an address on 24 July 1965, dedicating the DSSP's *Sealab II* seafloor habitat at the Long Beach Naval Shipyard, Robert W. Morse, the Assistant Secretary of the Navy for Research and Development, revealed that the "Navy is building at Mare Island, *Trieste III*, a successor to the bathyscaph that took two men nearly 36,000 feet down in the Mariana Trench. Eventually, the Navy wants to find the means to transfer men from one submarine to another, how to do search and rescue work down to 20,000 feet, and how to keep people on the ocean floor at depths of 600 feet for sustained periods."[15]

This was the second public disclosure of the new bathyscaph. Also in 1965, the San Francisco Bay Naval Shipyard published the unclassified report "Design of The Bathyscaph *Trieste II*," in which Herbert Graybeal revealed that "the next bathyscaph will have a more refined hull form with additional propulsion controls. Since the transit time from the surface to the bottom and back again can consume a large portion of available diving time, it is desirable to move vertically as fast as safety will permit. Special consideration will be given to this feature in future designs."

Graybeal's report's graphics showed a hull form similar to the ultimate *DSV-1* hull and a new sphere with vertical access through the float, a lower centerline main viewing window (similar to that in the Terni sphere), four smaller viewing ports to port and starboard, and one facing aft from which to see the rear shot tub ballast release valve. Graybeal also discussed the use of stronger materials for the sphere, including titanium—but evaluated it as not yet a sufficiently mature technology for 20,000-foot depths.[16]

For a supposedly "phantom," covertly built bathyscaph, an amazing amount of information about it was becoming available. However, it was not the Navy's intent to hide the construction of a replacement for the *Trieste II*, only to protect the highly classified mission for which it was being built—Winterwind.

In the autumn of 1965, a Navy officer from Special Projects in Washington arrived at Mare Island to "read in" John R. (Junior) Smyth, the test director for the new *Trieste*, to the Winterwind project. He advised Smyth that he would be commissioned as a naval officer and would shepherd the new *Trieste* through construction, testing, and into her operational phases. In addition, Smyth would play key roles in the new *Trieste*'s post-construction workup and its first operational employment.[17]

The Bathyscaph *DSV-1*

At Mare Island the shipyard team was extraordinarily skilled and inventive under the supervision of Philip A. Green.[18] When construction commenced, Lieutenant Commander "Buzz" Henifin directed that two or three enlisted *Trieste* crew members at a time rotate up to Mare Island for several weeks to become acquainted with the maintenance requirements of the new bathyscaph. As for all bathyscaphs and other deep divers, the sphere was critical. The Southwest Research Institute of San Antonio, Texas, designed the sphere for a three-man crew; like the Terni sphere, it consisted of two hemispheres with a flanged connection. The hemispheres were poured and spin-formed by the Lukens Steel Corporation of Coatesville, Pennsylvania. The first *DSV-1* sphere was machined by Hahn & Clay, of Houston, Texas. HY-120 steel was specified, for the designed operating depth of 20,000 feet plus a 50 percent safety factor. (The HY-100 steel used for spheres in the Navy's *Alvin, Sea Cliff,* and *Turtle* submersibles yielded an operating depth of 6,000 feet, with a safety factor significantly higher than 50 percent.)

In 1966, the Navy ordered at least two spheres from Lukens Steel, one for the Winterwind bathyscaph (which was referred to as T-III-1) and one for the Navy Electronics Laboratory's T-III-2. The sphere machined by Hahn & Clay was mounted in the *DSV-1*. Lukens' second sphere originally intended for the NEL bathyscaph, remained unmachined at Mare Island until 1974, when it was fitted in the *DSV-1*.[19]

One of the most substantial upgrades necessary for searching for and recovering objects from the deep ocean floor was, once again, the battery suite. Where the boat-shaped *Trieste II* had 90 amp-hours of 120-volt AC power in banks of automotive, lead-acid batteries for propulsion, the new bathyscaph would have 752 amp-hours of 120-volt propulsive power in silver-zinc batteries, good for 12 hours of bottom search—about three times the *Trieste II*'s limit. Auxiliary power from the forward battery bank—for the sonar, communications, lighting, TV, and other electronics—was similarly multiplied with 5,000 amp-hours at 24 volts from silver-zinc batteries. Like those of the

Trieste II, the new craft's propulsive and auxiliary batteries could be charged in place, but two sets of silver-zinc batteries for the sphere, to energize the magnets holding ejectable equipment and releasable ballast; these had to be removed and recharged on the support ship, one set recharging and the other in the *DSV-1*.[20]

Some of the highly classified aspects of the new craft's deep search and recovery abilities related to lifting objects from the ocean floor. The *Trieste* crewmen called these devices "kludges," which sometimes was represented as an acronym for "*k*lumsy, *l*ame, *u*gly, *d*umb, but *g*ood *e*nough."

Some kludges were ingenious, others quite straightforward. One of the more ingenious ones was the "sinking frame," a circular bowed frame about ten feet in diameter and weighing 500 pounds. It had pumps to drive it two or three feet into the bottom mud. Then a purse-net would be drawn through the mud to collect the remains of a target that had shattered on impact with the ocean's surface or while plunging to the bottom. When the purse net was lifted the sinking frame was left behind. Other devices were to recover torpedo-size targets, and some were intended to grab a truncated cone of specific dimensions, obviously a missile nose cone. The new bathyscaph could "graze" along the bottom, picking up treasures and placing them in special, trailing baskets. Another means of recovery was using a Dacron line, more than 26,000 feet long, that could be attached to an object by the bathyscaph and floated up to the surface with a flotation device. (The specific gravity of the Dacron line was 0.96, i.e., slightly lighter than seawater.) On the surface the floating end of the line would be led to a winch on the mother ship and hauled in. When the target was within a hundred feet of the surface, divers would rig an underhull transfer of the line from the winch to the dock's 35-ton-capacity crane, the dock-well gates in the stern would be opened and the ship ballasted down. Then the crane would swing the target—still underwater—through the gates, over the dock sill, and allow it to settle onto the well deck, all underwater. The Dacron line had a breaking strength of 100,000 pounds.

While none of these methods or devices involved high technology, they were the tools for underwater salvage work down to 20,000 feet. The genius of these devices was that they were developed covertly to be employed clandestinely.

The *ARD 20* Goes to Sea

In early November 1965, the Long Beach shipyard delivered the *ARD 20* to the fleet, and her crew took possession. The *ARD 20* was not a commissioned ship but rather a "yard" or "service" craft and rated an officer-in-charge vice a commanding officer. The first officer-in-charge was Lieutenant Commander James W. Watts. Unlike the *Trieste* crewmen, who were volunteers and hand-selected, the mother ship's crew had been assigned in the usual procedure. There soon would emerge a class structure separating the two crews—which would generate friction.

The first sea trials used a shipyard dive boat as the practice vehicle for launching and recovering through the *ARD 20*'s dock-well gates. Soon a practice shape called the "pig" was fabricated to substitute for the *DSV-1* for that purpose.

Winterwind would require mid-ocean operations, and during sea trials ways would have to be found of making the *ARD 20* safe to operate in the open ocean with the gates open. These trials would twice come perilously close to disaster. The first such incident occurred after shallow-water work and practice dockings in protected waters were completed. The *ARD 20* was towed into the open sea west of California's Channel Islands and ballasted down; the stern gates were then opened to allow waves and swells to enter the docking well. In accordance with the trial plans, Watts used the dock's newly installed bow and stern thrusters to hold its stern into the prevailing seas. The long swells swept the length of the docking well and then receded, in a safe and orderly rhythm.

Disaster loomed when the stern gates were closed.

The long-period swells, suddenly contained within the length of the ship, surged fore and aft within the well, and the *ARD 20* started to plunge and buck like an aggravated mule. Water in the well slamming against the forward bulkhead burst over the bridge area, and Watts ordered the stern gate reopened to save his ship. Master Chief Michel, serving as Dr. Craven's trial coordinator, was at the stern. He objected to the orders and raced up to confront Watts on the bridge wing:

Michel: What the hell are you doing?
Watts: What? I'm opening the stern gates.
Michel: You can't do that. We've got to see the effects of these tests.
Watts: We're in an unsafe condition, I'm opening the gates.

Michel: We're out here for one reason, and one reason only, to find out if this goddamn ship can perform the mission. So if we have to sink this goddamn ship, then we will sink this goddamn ship.

Watts later rescinded the order.[21]

The *ARD 20* survived, and Watts and her crew learned several lessons that day. First, the *ARD 20* would require yet more modification before she could safely support the bathyscaph in the open ocean. Second, the Deep Submergence Systems Project and the *Trieste* were the governing factors in the *ARD 20*'s operations. And, third, they learned that "black ops" can generate authorities and responsibilities not reflected in a person's rank or rate.

The dangerous vulnerability to a "free-surface effect" in the docking well in the open ocean had been perceived by the David Taylor Model Basin in tests on the ARD/*Trieste* pairing. Dr. Craven, who had earlier worked at the Model Basin, had commissioned the study but later objected to the costs and refused to make payment. The Model Basin consequently withheld the study results, and the *ARD 20* went to sea without warning of this critical potential danger.[22]

In the late summer of 1967, the *ARD 20* was towed to Mare Island for installation of a "wave break," called "the beach" by the dock's crew—a ribbed, perforated metal bulkhead sloped at about 45 degrees mounted in the forward part of the docking area to dissipate the energy of wave action.

Master Chief Michel departed the *Trieste* program in March 1967, and was presented the Navy Achievement Medal. The award's citation did not, of course, mention his key responsibilities in preparing the *Trieste* project for Winterwind.[23]

Submarine Development Group 1

On 12 August 1967, ceremonies were held in San Diego to celebrate the establishment of Submarine Development Group 1 and mark the assumption of command by Captain A. George Beutler.[24] He had been selected in part because of his experience with highly classified submarine intelligence projects. Beutler's previous assignment had been in the Pentagon in charge of the Winterwind project before its migration to the Office of Naval Intelligence.[25]

Prior to establishment of the Development Group the growing activities of the Deep Submergence Systems Project in the San Diego area needed oversight and organization. Accordingly, as previously noted, Dr. Craven had created the Deep Submergence Group, an informal activity reporting to the Commander, Submarine Squadron 3. The new group thus subsumed officially the function of the informal Development Group. Press reports the next day quoted the Commander, Submarine Force Pacific, Rear Admiral John H. Maurer:

> We need to explore, conquer, and develop the ocean to its fullest extent for the benefit of the United States and the U.S. Navy. We must develop the capability to operate anywhere, anytime, in the ocean. There is too much at stake here for us to neglect it. International law says that control of the deep ocean and the ocean floor is based on exploration and exploitation.
>
> Part of the job of this new command is to ensure that we can go farther and deeper and faster and safer than any adversary can. Congress will eventually have to appropriate as much money [for oceanographic programs], if not more, as it gives the space race.[26]

Hereafter, the diesel-electric, special-mission submarines *Baya* (AGSS 318), *Archerfish* (SS 311), and *Salmon* (SS 573), the bathyscaph *Trieste II,* and the floating dry dock *ARD 20* would report to Submarine Development Group 1; the Deep Submergence Group ceased to exist. Later, the Development Group would gain control of the special projects submarines *Halibut, Seawolf* (SSN 575), *Parche* (SSN 683), and others, as well as the future bathyscaph support ship *Point Loma* (AGDS 2), the deep submersibles *Turtle* (DSV 3) and *Sea Cliff* (DSV 4), and the Man-in-the-Sea/SEALAB program.

The ARD Swap

In the dark moonless Sunday night and small hours of Monday morning of the Labor Day weekend, 3/4 September 1967, the rectangular shape of *ARD 10*, the 450-foot floating dry dock normally at the Mare Island shipyard, sat immediately astern of the smaller *ARD 20* (to be named *White Sands* six months later), alongside a pier at the Hunters Point Naval Shipyard. Almost a week earlier the *ARD 10* had been towed down from Mare Island with the new *DSV-1* bathyscaph in her covered docking well.

On the pier was a small contingent of officials: Commander Charles D. Fletcher, the first chief staff officer of Submarine Development Group 1; Henifin (on the *ARD 20*); Lieutenant Commander Watts of the *ARD 20*; from the Mare Island yard, Paul Fossum and John R. Smyth; and several members of the *Trieste*'s crew.[27] Henifin's relief, Lieutenant Commander Robert F. Nevin, coincidentally had shown up on the pier two days earlier. He had been stopped by an armed guard at the head of the pier who refused to allow him to pass, even with his orders in hand. Nevin would report on board in January 1968, still completely bemused as to why a World War II–era floating dry dock carrying the well-publicized *Trieste II* would rate that kind of security.

Several weeks earlier the fleet tug *Apache* had towed the *ARD 20* up to Vallejo, where the shipyard installed the new wave-energy-dissipation "beach" and other features to support the bathyscaph. Subsequently the entire *Trieste* team had boarded the dry dock and set up shop. Without fanfare or celebration but in broad daylight and in view of civilians across the Napa River, the "repaired and reconditioned" boat-shaped *Trieste II* was lifted by a shipyard crane and gently positioned in the *ARD 20*'s docking well—the only time that it would ever be there. The *Trieste II*'s reported "shipyard availability" was obviously concluded and she was being housed in her new mother ship.

When the tug *Apache* towed the *ARD 20* out of the shipyard the team that was soon to be named the Integral Operating Unit (IOU), was just as obviously en route to southern California for post-overhaul testing and evaluation.

It was all theater.

The old, boat-shaped *Trieste II*, so publicly lifted into the *ARD 20*'s docking well, was *not* a recently repaired and updated bathyscaph. It was a shell, a decoy, the remnants of the once-operational float, cosmetically gussied up in paint and imagination. It would become a stalking horse whenever the Navy wanted to give the impression that the bathyscaph was undergoing repairs.

The *Apache* in fact towed the *ARD 20* with the stalking horse in her covered well to Hunters Point, mooring her there, astern of the *ARD 10*, that afternoon. Liberty was granted into San Francisco for most of the crew, and all who could departed for a long holiday weekend in the "City by the Bay." On Sunday night, with the pier darkened, a rail-traveling pier crane was positioned where it could reach both ARDs' docking wells, their covers now open. By the light of only flashlights and battle lanterns, the crane transferred the stalking horse directly from the *ARD 20* over the docking well of the *ARD 10* and down onto a cradle astern of the *DSV-1*. The bathyscaph then was lifted from the well of the *ARD 10* into the mother ship *ARD 20*. Both well covers were slid back into place. The swap took less than an hour. The *ARD 10* was then towed back to Mare Island, where it would keep the stalking horse completely out of sight for as long as necessary.

When Winterwind was unleashed, a press release would reveal that the Navy had reactivated a World War II–era floating dry dock that would be towed to either Guam or Yokosuka, Japan, the next summer. The stalking horse would then be made visible within the confines of the Mare Island shipyard, for all to see—including Soviet reconnaissance satellites.

"Black Ops" Preparations

After the Labor Day weekend the *Apache* towed the *ARD 20* with the *DSV-1* in her covered docking well directly to an operating area west of San Nicholas and San Clemente Islands, off Long Beach. There the *DSV-1*, *ARD 20*, and *Apache* would spend almost three months of shipyard trials and post-construction checks. When not under way, the *ARD 20* was anchored in Wilson Cove on San Clemente Island, where some support and air logistics were available. "Buzz" Henifin and his team were working up the new bathyscaph for Winterwind operations, but he ensured that the fact was not disclosed in the unclassified *DSV-1* dive logs (which commenced in April 1968, with Dive #1-68).[28]

In late December 1967, the *Apache* towed the *ARD 20,* with the *DSV-1* on board, to San Diego, where they would remain through the winter months. When the spring diving season arrived the bathyscaph and the *ARD 20* began their operational workup.

The Integral Operating Unit was now involved in classified operations. Most of the IOU officers knew the objective of the upcoming operation—recovery of a sensitive object on the bottom in mid-ocean—and they had been read into the Winterwind security system. With the exception of Master Chief Michel, who had detached in early 1967, and the *ARD 20*'s enlisted divers who would be handling recovered objects, none of the IOU's enlisted personnel were briefed into Winterwind.

The *Trieste* families, who had moved to Mare Island for eight weeks and had been window dressing for the shell game with the new mother ship and her stalking horse, never became aware of the role that they played in the security surrounding this covert bathyscaph. That was as intended.

When removed from service, the old bathyscaph's boat-shaped float had been slightly modified with a *V*-shaped tail and painted bright white with vertical black striping. The shipyard workers familiar with the *Trieste* knew something was amiss, as the black stripes did not match the tank bulkheads and stiffeners inside the float, but were evenly spaced along its length. Still, anyone monitoring the shipyard would report the continued presence of the *Trieste*. When not in plain sight it was in the covered docking well of *ARD 10* at Mare Island. It was moved outside, seemingly at random, and then returned to the floating dry dock. To an observer the moves would appear to reflect maintenance periods at Mare Island interspersed between operational periods at sea. However, anyone who could detect the errors in the paint scheme would see no rhyme or reason for the moves; it was simply a hulk, stripped of useful equipment.

After the *DSV-1* was brought out of hiding and conducted the *Scorpion* Phase II dives in 1969 (see chapters 15–17), the rationale for the stalking horse no longer existed. The cost of armed guards at the pier and at the *ARD 10* could no longer be justified.[29] Sometime in 1970 the boat-shaped *Trieste II* float—the stalking horse—was cut into sections at Mare Island and consigned to the deep. The Terni sphere was returned to San Diego and placed in storage at Point Loma, where it remained for several years.

Plans and Progress

Winterwind was moving forward with its DSSP-developed assets, under the management of Rear Admiral Philip A. Beshany in the Office of the Chief of Naval Operations in the Pentagon and the "operational command" of Submarine Development Group 1 in San Diego.

However, in mid-1968, Beshany lost control of both Winterwind and Underdog to the Naval Intelligence Command. The new "action officer" was Captain James Bradley, directing the small intelligence office sometimes called the command's Applications Division. This office would plan Holystone operations—the intelligence missions of fleet nuclear submarines penetrating territorial waters claimed by communist states—and plan and direct Sand Dollar programs, including Winterwind.

In December 1969, the Applications Division was split into two sections: Captain Edward A. (Al) Burkhalter relieved Bradley of his Pentagon position (including responsibility for Holystone) and retained the majority of the Application Division's activities. Bradley then became a founding member of a new national intelligence agency—the National Underwater Reconnaissance Office (NURO).

The newly formed NURO would supervise *Trieste* operations in 1971–1972. Captain Bradley would be in charge of the Navy programs within NURO; John Parangosky of the CIA was in charge of Project Azorian, the salvage (by the CIA) of the Soviet ballistic missile submarine *K-129* in the North Pacific.[30] There were several other "projects" within NURO, whose Advanced Technology division was directed by former *Trieste* program manager Dr. Andres B. Rechnitzer, whose deputy was to have been the first *Trieste* officer-in-charge, Commander Richard B. Davey.

The first NURO staff director was Rear Admiral Maurice H. (Mike) Rindskopf, relieved two years later by Rear Admiral Walter N. (Buck) Dietzen. Both Rindskopf and Dietzen were double-hatted, simultaneously holding the classified position of NURO staff director and the unclassified billet of director of the Deep Submergence Systems Division in the Office of the Chief of Naval Operations.

NURO was disestablished in 1976 at the behest of the Navy; its successor incarnation in the unbroken evolution of the National Underwater Reconnaissance Program was the Special Naval Collection Program, founded that year and operating in the same office space—behind the "blue door"—in the Pentagon earlier occupied by NURO, with many of the same personnel.

Then, despite Dr. Craven's broad authority and boundless energy, he was not managing Winterwind operations, but supporting them. From that point forward the *DSV-1* was a covert operational asset of the Navy; the *Trieste*'s days of research and development under DSSP control had ended.

The operational workup of the *DSV-1* was receiving the highest national priority. The bathyscaph's handpicked crew was preparing for a recovery mission as soon as the first Soviet re-entry vehicle could be located.

14 Mission Lost

I christen thee DSV-1.

—Lieutenant Commander, Edward E. (Buzz) Henifin, Officer-in-Charge, *Trieste*

The year 1968 was a very bad year for submarines. In January 1968, the Israeli diesel-electric submarine *Dakar* with her crew of 69 men disappeared in the eastern Mediterranean while en route from Portsmouth, England, to Haifa. Despite massive search efforts conducted by ships of several nations, the submarine was not located for more than three decades.[1]

Two days after the *Dakar* was lost in the Mediterranean, the French submarine *Minerve* was lost at the other end of the Mediterranean with her crew of 52. Despite immediate local searches and later efforts by the French bathyscaph *Archimède* and the U.S. Navy, the *Minerve*'s remains were not located for 50 years.

In the Pacific, the Soviet Golf II ballistic missile submarine *K-129,* en route from her homeport of Petropavlovsk on the Kamchatka Peninsula to her North Pacific patrol station, sank during the night of 11/12 March. There were no survivors from her crew of 98. Her location and time of loss were not known at that time to the Soviet Navy.

A little more than two months later, on 22 May, the U.S. nuclear-propelled submarine *Scorpion* (SSN 589), returning to her homeport of Norfolk, Virginia, from a deployment in the Mediterranean, sank with 99 men on board. Despite subsequent location information derived from the U.S. acoustic sensors in the North Atlantic, the rough ocean floor hid the *Scorpion*'s wreckage from towed cameras and sonars; it would be more than five months before the *Scorpion*'s remains were located and photographed by the venerable *Mizar.*

Finally, on 16 October 1968, the U.S. Navy deep submersible *Alvin,* operated by Woods Hole Oceanographic Institution, sank in 5,000 feet of water east of Wood's Hole, Massachusetts. The *Alvin*, with three men on board, was being hoisted into the water by the support ship *Lulu* when two supporting chains snapped, dropping the submersible into the water with a hatch open. It immediately started to take on water and sink; the three men were able to escape as the submersible went down. (In June 1969 the *Mizar* located and marked the *Alvin*'s location with deep-ocean transponders; on 27 August the submersible *Aluminaut* attached lines to the sunken *Alvin*; the *Mizar* hauled the craft off the bottom and towed it back to Woods Hole.)

A black year indeed for undersea craft—four submarines and one submersible sank, with no survivors from any of the submarine crews.

The *DSV-1* Workup

Just prior to the first "wetting" of the third version of the *Trieste*, the officer-in-charge, Lieutenant Commander "Buzz" Henifin, arranged to have a bottle of champaign available on the ARD for a covert christening. The *ARD 20* ballasted down slightly to allow a few feet of water into the docking well; he paddled a punt to the bathyscaph, shattered the bottle against the support structure at the bow, and intoned, "I christen thee *DSV-1*."[2] During its design and construction, and for more than two years after the christening, its most common designation was simply "*DSV-1*."

The first undocking from the *ARD 20,* in calm waters, went without incident. Filling the new float with 66,000 gallons of 115-octane avgas and "shotting" her with 22-plus tons of BB-sized pellets for ballast also went well. The batteries were charged, and the high-pressure air flasks had been filled while en route in the mother ship from San Francisco.

The first dive for this new bathyscaph was a weight-and-trim dive in late September 1967. Topside on the *ARD 20* were Henifin and naval architect George Childs, lead architect of the *DSV-1* program at Mare Island; in the water with the bathyscaph were Navy divers. Henifin and Childs oversaw the divers' activities, communicating with the bathyscaph by UQC underwater telephone. On board the *DSV-1* for this first dive were Lieutenant Mike Staehle piloting, John R. Smyth, who had acted as the DSV-1 program manager through its construction, and Electronics Technician 1st Class Robert D. Bowen, a diver from the *Trieste* crew who was communicating with the divers.[3]

Staehle piloted the first several dives and was pilot, co-pilot, or training officer on all but one of the dives of the new *DSV-1* prior to her departure for the first operational mission in early 1969. He had shepherded the new bathyscaph through construction at Mare Island and possessed an intimate knowledge of the craft—every nut and bolt in it. Although he was the most junior officer in the crew and never before had piloted a deep submersible other than the *Trieste* trainer at Charlottesville, it was largely in Staehle's hands to develop the diving procedures for this new craft and to train the new officers in techniques and practices.

The first three test dives were made to conduct checks and tests for the shipyard inspectors. The tests and dives developed a "punch list" of deficiencies and problems for shipyard attention.

First Docking

Upon completion of these first dives the *DSV-1* was re-docked in the ARD, and that evolution exposed a major problem with the avgas system. Necessarily, the structure of the float was not massive enough to support the weight of the 66,000 gallons of aviation gasoline unless the bathyscaph was in the water. To lighten the craft for re-docking at least 63,000 gallons had to be transferred into the ARD's tanks through a floating hose. And, each time that the bathyscaph was docked in the ARD more than 8,000 cubic feet of nitrogen would be forced into the bathyscaph's buoyancy tanks to render them explosion-proof. Following de-fueling, a compressed-air flush would vent the nitrogen to the atmosphere. There was normally enough nitrogen on the ARD *for only one re-docking.*

In this first docking, the avgas pumping system could not de-fuel the *Trieste* completely due to an improper design of the fuel transfer pump. To force out more avgas, the nitrogen was pressurized to about five pounds per square inch over atmospheric pressure, exhausting the ARD's nitrogen supply prematurely.[4] When docked the bathyscaph still had about three feet of fuel in each of its buoyancy tanks; the docking plan called for no more than two inches of residual gasoline. The combination of a large excess of avgas under pressure in the bathyscaph and an inoperative transfer pump caused the spilling of several hundred gallons of highly volatile avgas into the ARD's docking well. There was imminent danger of fire in the docking well, a fire that would likely be uncontainable. The situation was so dire that the ARD's safety officer "suggested" that all crewmen move to the questionable security of the forward part of the ship.

The docking officer reversed the ARD's "scupper" pumps, sending avgas-contaminated seawater into the dock's ballast tanks. This presented a less immediate, although possibly an even more dangerous situation. The ARD's ballast tanks extended some 100 feet along each side of the docking well; should gasoline vapors collect there to explosive levels and be ignited the ship would be blown in half and sink. The ship was one spark away from destruction.

The ship's engineers quickly re-aligned the ballast pumps and began pumping the ARD sidewall tanks overboard, which was achieved without incident. Compressed air was pumped through the sidewall tanks until they registered safe, and the ARD's officer-in-charge, Lieutenant Commander James Watts, personally crawled, at great personal risk, through every tank to check for any residual avgas pockets.

To ensure his safety (or at least his rescue), Watts had a man topside call out through an open tank top every 30 seconds, "You OK?"—to which he responded, "OK." If he had been rendered unconscious a team stood ready to find and retrieve him. The *ARD 20* dodged a bullet that day.[5] Earlier, Dr. John Craven had rejected a foam firefighting system for the ARD as too expensive. Following this incident a foam firefighting system was installed.

The *Apache* towed the *ARD 20* and the new *Trieste* into Wilson Cove on San Clemente Island for a several-week period to correct the shipyard "punch list." This work was done by *Trieste* crewmen under the supervision of "Junior" Smyth, the Mare Island quality reliability officer, and shipyard engineers and technicians who were riding the *ARD 20.*

The Integral Operating Unit left Wilson Cove in late October for a second extended underway period west of the San Nicholas and San Clemente Islands. A program of four dives tested sonars, bottom maneuverability, buoyancy controls, and responsiveness. There was no attempt to exercise the several capture systems (kludges), which would ultimately be required in an upcoming Winterwind mission.

Throughout this dive series the *ARD 20* provided fueling and ballasting, while the tug *Apache* handled towing and area security. When the *DSV-1* was not diving, the *Apache* would assemble a "tandem," or three-part, tow: the ARD with a towline from its stern to the bathyscaph, which in turn pulled a 35-foot hawser-handling boat. The boat was there to steady the *Trieste* under tow and to tow the bathyscaph itself in an emergency. Until it was time for the next dive, the *Apache* led this little nautical parade in a long north–south racetrack pattern to the west of San Nicholas Island, day and night, at a constant two or three knots.

One night, when the *Apache* had reached the northern end of the circuit and was making a long, sweeping starboard turn to the south, the towline to the 35-foot boat jumped its towing cleat and turned the boat sideways to the direction of movement, swamping it and throwing its crew of two sleeping ARD sailors into the sea. Mike Staehle, on the bathyscaph, heard the sudden "jump" of the tow line and felt the change to the *DSV-1*'s motion through the water. He raced to the stern of the float and with a touch from his pocketknife severed the towline to the boat. He then radioed the emergency to the *ARD 20*.

Interior Communications Technician 1st Class Peter A. Harnly and two other *Trieste* sailors jumped into a 15-foot Boston Whaler at the starboard quarter of the ARD and raced back to find a darkened sea full of debris and floating life jackets. It took more than an hour to locate and recover the sailors, both alive. By that time that the *Apache* and her caravan were nowhere to be seen. Upon consultation with Henifin over a walkie-talkie, Harnly chose to take his five-man "detachment" to Wilson Cove for the night. The 35-foot hawser boat was not recovered. (Almost a year later a Mexican fisherman returned the boat to San Diego for a financial reward.)

The next day Harnly brought his boatload of five back to the ARD. While the Boston Whaler was being hoisted on board at the starboard quarter, with Harnly and another man still in it, the crane rigging parted, dropping the whaler into the water from a height of 30 feet and spilling both occupants into the ocean. They were quickly recovered, but thereafter Harnly was considered a "marked man."[6]

Spring 1968 Diving

The second series of classified dives concluded in late December 1967, and the IOU retired to San Diego for year-end holidays.[7] The *ARD 20* was moored to a buoy off Ballast Point, where she was serviced by small boats. Her docking well remained covered at all times, and only personnel attached to the bathyscaph, ARD, the *Apache*, or the staff of Submarine Development Group 1 were allowed to board. At that time the only people in San Diego "read into" the Winterwind program were the officers-in-charge of the ARD and the *Trieste*, the commanding officer of the *Apache*, the *Trieste* pilots, the divers, and selected officers on the DevGru staff—perhaps no more than 25 people. Everyone else was simply told that they were involved in classified work—"period."[8]

In unclassified documentation distinguishing the new bathyscaph from the old, boat-hull *Trieste II,* it occasionally was identified as the "*Trieste III,*" at least once as the "*Bathyscaph III.*" In Deep Submergence Systems Project communications it was always simply the *DSV-1.* Prior to September 1970, there were official public statements using "*Trieste II*," as if the third incarnation of the bathyscaph was a minor modification of the second. It would not be until late 1970 that the Navy would officially apply the (perhaps deliberately confusing) name and hull number "*Trieste II* (DSV 1)" to this new craft.

Lieutenant Commander Robert F. Nevin reported on board as Henifin's prospective relief, the change in command taking place in May 1968. Nevin had been executive officer of the submarine *Spinax* (SS 489) and was the third consecutive Naval Academy graduate to command the high-visibility *Trieste* program.

Lieutenant Anthony Dunn, who had reported in September 1967, first as hull officer, later as chief engineer, finally in March 1968 as assistant officer-in-charge. Dunn had served in the Navy for 14 years as an enlisted man, advancing to senior chief electronics technician prior to being commissioned in 1961.

On 9 March 1968 the *ARD 20* was officially named the *White Sands* (hull number ARD 20), although she remained a "yard craft" as opposed to a ship in commission. And on 1 April, the *Apache* towed the *White Sands* with the bathyscaph in her covered docking well to the area that had been used for the 1967 classified workup dives. Of all the officially certified pilots and trainees from the earlier *Trieste II* only Henifin and Staehle remained, and only Henifin, about to be relieved, was *officially* qualified. Staehle was effectively the chief pilot, although he had not been officially certified, and was training newly arriving officers as pilots.

Dives April–July 1968

Four dives in April were primarily for equipment testing and pilot training. The first and second dives were to a depth of 50 feet. As a covert vehicle under intense pressure to become able to respond to the first Winterwind target, whenever it should arise, the *Trieste* was exempted from the formal requirement that pilots have five dives in training plus one for qualification. For the *Trieste* the student-officer qualification occurred on the date that the officer-in-charge decided he was ready. This commonly occurred on his third dive. Nevin was a hands-on manager and made 10 of the 13 workup dives from April to December 1968, personally monitoring each officer. Staehle made 12 of these 13 dives, to train the new officers.

After the two 50-foot dips came an ocean dive to 450 feet, with Staehle piloting, Nevin in training, and Henifin in the third seat providing his expertise and experience. That dive lasted three hours, 45 minutes submerged. The fourth and final dive for this series was to 2,312 feet on 13 April, again with Staehle, Nevin, and Henifin on board. Staehle considered this his qualification dive although it was his tenth dive piloting the *Trieste*. It was Henifin's last bathyscaph dive. He has the distinction of being the only man to record dives in all three variants of the *Trieste*.

In May 1968 Buzz Henifin was detached, going on to command the submarine *Promfret* (SS 391), homeported in San Diego. Henifin's superior performance in readying the bathyscaph for Winterwind brought him to the Deep Submergence Systems Division office in the Pentagon only 18 months after reporting to the *Promfret*. He would spend five years in the Office of the Chief of Naval Operations, from 1971 to 1975, involved in both "black" and "white" deep submergence activities, and in 1978 he would be ordered to the Naval Research Laboratory as commanding officer. As his "twilight" tour in the Navy, in 1981 he became coordinator of all research programs at the Naval Sea Systems Command. It was a distinguished naval career.

The new *Trieste* re-docked into the *White Sands* on 14 April, and the *Apache* towed the ARD back to San Diego. The *White Sands* once again moored to a buoy off Ballast Point and remained there until the next series of dives, in mid-May 1968.

This dive series, from 13 May to 18 May, included the first deep dive, Dive #7-68 on 18 May to 6,228 feet, with Nevin piloting, Staehle as co-pilot, and Naval Ship System Command (NavShips) representative Anthony J. Taromino observing. Taromino was the NavShips project engineer for the *DSV-1* and as such was required by custom and regulation to personally participate in this first deep dive. Taromino later said, "It kept us honest."[9]

The July series of four dives was to work up the *Trieste*'s sensor, navigation, and control systems. On the dive of the 19th, Lieutenant Field piloted under the instruction of Nevin, and Homer Hemming, the Westinghouse techrep, observing sonar operation. On this dive Field was skiing along the bottom with the skegs a few inches into the silt when the bathyscaph ran head-on into a sunken Mark 14 torpedo. The torpedo slid between the port skeg and the sphere, striking the Doppler sonar transducer and bending it upward. It seemed that sonar sensitivity would need to be improved to give timely warning of such obstacles ahead of the craft.

Dive #11-68 was a deep dive made at the direction of Captain George A. Beutler, Commander, Submarine Development Group 1, to 12,720 feet on 29 July. The depth was specified for an operational reason that was very closely held. Nevin later said, "They just told us to dive as deep as we could, the 12,000-foot dive was the deepest spot we could find."[10] The *White Sands* and *DSV-1* were towed back to San Diego on 29 July; there they remained, moored off Ballast Point, until the start of the December dives four months later.

The Impact of World Events

World events during this 1968 workup would greatly affect the *Trieste* program and in fact changed the destinies of those involved with the craft. The first major event was the seizure of the U.S. intelligence ship *Pueblo* (AGER 2) by North Koreans in the Sea of Japan on 23 January 1968. American reaction included the deployment of three aircraft carrier groups into the waters off the Korean Peninsula. U.S. submarines were surged into the Sea of Japan, and the American armed forces girded for a re-ignition of the Korean War. Ultimately, neither side took a step too far, the United States choosing to continue to fight the Vietnam War without opening a "second front" in Korea.[11]

The second event was known to very few persons at the time, but was to have a significant impact upon the *Trieste* and Project Winterwind. In the northern Pacific, unusual Soviet air, surface, and submarine deployments out of Petropavlovsk and Vladivostok began on 11 March. Some of this activity was in response to a diesel-electric Zulu IV ballistic missile submarine in distress north of Midway Island—but not all of it. In late March the unique spectacle of four Foxtrot-class diesel attack submarines out of Petropavlovsk searching in line abreast on the surface directly southward to the 40th parallel, thence due east to the International Date Line, was especially revealing. They were searching for the submarine *K-129*, a Golf II–class diesel-electric submarine carrying three one-megaton nuclear missiles and two nuclear torpedoes.[12]

The Soviet submarine loss in the Pacific had, paradoxically, the effect of killing the *Trieste*'s participation in Winterwind. The Soviets did not appear to know when, where, or why they had lost the *K-129*. Thus they could be expected to show special sensitivity to operations by the *Trieste* anywhere in the northern Pacific.

The *Pueblo* seizure, meanwhile, made clear that the rules of the nautical intelligence-gathering game were changing. Prior to the *Pueblo*'s seizure, the "rules" of that game were that intelligence collection ships—of any nation—would not be harassed or seized. Until the assault on the *Pueblo*. Soviet and U.S. intelligence collection ships had rarely if ever been "bothered" so long as they remained beyond claimed territorial waters.[13] The ongoing *Pueblo* crisis made it appear possible that this unspoken agreement could be broken with impunity should the Soviets feel it was in their interests. Such a perception could be triggered by a U.S. Navy ship loitering in the northern Pacific carrying a deep-diving search and recovery vehicle. And, with the *Trieste* being surface supported no clandestine operation was feasible.

The U.S. Chief of Naval Operations, now Admiral Thomas H. Moorer, believed that any attempted surface-based deep recovery in the northern Pacific could lead to an unpredictable and possibly aggressive Soviet reaction. Accordingly, Moorer personally decided to remove the *Trieste* from that potential danger. He had the Director of Naval Intelligence modify the Winterwind project by deleting plans for surface ship–based open-ocean recovery of Soviet missile re-entry vehicles. If Winterwind was to continue as a viable project its planners would have to find a way to conduct a recovery operation from a submerged platform.[14]

That decision left the new *Trieste* without a mission. Suddenly an orphan, the *Trieste* program embodied an expensive system with a unique and very narrow capability that might never find another mission that could justify the cost of its upkeep. It would take some time before the mission loss would percolate down to the *Trieste*'s officers and enlisted crew, but their lives and their careers had been changed.

However, a third major historical event would soon give the *Trieste* program a temporary reprieve—the loss of the U.S. nuclear-propelled submarine *Scorpion* in the Atlantic on 22 May 1968.

15 *Scorpion* Preparations

The termination of your assignment brings to a conclusion one of the most complex and professionally demanding missions in the history of search and salvage operations. You and your entire task unit are to be congratulated for the determination and professional competence demonstrated.

—Admiral E. P. Holmes, Commander in Chief Atlantic Fleet

The loss of the *Scorpion* on 22 May 1968, immediately put the *Trieste III*—known at the time as the *DSV-1*—on notice for deployment to the Atlantic to examine the submarine's wreckage. However, until the submarine was located no changes to the *Trieste*'s operational schedule would occur because the covert submersible was awaiting a Winterwind mission.

The Naval Research Laboratory's research ship *Mizar* sailed from Norfolk on 2 June 1968, six days after the *Scorpion* was reported missing, sailing toward the best estimated location of her loss. The *Mizar*'s towed search system, four years after its debut off Boston during the 1964 *Thresher* search, was the best deep-ocean search system then in existence.[1]

On 26 June, during the ninth run of the *Mizar*'s fish, Chester (Buck) Buchanan, heading the *Mizar*'s search team, obtained a photo of a 30-inch piece of sheet metal that appeared to be newly deposited on the ocean floor. It was potentially a breakthrough moment, but subsequent days of maneuvering in the area failed to generate additional clues. It was not until four months later, on 28 October, that the main hull sections of the *Scorpion* were found—within 200 feet of the piece of bent sheet metal photographed on 26 June.[2]

Admiral Thomas H. Moorer, the Chief of Naval Operations, announced on 31 October that the *Scorpion*'s remains had been found some 385 nautical miles southwest of the Azores in 10,000 feet of water. The next day a United

Press International story quoted officers at Submarine Development Group 1 as considering the possibility of using the *Trieste* to investigate the *Scorpion* wreckage and possibly retrieve parts of the wreckage.[3] A very quick renunciation of the article came from the Eleventh Naval District in San Diego, where a spokesman insisted that the Navy "had no plans to use the bathyscaph *Trieste II* [*sic*] for photography or salvage work in connection with the discovery of the sunken nuclear submarine *Scorpion*."[4] It appeared that some Navy activities had not been notified that the *Trieste*'s participation in Winterwind had been terminated and that the underlying reason for covertness had ended.

In an act that appears to reflect the CNO's decision to remove the surface recovery option from Winterwind, Lieutenant Commander Bob Nevin and Lieutenant Tony Dunn, the *Trieste*'s officer-in-charge and his assistant, were summoned by the DevGru commander in early August. When the door was closed, Captain George Beutler informed them that the *Trieste* would in fact be deploying to the Atlantic for *Scorpion* operations. They had four months to prepare—and the crew was *not* to be informed. Dunn later related that this order forced him to work everyone long hours without explanation to squeeze nine months of workup into the four months allocated. Even without "official" notification, the crew easily guessed the forthcoming target.

One of the earliest indications that the *Trieste* would be sent to the Atlantic for *Scorpion* dives had been the 19 July dive to 12,720 feet off San Diego. Mare Island's John (Junior) Smyth was on that dive, sent by the Naval Ship System Command to certify the *Trieste* to that depth.[5]

The *Scorpion* Mission

Commander Brad Mooney, a former *Trieste* officer-in-charge, received orders to the Office of the Chief of Naval Operations on 9 October 1968 as the Deep Submergence Plans and Programs Officer. This newly established billet was charged with planning and developing missions for deep submergence vehicles, including search and rescue for disabled or sunken submarines, and for locating and recovering small objects from the deep ocean floor.[6] Mooney would serve as the OpNav action officer for manned deep submergence operations, both classified and unclassified, from 1968 to 1971.

In November Captain Beutler gave Nevin permission to brief the *Trieste* crew, Lieutenant Commander Charles Poarch to inform his crew on the *White Sands,* and Lieutenant Lawrence W. (Larry) Lonnon, commanding the *Apache,* to reveal to his crew, that they all about to deploy to the Atlantic for the *Scorpion* investigation. The *Apache* towed the *White Sands* with the *Trieste* embarked back to the San Clemente Island dive area for two final Pacific dives, on 12 and 13 December.

Lieutenant John Field received his qualification dive on the 12th, descending to 4,180 feet, with Lieutenant Mike Staehle evaluating and Junior Smyth along. On the 13th, Dunn received his qualification dive to 1,404 feet, with Nevin and Staehle evaluating. Thus, when Staehle would depart in March 1969, the *Trieste* program would have three qualified pilots (Nevin, Dunn, and Field) and four men at various stages of training or en route to the program: William J. (Stretch) Leonard and Ross E. Saxon (both lieutenants), Lieutenant (Junior Grade) David T. Byrnes, and Chief Warrant Officer Larry E. Hawks. Byrnes would join just before the *Scorpion* operation.

The final approval for the *Trieste* to deploy to the Atlantic for dives on the *Scorpion* wreckage was received from the CNO on 28 December.[7]

Captain Beutler was relieved as Commander, Submarine Development Group 1 by Captain Robert H. Gautier on 22 January 1969. Beutler was awarded the Legion of Merit for exceptionally meritorious service as the first commander of the DevGru. Gautier had commanded two diesel-electric submarines as well as an attack transport and a submarine division.

Shortly before departure for the Atlantic, the *Trieste* team received and installed an improved propulsion battery suite. It would provide 962 amp-hours of 120-volt AC power for propulsion—enough for 20 hours of bottom search with the centerline 6.5-horsepower thruster. This was almost a 12-fold increase in propulsive power and quintupled the bottom-search time possible for the *Trieste* when searching for the *Thresher* in 1963. Mare Island's test director, Junior Smyth, rode the *White Sands* as far as the Panama Canal to check out the new battery system.[8]

On 3 February 1969, the venerable fleet tug *Apache* with the *White Sands* in tow (carrying the *DSV-1*) slipped out of San Diego Harbor without ceremony or announcement. Lieutenant Lonnon's *Apache* was embarking on the longest tow in U.S. Navy history: 14,000 nautical miles, requiring more than five months of towing time. (Probably the longest U.S. Navy tow prior to this *Trieste* odyssey occurred in 1905–1906, when the floating dry dock *Dewey* was towed from Chesapeake Bay to the Philippines—a 12,000-mile tow that took six months, 12 days.)

Lieutenant Poarch, officer-in-charge of the *White Sands*, was not on board, having received the horrible news that his teenage son had been killed; he never returned to the *White Sands*. Three days later, at sea en route to Panama, Nevin, officer-in-charge of the *Trieste,* assumed the additional duty of commanding the *White Sands*. For the next three years command of the *White Sands* was an additional responsibility of either the officer-in-charge of the *Trieste* or the ARD's executive officer, Lieutenant Beauford E. (Beau) Myers.[9]

The *Apache* could tow the *White Sands* at between three and four knots, depending greatly on currents, tides, and weather. To make the best time possible the engineers on the *White Sands* activated the dock's two stern-mounted 635-horsepower thrusters and ran them at full power throughout the tow. The dry dock's twin Detroit diesel engines thrived on running constantly at full power.[10]

The draft of the old *Apache*—completed in 1942—would of course decrease as she burned fuel; eventually her single screw would be halfway out of the water, at which time she would be no longer able to tow. To prevent that happening the *White Sands* refueled her while under way. The *White Sands* carried enough diesel fuel for her own thrusters and to refuel the *Apache* as required.

As an example of the frustrations of the long tow, the *Apache* and the *White Sands* had the resort city of Acapulco in sight for four days without liberty ashore for the crew. The tow later ran into gale-force winds—called the Tehuano—off the Gulf of Tehuantepec and lost 12 miles of progress over a 24-hour period. The tow to Balboa, Panama, took more than a month. The *Apache* and *White Sands* moored at the Rodman Naval Station from 6 to 10 March, and their crews had liberty in Balboa. There Mike Staehle departed for his next duty station after 3½ years with the *Trieste.* With his detachment Nevin lost a most knowledgeable pilot and engineer.

In Panama the *White Sands* received an unannounced visit from General Robert W. Porter, U.S. Army, commander of the U.S. Southern Command, and Rear Admiral George B. Koch, commandant of the Fifteenth Naval District

(Panama Canal Zone). They were held up at the gangway, with Nevin's apologies, until permission could be obtained from the DevGru in San Diego to allow them on board for a tour of the bathyscaph. This visit by two senior officers was the triggering event that brought the *Trieste* out of the "black." Shortly thereafter, Nevin was directed by message to uncover the *White Sands'* docking well and to present the *Trieste* as an unclassified deep submergence vehicle. Still, the radically improved capabilities of this third version of the *Trieste* remained classified.

Lieutenant Lonnon in the *Apache* led his tow along the Central American east coast until north of Cuba, then eastward off the northern coast of Cuba, and from there north along the eastern coast of Florida to Mayport. Along the Florida coast, Lonnon encountered the Gulf Stream, which boosted the pair's speed of advance to almost nine knots.

Preparations at Mayport

The *Apache*, *White Sands*, and *Trieste* remained in Mayport from 26 March to 21 April, making final preparations for the upcoming dives. During the Mayport period Admiral Bernard A. Clarey, the Vice Chief of Naval Operations, and other flag officers boarded the *White Sands* to inspect the *Trieste*. Dr. John Craven, still the DSSP's chief scientist, and Captain Harry A. Jackson, senior member of the *Scorpion* Technical Advisory Group, also came on board.

The equipment installed at Mayport included one of the first satellite navigation sets in the Navy's inventory. An experimental model, the Magnavox 70-HP, had just been on board the battleship *New Jersey* (BB 62) when, conducting naval gunfire missions off South Vietnam. The battleship's commanding officer had been directed to offload the set and forward it to the bathyscaph at Mayport.[11]

The Naval Reconnaissance and Technical Support Center (NRTSC), in Suitland, Maryland, sent automatic photographic-developing equipment and enlargers to be housed on the *White Sands*. This hardware could take the 35-mm film from the *Trieste*'s cameras and blow up the exposures to eight-by-ten positives.

One of DSSP's contributions to the *Trieste*'s investigation of the *Scorpion* site was the first remotely operated deep submergence vehicle, for the *Trieste* to carry down to the ocean floor. Dr. Craven's idea was that a vehicle small enough to enter one of the *Scorpion*'s 21-inch torpedo tube might help to determine if her loss had been caused by an internal or external torpedo explosion. A $100,000, 40-day crash program produced the vehicle at the Naval Undersea Center in San Diego; two of these "flying eyeballs" were built by the firm HydroProducts and delivered to Mayport. To the underside of the bathyscaph's float were added two horizontal, open-end pipes to house the vehicles.[12] The San Diego center supplied a technical representative on the *White Sands* for the entire *Scorpion* operation.

Several technical representatives boarded the *White Sands* prior to departing Mayport. Others would join at Terceira in the Azores. Among the individuals boarding at Mayport were Lieutenant Ross and Lieutenant (jg) David Byrnes, the sixth and seventh *Trieste* pilots. Junior Smyth from Mare Island also boarded at Mayport, potentially providing another pilot, as he had qualified a year earlier during the post-overhaul, classified operations off San Clemente.[13]

One day out of Mayport, before sunset on 15 April, the bathyscaph was floated out of the *White Sands* and taken in tandem tow behind the dry dock, which was itself being towed by the *Apache*. Nevin was under orders to conduct a test dive and to do so while within support range of Naval Station Mayport in the event that a problem was encountered. The dive was planned for the next morning.

The bathyscaph was riding high, with no fuel or ballast. A storm struck during the night, accompanied by thunder and lightning, and the *White Sands*' stern lookout was unable to keep the *Trieste* in sight. At some point prior to 0400 on 16 April the bathyscaph parted its towline and disappeared into the dark. The *White Sands*' executive officer, Lieutenant Beau Myers, on the ARD recalled:

> I had turned in and was sound asleep. The bridge messenger woke me and told me the captain wanted me on the bridge. As I quickly donned my trousers and shoes, I noticed the quick movements of the ship. They were not there when I turned in. I made my way to the bridge.
>
> Nevin and other DSV pilots were there. I detected a sense of panic. One fear expressed was that the Soviets (there were constant rumors that they were monitoring us) might find the *Trieste* before we could get to her. I had quit a twenty-year smoking habit seven months earlier and I found myself reaching for my left shirt pocket as if to pull out a non-existent pack of cigarettes.[14]

Notified by Nevin, Lonnon reversed the *Apache*'s course at 0420 and backtracked with the dry dock in tow. He sighted the errant bathyscaph's running lights at 0525, about two miles distant. Lonnon placed the *White Sands* about 1,500 yards from the *Trieste*, cast off the tow, and proceeded to the bathyscaph's location, meeting there with two motor launches from the *White Sands.* These were manned by Dunn, Leonard, and several divers. Dunn recalled,

> Stretch [Leonard] and two [other] divers dove down to check out the underside of the *Trieste*, while a couple [of] men and I secured the tow and checked out the topside for damage. While under the bathyscaph securing the two "flying eyeballs" ROVs (which were dangling into the depths from their control cables), Leonard had his weight belt stripped away by wave and water action, which then thrust him upward, bashing his skull into a bottom beam on the 'scaph. Some guardian angel guided his nearly unconscious body to the surface where he was flopping around in distress. We got him on deck of the *Trieste*, into a boat and took him back to the ARD to bandage up his head.[15]

The remaining *Trieste* divers passed the towline to the *Apache* at 0656, and the *Apache* towed the *Trieste* back to the dry dock. At 0722 the *White Sands* began the process of docking the bathyscaph (there being no fuel in the float to remove). Once safely docked the *Trieste* was inspected for damage. Both remote vehicles were in need of refurbishment; the other storm damage was minor and easily handled by the *Trieste*'s crew. The towline apparently had been severed by rubbing against the boom structure installed on the *Trieste*'s bow in Mayport to mount cameras and lights. Following this incident the bathyscaph would be connected to the tow rope via a short chain bridle.

The Naval Supply Center at Mayport, gleeful at the departure of the *Trieste* group, soon had one last bathyscaph-related crisis to handle. As the *Trieste* was being recaptured and brought back into the safety of the ARD's docking well, the *White Sands*' engineers radioed a Brickbat/Priority-1 casualty report and requisition for an armature assembly for one of the ship's electrical motors. By the time the part could be located and dispatched by helicopter the *White Sands* was almost 200 miles offshore and heading eastward. An SH-3 Sea King flew more than 400 nautical miles round-trip to make the delivery, with a fixed-wing aircraft as escort. The *White Sands* officers had been stunned by the priority and precedence the *Scorpion* search mission was carrying, and now they were stunned again by the power of a requisition related to that mission.

While the *Trieste* was no longer a part of Winterwind, its Brickbat/Priority-1 authority melded seamlessly into the *Scorpion* operation. The *Trieste* would receive almost unlimited and unquestioned support during the investigation.

On 22 April, at 0830, the *Apache* experienced a casualty affecting her diesel engines and went dead in the water. Nevin, concerned that she might collide with the *White Sands*, ordered the dock's tow cast off, and the *White Sands* soon was blown several miles away from the tug. Junior Smyth was sent by small boat to the *Apache*, where he devised and completed an emergency repair to the tug's engines, using material available on board.

The *Apache* was under power by 1350. Less than 12 hours later the "convoy" hit a storm with seas six to ten feet high and winds up to 50 knots. On 26 April, at 1415, the fleet tug *Salinan* (ATF 161) rendezvoused with the group to deliver the parts the *Apache* to effect a permanent repair.[16] (The Mare Island shipyard later received a letter of appreciation from the *Apache*'s commanding officer praising Smyth's emergency repairs and acknowledging that if he had not succeeded so expeditiously, "all the units would have been at the mercy of heavy seas and high winds.")

About two days later the *White Sands*' head cook came down with severe abdominal distress. Chief Hospital Corpsman Troy W. Brown on the ARD quickly diagnosed appendicitis, established an intravenous penicillin drip, and had a radio message sent requesting assistance.[17] The Atlantic Fleet submarine command arranged for the submarine tender *Canopus* (AS 34), en route from Rota, Spain, to the Panama Canal, to rendezvous with the *Apache* and *White Sands* for an at-sea transfer. The cook was trussed up into a stretcher and swung out by crane into a ship's boat from the *Canopus*. A surgical team on the tender removed his appendix. It had ruptured. His life was saved.

On 3 May, the *Apache* cast off her tow and cozied up to the *White Sands* to refuel from the ARD. It was a half-day evolution, and when it was completed the *Apache* reestablished the tow and continued on to the Azores.

A nearly month-long transit to the Azores made for a lot of free time. In addition to training new pilots and reviewing the *Mizar*'s photos and bottom data, Leonard and Byrnes had time to design and draft a proposed deep submergence pin to be worn by certified deep submergence pilots, similar to aviators' wings, submariners' dolphins, and the later surface warfare insignia. Nevin agreed to forward the proposal to the Pentagon, through channels; Commodore Gautier would be the first "chop" on its long journey through the Navy's bureaucracy. It was to have a 12-year gestation period, finally being approved 6 April 1981![18]

Arrival in the Azores

The *Apache* towing the *White Sands* carrying the *Trieste* arrived at Praia da Vitória, on the island of Terceira, on 21 May. There Commodore Gautier, Captain Harry Jackson and his Technical Advisory Group, a medical officer from Submarine Development Group 1, and some late-arriving technical representatives boarded the *White Sands.* The period 21–26 May offered liberty for the crews while the *Apache* and the *White Sands* were taking on stores and the bathyscaph was made ready for diving.[19] While in port, a Johns Hopkins techrep was flown out especially to repair the satellite navigation equipment, which had failed on the way to the Azores.[20]

The *Apache* got under way from Praia da Vitória on the morning of 26 May, with the *White Sands* in tow, carrying the *Trieste*. Left behind with U.S. Navy and Air Force personnel at Lajes Field on Terceira was Commodore Gautier's staff supply officer.[21] He would coordinate shore-based logistics support for operation. Nevins later described him as a "magician" at getting badly needed repair parts and supplies air-dropped to the ships.

A day's voyage took them to São Miguel Island, where the Azores' largest city, Ponta Delgada, could provide fuel and a day of liberty for the crews. There Gautier and Junior Smyth accepted an invitation to tour Jacques Cousteau's ship *Calypso*. Cousteau graciously offered lunch and a tour of his research submersible, the "diving saucer" *Denise*.[22] On the 29th, after a day spent refuel, the *Trieste* group sailed for the *Scorpion* dive site.

Scorpion Operations, Phase I, in 1968, had involved the research/oceanographic ships *Mizar* and *Bowditch* and other ships. Phase II was about to begin with a close examination of the *Scorpion* wreckage by the bathyscaph *Trieste*.

16 At the *Scorpion* Site

The officers and men of this Integral Operating Unit, through their superb teamwork, untiring efforts and dedication, merged their diverse elements into a unit which provided a deep-ocean search capability for the United States.

—John W. Warner, Secretary of the Navy

Two days after the *Trieste* Integral Operating Unit arrived in Praia, the operation's support ship, the high-speed transport *Ruchamkin* (APD 89), entered port with damage to both propellers. Originally built as a destroyer escort (DE 228), the *Ruchamkin* had been converted to a high-speed transport, intended to carry 160 commandos and raiders and equipped with four small, LCVP-type landing craft. The *Ruchamkin* would provide both on-site security and a "taxi service" into Praia from the *Scorpion* search site on an as-required basis to transport arriving and departing personnel, deliver mail, bring out fresh food, and fulfill special logistic requirements. She would develop an approximately ten-day cycle: four days with the IOU, one day to Praia, a few days in port, and one day returning to the dive site.

Prior to entering Praia the transport had steamed through a field of wire-tethered fishing traps or nets that chewed up both of her propellers. The damaged propellers caused severe vibration. Lieutenant William Leonard and Lieutenant Ross Saxon, who were on board the *White Sands,* dived on the *Ruchamkin*'s propellers to make whatever repairs could be manually effected. The *Trieste* group departed from Praia more than a week before repairs could be concluded, thus the *Ruchamkin* was late to rendezvous at the dive site.[1]

The *Apache* had all radio communications responsibility for the IOU, passing messages by signal light or semaphore to the *White Sands*, thus keeping the signalmen on both ships very busy. Coordination between ships, boats, and the bathyscaph was made with commercial walkie-talkies, which proved both reliable and effective. The *Ruchamkin* provided a post office and a paymaster. Once a month the *Ruchamkin* sent her paymaster and postal clerk to pay the men on the *White Sands* and *Apache*—in cash, so they could buy money orders to send home. Most men who did not have to send money home allowed their pay to accrue until just before hitting liberty ports.[2]

The trip from Ponta Delgada to the dive site was just under 370 nautical miles at a tow speed of four knots; the transit took until 1 June. The *Apache* handled the navigation, as she did throughout the eight months away from San Diego, and almost exclusively the old-fashioned way—by star fixes and sun lines. The tug's LORAN set provided only sporadic information.[3]

The *White Sands* had been equipped with both an SRN-12 Omega radio navigation receiver and a Bendix acoustic interrogator while in Mayport. The former was to augment the satellite navigation system (also installed at Mayport), the latter to interrogate deep-ocean transponders left at the *Scorpion* site by the *Mizar* seven months before. Locating a specific, 300-yard-diameter site in the trackless ocean would have been nearly impossible but for satellite navigation and the transponders. Saxon recalled that the *Mizar* had left six transponders at the site in November 1968, but in June 1969 only three responded to their "wakeup" calls from the *White Sands*.[4]

The First *Scorpion* Dives

The plan was to establish a long-baseline underwater acoustic navigation network at the *Scorpion* site by placing three Deep Ocean Transponders (DOT) in an equilateral triangle, a transponder at each apex and the *Scorpion* wreck in the middle. There had been lengthy discussions in Washington on exactly how to ensure that each DOT was two to three miles from the other two. At one meeting, after silently absorbing all of the carefully thought-out and mathematically rigorous solutions being proposed, Lieutenant Tony Dunn, then the *Trieste*'s assistant officer-in-charge, asked innocently, "We'll have three ships. Why don't we just maneuver so the ships are spaced as we desire, then all three ships drop a transponder at the same time? Won't that result in the DOTs deploying in a proper pattern at the desired locations?" No one said a word.[5]

Dunn's suggestion was exactly how Commodore Robert Gautier would handle the task. On 4 June, using satellite navigation, the *Mizar*'s DOTs, and radar, he positioned the *Apache* and *Ruchamkin* at distances of about 4,200 yards from the *White Sands* and from each other. Transponders DOT-4, -6, and -8 were dropped nearly simultaneously. The *White Sands* carried ten DOTs, which would be expended throughout the summer to reseed the transponder field whenever a DOT failed. On 8 June, at 2100 the *Ruchamkin* was released for her first shuttle run into Praia da Vitória.

The *Trieste* was floated out of the *White Sands*' docking well on 11 June for *Scorpion* Dive #1. This dive was not intended to reach the ocean floor, only to check the bathyscaph's controls, electronics, and transponder and navigation systems and to put all systems through their paces. Also, a primary purpose of that and the next dive would be to establish the long-baseline navigational grid. Lieutenant Saxon described a typical descent:

> We would load the sphere with carrot sticks, celery, tuna fish sandwiches, and drinking water. The descent and ascent were times when we could relax, joke around, laugh, and generally shake off the previous period of stress. Even though it could be two hours or more in vertical transit, it always seemed to go by quickly. For bodily functions, we carried a "wiki-bottle" should anyone need to urinate. Usually the first user of the wiki-bottle would lose a bet—the bet might require him to buy a round of drinks on the next liberty. We would eat very lightly for the day prior to a long dive and attempt to void our bowels prior to the dive. It would be *very bad form* to void your bowels in the sphere, and I don't know if it ever happened. Intestinal gas could sometimes burst forth, and everyone would point toward someone else as the offender. If someone *could* be identified as the source of the odor, he might lose whatever was bet on the dive.[6]

A large floral wreath was purchased in Mayport with the Welfare and Recreation fund of the *White Sands*. The wreath was dropped overboard as the *Trieste* made the first dive at the *Scorpion* site.

The dive was made on Friday, 13 June 1969. The descent began at 1338. Six months earlier, the last dive off San Clemente had been on Friday, 13 December 1968. It would appear that neither Gautier nor Nevin was superstitious.

Assistant officer-in-charge Dunn piloted, with John Field acting as co-pilot and Nevin in the jump seat as the senior officer on the dive.

The dive was programmed to resurface in the early evening while it was still light. Dunn took the bathyscaph down to within 600 feet of the bottom. The plan was to maneuver to cross each baseline at approximately its mid-point. To ensure accuracy, the goal was to cross each baseline twice. It was known that DOT-6 and -8 were performing but that DOT-4 was not responding normally. Total dive time was eight hours, 27 minutes.

On this first survey the NAVNET system generated an excessive percentage of false returns and echoes.[7] These increased over the next two months, eventually making the Mark 15 navigational computer on board the *Trieste* nearly inoperable. The bathyscaph pilots were reduced to looking for repeatable transponder returns and plotting positions manually. On Dive #1 they were successful only in running the baseline between DOT-6 and -8. They made no attempt either to locate the debris field or photograph the bottom. During their ascent, the bathyscaph's electrical system started shorting, arcing, and sizzling. The problem was traced to the on-deck electrical junction boxes, which at the next opportunity were cleaned and refilled with silicon oil to prevent saltwater coming in.

On the Surface, 14–17 June

Following Dive #1, Nevin established a unique time system, moving the ships' clocks three hours ahead of local time. Called "bathyscaph time," it allowed the *Trieste* crew to board the bathyscaph about sunset (about 2300 bathyscaph time), conduct the dive through the night hours, and surface well before local noon the next day (which would be after noon bathyscaph time). The maintenance crew thus would have a full night's rest, then daylight to re-ballast, replenish consumables, and make any necessary repairs to the *Trieste*. The maintenance people found this time scheme to be much easier on their biological clocks than using the "real" local time.

The *Ruchamkin* returned to the IOU from Praia just before midnight on 14/15 June. Her second shuttle run would begin on the afternoon of 18 June.

Due to the problems with DOT-4, the NAVNET team decided to "seed" a new triangle. The *Apache*, vectored using radar to the correct drop site by the *White Sands*, which had taken a satellite fix, launched DOT-1 on 17 June.

The bathyscaph pilots used the transit to the dive site to review the paper-strip recordings of the acoustics marking the *Scorpion*'s demise, shown to them by Captain Harry Jackson. The photos of the complete debris field taken by "Buck" Buchanan on the *Mizar* more than eight months earlier also were available. However, the pilots were denied access on grounds of classification to the blueprints and construction drawings of the *Scorpion*, which could have helped them recognize the debris that they might encounter on the ocean floor. It was a peculiar decision.

The automatic film developer and the enlarger brought on board the *White Sands* at Mayport were supervised by specialists from the Naval Reconnaissance and Technical Support Center. For *Scorpion* Dives #1 through #3, Richard K. Sibley, the civilian director of NRTSC was on the dry dock; for Dive #4 though the end of operations, his deputy, James Cathro, provided his expertise in photographic interpretation. At the end of each dive, Sibley or Cathro would receive the exposed film and develop negative strips for visual scan. Views worth developing would be marked for positives, usually printed in eight-by-ten format.

Prior to a dive the pilots would be briefed on the findings of previous dives and review the photos of the debris field. By the end of the operation NRTSC had processed about 10,000 photos and had a good understanding of the wreck site. The *Trieste*'s 35-mm camera pair had provided stereo-pair photography, allowing three-dimensional photo analysis of the wreck. NRTSC ultimately produced a three-dimensional model of its two major sections, the bow and the control compartment, for the *Scorpion* court of inquiry. (NRTSC was awarded the Navy Unit Commendation for its contributions to the *Scorpion* search.)

Dive #2, 18–19 June

On 18 June, Warrant Officer Larry Hawks submerged about sunset on *Scorpion* Dive #2, descended to a depth about 600 feet above the bottom, and again ran baselines for the NAVNET. This was the first night dive, setting the pattern for all succeeding dives at the *Scorpion* site on bathyscaph time. Nevin and Dunn with Hawks tried to make the automatic navigation system—the Sperry Mark 15 computer—perform as specified but without success. Dunn's theory was that the device, originally designed for the Boeing 707 aircraft, could not deal with speeds over the ground of less than about 100 knots. Since the bathyscaph operated in the one-knot regime, that meant the computer was completely "out of its depth"—Dunn's pun. In reality, the computer's problems were false returns, echoes, and reverberation of the DOT acoustic replies to interrogations. The navigation computer could never be coaxed into cooperating, but the crewmen transitioned with little problem to dead-reckoning plotting boards on their laps. All *Trieste* divers were qualified submariners and could do the basic plotting in their sleep.

On this dive, DOT-1, -6, and -8 were responding, and suddenly -4 joined in. Hawks was able to run all baselines except the 1-to-4 line. Observation discerned that the acoustic navigation system worked best with the *Trieste* at about 600 feet above the bottom; useable readings were obtainable as high as 3,000 feet above the bottom.[8]

Dive #2 ended after 12 hours, nine minutes submerged, surfacing on the morning of 19 June with the main propulsion electrical system again sizzling and arcing. Investigation disclosed that seawater again had entered the on-deck junction boxes.

Dive site security was maintained by either or both the *Apache* and the *Ruchamkin*. Whenever the *Trieste* was submerged the area within two miles of the dive site was a "protected" area. During the day the *White Sands* flew the international signal for "inability to maneuver," a red ball over white ball over red ball. At night, the signal was expressed with lights, red over white over red. The presence of a nearby commercial shipping lane between Europe and South America meant that transiting freighters continually posed a risk. With Nevin's program of night diving, any confrontation between a passing freighter and one of the dive security ships was likely to occur in darkness.

Lieutenant (Junior Grade) Larry Hagerty, executive officer of the *Apache*, remembered one incident when a radar contact on a steady bearing, decreasing range and looked to be about to pass through the protected area. The *Apache*'s captain, Larry Lonnon, moved to intercept at the greatest feasible range and divert her, repeatedly sounding five short blasts on the fleet tug's whistle to warn the freighter that she was standing into danger. Seeing no change of course, Lonnon lit up the ship with the largest of the *Apache*'s signal lights and saw a merchant sailor racing up several ladders to the bridge to take the ship off of autopilot. Lonnon later reported that he had been prepared to fire a shot across her bow had she continued to ignore his warnings. He had mustered the tug's 3-inch/50-caliber gun crew and later seemed disappointed that he had not had an opportunity to fire a shot.[9]

The *Ruchamkin* returned to the dive site late in the evening of 21 June.

Another concern at the dive site was the disposal of trash and garbage. All ships were under orders not to dump trash at or near the *Scorpion* wreck site. On the *White Sands* the crew used empty 55-gallon drums, which had earlier contained cloth bags of steel shot for bathyscaph ballast, to hold trash and garbage. When filled, they were stored in the docking well. Every few days the *White Sands* would move five or ten miles away from the dive site using her thrusters and dump trash, puncturing the steel barrels to ensure that they sank. The *Apache* and *Ruchamkin* also were careful not to contaminate the dive site.

Dive #3, 22–23 June

On 22 June, *Scorpion* Dive #3 took the *Trieste* to the ocean floor, 11,100 feet down, piloted by William (Stretch) Leonard, with Hawks and Field rounding out the crew. Using the NAVNET, Dive #3 and later dives landed within the target area, although not necessarily within the debris field. Field, handling communications, upon reaching the bottom reported by underwater telephone that they were sitting on top of "a large carrot."

Leonard realized that they was resting on some portion of the *Scorpion* wreckage and took the cautious decision to ascend directly upward, off the "carrot," and find a less precarious place to set down. He was able to arrest his ascent at about 500 feet up. Leonard then returned to the bottom but now found the acoustic navigation system unresponsive. The bathyscaph was within some kind of acoustic shadow and could not establish a "handshake" with any of the DOTs. Without an accurate fix on his location, Leonard was forced to strike off on a heading he had computed when querying the DOTs at 500 feet off the bottom. The dive terminated without visual contact with the debris field, surfacing in the morning hours of 23 June. Once again the *Trieste* surfaced to the sizzling sounds of the main electrical system shorting out the battery cables and junction boxes.

Later, after review of the photos taken on Dive #3, Captain Jackson's on-scene advisory group determined that Field had seen the shaft and hub of the *Scorpion*'s propeller, carrot-looking in their red-lead primer. Leonard had dropped directly onto the shaft, the submarine's propeller blades just inches from the 3/16-inch sheet metal skin of his avgas tanks. His only safe option had been to drop ballast and ascend out of possible danger, and he had done exactly that.

The Monkey-Shit Incident

These first three *Scorpion* dives all had experienced electrical shorts and battery fires near the end of their dives. On 23 June, a major review, involving everyone from divers to Commodore Gautier, of the electrical problems explored different ideas as to the cause and solution. Chief Electrician's Mate Royal G. (Roy) Fort would later confess that at the meeting he brought up solving the problem with "monkey shit" more as a joke than as a real suggestion, but the idea took on a life of its own. Gautier turned to Nevin and told him to get some of this "monkey-shit stuff," and Nevin turned to the *Trieste*'s supply officer, Dave Byrnes.

Byrnes, the junior commissioned officer in the *Trieste* crew, had been assigned supply as a collateral duty. He was extremely uncomfortable ordering something by that name on a Priority-1 requisition. No one in the IOU could determine the official name of the "gunk" or had any idea as to the federal stock number required for ordering parts and material. In innocence or desperation, he just cut a message for Gautier's signature, in casualty report format, identifying the casualty priority as Brickbat/Priority-1, the material needed as "monkey shit to seal the electrical junction boxes and fuse box from seawater infiltration," and the quantity needed—a case. It was addressed to the Mare Island Naval Shipyard for action. The message went out as "operational immediate" the highest message precedence short of a report of sighting an enemy force.

The message was received at Mare Island and because of its Brickbat priority got immediate action. An officer walked the message to the electrical shop and found electricians who had worked on the *Trieste*. One of them said, "Oh, I know what they want." He went to his locker and came back with a can of "Duct Seal," with the essential federal stock number on the label. With that information, the supply center sent a case of Duct Seal by air to the Lajes airfield on Terceira. There it was placed on board a P-3B Orion maritime patrol aircraft, which in turn arrived over the IOU two hours later and radioed that it had an airdrop for the ships. The 50-pound case of monkey shit, which had been requested less than 48 hours previously, came down by parachute. The *Apache*'s motor whaleboat fished the package out of the water and brought it to the *White Sands*. Commodore Gautier had his monkey shit, and Chief Fort was praying that it would work.

At 2233 on 24 June, the *Ruchamkin* began another shuttle trip to Praia da Vitória. Outgoing personnel on board included NRTSC director Richard Sibley and Lieutenant Commander Philip L. Hendricks, a physician and diver from Gautier's staff in San Diego. The *White Sands* had a decompression chamber on board for the operation, and Gautier had made certain that a diver-qualified medical doctor was on the dock throughout the program to handle diving accidents.[10] Additionally on board the ARD were two *White Sands* sailors bound for the U.S. naval facility at Rota, Spain. The *Ruchamkin* would remain in port for a few days to await the return of the two sailors from Rota and the arrival of reliefs for Sibley and Hendricks.

The *Ruchamkin* again rendezvoused with the IOU on 1 July, bringing back the two sailors from Rota. They had made their trip in response to the mounting discontent over the movies available in the IOU. Movie call had become more and more raucous, as the crews loudly expressed their disdain for third or fourth viewings of films. Commodore Gautier had taken action. The two sailors had flown out of Lajes to Rota to exchange their old movies (not as DVDs but in heavy, beat-up boxes of several 16-mm reels) for a completely new set. Arrival of the new movies was greeted with rejoicing by all in the IOU. Not only were they fresh but submarines always received the best films that the Navy had to offer. It was a good day to be in the Navy and in the *Trieste* group.

Also on board the *Ruchamkin* when she returned to the site were NRTSC deputy Jim Cathro and Lieutenant Commander Edward A. Beron, a medical doctor and diver.

After three preparatory dives establishing a bottom navigation grid and testing the *Trieste* at 11,100 feet, Commodore Gautier was ready to commence investigations of the site.

17 *Scorpion* Site Investigation

We were young and bullet-proof. One could say that the good Lord had taken a liking to us and at times we had His full and undivided attention.

—Lieutenant Anthony Dunn, Assistant Officer-in-Charge, *Trieste*

With the preliminary preparations, surveys, and test dives completed, the detailed investigation of the *Scorpion* wreckage began with *Scorpion* Dive #4 on 2 July 1969. Lieutenant "Stretch" Leonard once again piloted, with Lieutenant Ross Saxon on his first bathyscaph dive, and Lieutenant Tony Dunn in the third seat.

The black-and-white TV monitors inside the sphere provided a view forward from the *Trieste*; the forward window was only manned when attempting a capture with the manipulating arm. External lighting was very bright, being designed for both TV and eyeball. With very clear water, people in the *Trieste* often could see more than 50 feet. High-intensity strobes enabled film photography of objects as far away as 200 feet. However, such visibility was rare; when, as often happened, the bottom sediment was disturbed by the bathyscaph's trail ball or thrusters, it immediately went into suspension, blocking forward visibility for up to 20 minutes.

Scorpion Dive #4 commenced about sunset on 2 July and would surface after 14 hours, 17 minutes submerged, with almost 11 hours of that time on the bottom. With all systems functioning—except, of course, the Boeing 707 computer—the *Trieste* dropped into the debris field and set about photographing the small items of debris within the illumination zone. Dunn noticed a looming bulk at the very edge of the illumination field that resolved into the *Scorpion*'s bow section.

Dunn later related, "We had a head-on view and it was a big surprise."[1] They deployed a Mini-DOT homing transponder to mark the location of the bow section, but it failed to activate. (A photograph taken on a later dive revealed that the Mini-DOT had not properly deployed.) He then undertook a close examination of the bow section and for the next several hours took still photos and videotape of the section. His surprise was at the evident violence of the hull's destruction: "When she imploded it created such a hammer effect it blew the [torpedo room] hatch off."[2]

Near the end of the *Trieste*'s battery life the pilots began to experience lighting problems and to hear again electrical shorts and sizzling cables. Dunn noticed that they were pulling 50 amps even with the circuit breakers off. When yet again "topside" inspection found the electrical junction boxes full of seawater, Dunn explained, "The design people had not taken into account that the power cables would be compressed, allowing salt water to enter the junction box. We also found that the fuses had become buss bars because of their design, conducting high amperages rather than interrupting the circuit."[3]

These electrical problems caused consternation among the *Trieste*'s maintenance team. The Duct Seal ("monkey shit") was supposed to have solved them, yet there they were again. In addition to the shorting, arcing, and sizzling that had been experienced for the fourth time, an entire battery pack had reversed polarity and would have to be replaced.

The bathyscaph was placed in the *White Sands* for a major inspection of the electrical system. The nitrogen and compressed-air flushes of the *Trieste*'s buoyancy tanks to purge gasoline fumes could take up to 15 hours, until an explosimeter registered "safe." All machinery on the dry dock, except the diesel-powered thrusters and electrical generators, were shut down during that process. Saltwater sprinklers in the dock's diesel engine exhaust stacks were kept on to douse possible sparks. Firefighting crews in protective clothing were strategically stationed, and the ship was completely buttoned up, with all air supply and exhaust fans secured. The odor of avgas fumes permeated the air.

During the long hours of de-gassing the dock's galley prepared cold sandwiches, and messcooks delivered them throughout the ship. It was a relief when the *White Sands* could return to normal conditions and turn back on the air conditioning systems and galley stoves.[4]

The reversed battery cell was a potentially mission-ending casualty, as the cell was designed to be replaced only with shipyard-level equipment and expertise. The *Trieste*'s electrical officer, Saxon, instead developed a workaround, planned and supervised the battery-cell replacement with only onboard resources, and thus prevented cancellation of the *Scorpion* Phase II operations. (For his initiative and his superior professional performance, he was awarded the Navy Commendation Medal.)

The battery-cell polarity reversal was traced to a mixture of silicon and carbon-based oils in the fuse box, an error made in California. Pure silicon oil was on hand, but new fuses had to be fabricated in California and flown out to the *White Sands*. The fuses were air-dropped and installed in time for Dive #5—another Brickbat/Priority-1 triumph.[5]

But more bad news followed. Inspection of the *Trieste*'s hull and safety systems revealed a critical problem: the shot containers could not be dropped. They were frozen into place by rust. The jettisoning of that weight in an emergency was a primary method by which Auguste Piccard had rendered the bathyscaph fail-safe. Commodore Robert Gautier dispatched a casualty report to the Naval Ship Systems Command, which recommended immediate termination of the operation and return to port if the *White Sands* and the *Trieste* crew could not fix the problem at sea, or if NavShips could not send a team to Praia to handle it.

Gautier leaned on Nevin, and Nevin leaned on Field, the *Trieste*'s engineer officer. After considering a number of ideas, however, Field and Dunn struck out. They could not solve the problem at sea. The choice was returning to port or diving without the ability to jettison the shot containers.

Nevin called a meeting of the *Trieste* officers and asked if anyone would refuse to dive with the shot containers frozen in place. All agreed to continue the dives. Gautier agreed and sent messages to both the Atlantic and Pacific submarine force commanders stating that, having received the advice and consent of all concerned officers and officials on scene, he would—unless otherwise directed—continue the dives. No objection was raised.

Later, of this decision Dunn said, "We were young and bullet-proof. We were very mission oriented. One could say that the good Lord had taken a liking to us, and at times, we had His full and undivided attention."[6]

On 5 July the *White Sands* refueled the *Apache*. This was the first refueling at the *Scorpion* site. On this refueling, Lieutenant "Beau" Myers, executive officer of the *White Sands*, boarded the *Apache* to clean her "windshields," the pilothouse windows—"service with a smile." About a day later *Ruchamkin* departed for her shuttle run into Praia. On the afternoon of 8 July the *Apache* dropped DOT-7 into the target area to replace failing DOT-6 but closer to the debris field.

Scorpion Dive #5 on 10 July took the *Trieste* slightly deeper than the earlier dive. Piloted by Saxon—this being his qualification dives—with David Byrnes on his first dive and Nevin in command, they descended to 11,687 feet. The NAVNET transponders were showing their age: only one useable return was received from DOT-4, after which it was never heard from again. However, DOT-1, -7, and -8 still were active and useable.

The *Scorpion*'s bow section was again sighted. It was bow-down in an elliptical depression about 100 feet in diameter and in depth. Nevin instructed Saxon to put the *Trieste*'s bow into the depression to get a boom-mounted camera into position. The pilots had been instructed not to photograph human remains, but Nevin now inadvertently took a photo of what appeared to be a skull on the lip of that hole. He later destroyed that image, as required.[7]

The dive was cut short after 12 hours, 28 minutes, and the *Trieste* surfaced because of, again, battery shorts and an electrical fire. The maintenance crew discovered that the fuses recently air-dropped had a plastic case, which at depth caused a short circuit. Saxon, the *Trieste*'s electrical and diving officer, discussed with Captain Harry Jackson, the senior member of Technical Advisory Group, the idea of modifying the fuses. While what they agreed on and carried out was not an ideal fix, it seemed to eliminate the problem of cables frying (rather than fuses blowing, as they should have) whenever water intruded into the junction boxes. By 15 July the *Trieste*'s crew had the rebuilt fuses installed, tested, and ready for the next dive.[8]

The *Ruchamkin* returned to the dive site late on 12 July, with Dr. John Craven on board. Craven had been appointed coordinator for this effort to investigate the *Scorpion* wreck. He would remain at the dive site for Dives #6 through #9.

Bottom Investigation

The *Scorpion* Dive #6 on 15 July took Byrnes, newly promoted Chief Warrant Officer Larry Hawks, and Nevin into the depths. The *Trieste* landed farther from the *Scorpion* than on previous dives because of a SATNAV delay and as a result had problems locating the debris field. No major hull sections were sighted.

During this search Byrnes reported sighting a cruise ship's deck chair lying on the bottom. There has since been some discussion of this sighting. It seems wildly improbable that a deck chair would be found on an attack submarine, and the chance of an artifact from another ship showing up in *Scorpion*'s debris field was almost nil. The answer may be suggested by the case of a photo taken by the *Mizar* of what had been initially reported as an umbrella (and reported once again, below) but later was identified by biologists as a deep-sea creature.

If the "umbrella" was misidentified, perhaps the "deck chair" too was a denizen of the deep, unrecognized because of its alien form. Byrnes, nevertheless, thereafter remained adamant about the accuracy of his report. It is one of many completely inexplicable findings of the investigation at the *Scorpion* site. Dive #6 terminated after 13 hours, 54 minutes submerged. This was the first *Trieste* dive on the *Scorpion* site that did not experience electrical arcing, shorting, sizzling cables, or an electrical fire.

On 17 July the task unit readied for its first (and only) replenishment at the dive site. The Navy cargo/replenishment ship *Victoria* (T-AK 281) provided a mid-operation restocking of the storerooms and freezers of the *White Sands*, *Apache*, and *Ruchamkin*. After more than 60 days out of Mayport, the timing could not have been better. The provisions were shuttled from the *Victoria* by one of the *Ruchamkin*'s landing craft. There was joy in the crewmen's hearts as hundreds of cases of Pepsi and other soft drinks passed through their hands to fill the two vending machines on the dry dock. ("Profit" from the machines went into the ARD's Welfare and Recreation Fund.)

The *Victoria* delivered *submarine* rations, undoubtedly the best chow in the Navy: steak and lobster tails for Sunday dinner, with strawberry shortcake for dessert. The "retail stores" (ship's stores) on all three ships were replenished with snacks, shaving gear, and a new selection of paperback books. There also was a major mail call, with "Care packages" from home, complete with photos of loved ones, cookies, brownies, and other gifts that conveyed the message "I love you and I miss you."

Back Down Again

Scorpion Dive #7 on 18 July gave Larry Hawks his second opportunity as pilot and became his qualification dive. The *Trieste* submerged at 1953 with Dunn in command and Leonard with him. During the descent navigation was aided by contact on DOT-8, which allowed Dunn to take a direct path to the area of highest-density debris.

By that time the transponders were becoming less and less effective, and not only from age. One reason was that at least one DOT was in an acoustic shadow and available only if the *Trieste* was several hundred feet above the bottom. When the bathyscaph was actually on the bottom they provided little or no information.

Dunn recalls that they sometimes navigated by following trail-ball grooves produced by previous dives. On this dive they photographed the bow and then worked around to its open, after end to peer into the hull. Dunn ordered Hawks into the hole to allow better photographs of the interior with fixed and TV cameras. It seemed a perfect opportunity to test the remotely operated submersible "flying eye" as well, but Dunn passed on that idea as the interior of the section was filled with dangling pipes and wires.

The bathyscaph team also took photos of random debris near the bow section, including damaged main battery storage cells. Dive #7 terminated after 15 hours, 55 minutes submerged. Upon surfacing it was confirmed that between the monkey shit brought in under such high priority and the subsequent repairs, the electrical problems had been solved. Shortly after Dive #7, Hawks celebrated his qualification as a bathyscaph pilot with a letter to his mother:

> I am now a Chief Warrant Officer, a promotion. I am also a qualified deep submergence vehicle pilot and copilot/navigator and I have been to the bottom of the sea where no living man has been. I have been down there many times and viewed all the wonders of the deep, including the sunken submarine USS *Scorpion*—a very awesome and somewhat forbidding sight—and all this still suspended in a time and place that will never change. *Who would have ever thought* that I might be lucky enough to have come to this? In fact, I just returned from

> 14 hours on the bottom here, and would have talked directly to the astronauts on the moon (or orbiting) had we gotten the right people involved to make arrangements [emphasis original].[9]

On the Surface—*Not* of the Moon

On 20 July 1968, Neil Armstrong became the first human to set foot on the Moon. Saxon and Nevin had proposed that a dive be coordinated with the Moon landing and that a special communications hookup be established to enable the deepest divers on the planet to communicate with the men on the Moon. It was not to be. Even though it had been published in many newspapers that the *Trieste* was diving on the *Scorpion* site, the location and depth of the *Scorpion* wreck and the improved operational capabilities of the newest version of the *Trieste* were still highly classified. Such a link would have exposed the *Trieste* operation to unwanted attention.

There was no dive that day: everyone in the little flotilla listened to the Voice of America account of the Moon landing and felt the similarities between the nation's achievements in outer space and in inner space. Dr. Andreas Rechnitzer later observed, "Strange, isn't it, that the initial conquest of inner space—our last geographic frontier—and outer space should occur within a single decade?"[10]

Dr. Craven recorded that when Armstrong placed his foot onto the Moon's surface everyone in the wardroom thrust their hands above their head—a sporting celebration for "score!" An anonymous voice shouted out, "No, dammit, no! *Two* small steps!," referring to the *Trieste*'s simultaneous work at a depth of 11,100 feet.[11]

There was a felt need to document the *Trieste*'s accomplishments. Prior to sailing from San Diego, "Beau" Myers on the *White Sands* had contacted the Allen Company of Anaheim and had received a do-it-yourself cruise-book kit. He asked for volunteers to undertake the project; two men, Storekeeper 2nd Class Peter A. Tyrrash on the *Apache* and Seaman Eric L. Swenson on the *White Sands,* stepped forward. The pair did an outstanding job, and the cruise book, titled Scorpion *Operations, Phase II,* became not only a prized possession for all who participated in the operation, but a major source of information about the dives.[12]

Bottom Investigation, 21–22 July

Scorpion Dive #8 commenced at 2239 on 21 July and produced one of the most contentious sightings of the operation. Saxon piloted, assisted by Field in the co-pilot position, and Byrnes in the jump seat controlling the cameras.

Upon reaching the bottom they found that DOT-8 was not providing useable information but DOT-1 and -7 were. Saxon headed toward the *Scorpion*'s bow section and then turned toward the aft section. There is some disagreement about what happened next. Stephen Johnson relates in his *Silent Steel* that Field sighted the corpse of a *Scorpion* sailor, in the attire of a submarine crewman and wearing a life jacket. He quoted Field: "Although it doesn't seem scientifically possible, it appeared to me that the flesh was still on the face, and hair—dark and about 3 inches long—was still on the head. The corpse was wearing dark (possibly blue) coverall-type clothes. The clothes seemed to be 'puffed' out, as if filled with a gas. The trousers appeared to have a 'tight' elastic type of cuff. The corpse also had on what appeared to be a light-colored . . . yellow vest."[13]

Saxon and Byrnes have verified some of this information, but not all. They stated that there was a brief, perhaps 25-second glimpse of a pile of clothing on the sea floor. Byrnes, who was on his stomach looking out of the forward window, had the best view of it. Saxon observed the scene on a small, monochrome TV monitor, as did Field. Byrnes agreed that the clothing was arranged as if it had once covered a human form and that there a yellow or orange vest or life jacket visible but insisted that there was no sign of skull, bones, hair, or flesh. Saxon too saw the clothing, which gave him the impression of a "sailor, wearing a pair of 'nuclear power type' overalls, lying on his stomach with legs askew."

All three men described a light-colored vest draped across the shirt area, Saxon and Byrnes called it a Kapok-type life jacket. Later, discussing the sighting, both Saxon and Byrnes felt that Field's grisly description of flesh and hair was not plausible and attempted to dissuade him from including it in his de-brief.[14]

Six months later, all three men prepared written descriptions of the sighting for the *Scorpion* court of inquiry.[15] Nearly 50 years later, in interviews and discussions with the authors of this book, both Saxon and Byrnes registered discomfort with their February 1970 descriptions of the sighting and acknowledged the low probability of a corpse surviving intact between May 1968 and their sighting in July 1969. This sighting was not used by the *Scorpion* Special Advisory Group as evidence for any theory of the cause of the *Scorpion* loss, and the clothing (as well as what Field told Steven Johnson he saw in it) remains yet another completely inexplicable sightings reported during the *Trieste*'s investigation of the wreck site. Attempts by Saxon to return the *Trieste* to the clothing were unsuccessful. Attempting to back down to it, he stirred up the silt with the reversed motors and trail ball. As it settled, the three men sighted instead the submarine's operations compartment. Investigating that section became the higher priority.[16] Saxon related:

> When the silt cleared, we were able to see into the operations compartment, which was a tangle of wires, piping, and other items. John Field suggested we lower the boom camera into the compartment. We discussed the pros and cons, the desire for better photography vs. the possibility of the boom becoming snagged inside the compartment, an especially significant problem since we did not have the ability to drop the shot pans. I did not think it prudent to jeopardize the operation by perhaps becoming entangled, so advised "Topside" that, as the senior officer and in command, I had decided *not* to lower the camera boom due to safety concerns.[17]

Turning their attention to the remainder the operations compartment, Byrnes noted big wrinkles in the hull where pressure had buckled the hull around the frames. They received instructions from the surface to turn off all lights and look for a green glow near the reactor compartment. There was no green glow. Saxon terminated the dive upon seeing a low-battery indication and surfaced in the morning hours of 22 July after a total of 14 hours, 56 minutes submerged.

The *Ruchamkin* returned to the dive site from another shuttle run the next day. The task unit had been at the *Scorpion* site almost eight weeks by the end of Dive #8. The weather had remained favorable throughout, and the ships and crews had settled into a comfortable pattern of watch standing, workdays, and free time.

The Final Dive

More than a week later, on 30 July, Dive #9—the final dive on the *Scorpion* wreck—began late in the day and ended on the surface 15 hours, 38 minutes later.[18] Nevin was pilot, with Byrnes and Captain Jackson with him in the *Trieste*.

Byrnes provided a firsthand narrative of the dive:

> A moderate swell made me a little queasy as I stepped aboard the bathyscaph. I saluted several U.S. flags mounted on-deck, and took a moment to tie a nylon bag of Styrofoam cups onto the handrail. The cups would be squeezed to about the size of a thimble at 11,000 feet depth. The flags and the cups would become *Scorpion* mementos.
>
> I followed Bob [Nevin] and Harry [Jackson] into the sphere, climbing down the trunk providing passage through the surrounding gasoline tank. A diver followed me down the trunk, and when I signaled, we coordinated to close the hatch.
>
> We could watch the divers as they made final adjustments to the TV cameras and the manipulator and removed the ballast tank covers. Bob instructed the divers via a sound-powered telephone to flood the trunk and confirm all divers were clear. The lead diver responded, "All systems go." Bob flipped a switch to release the air from the fore-most and rear-most ballast tanks, flooding them. For several minutes, nothing happened. The scuba divers topside started to jump up and down to help break the ocean's surface tension. The depth gauge quivered then started to move and the sapphire blue water seen through the lower window darkened.[19]

Descent took more than 90 minutes, and the three men in the sphere had time to relax and discuss events. Byrnes continued his narrative:

> Once the fathometer indicates that we are approaching the bottom, we turn on the exterior lighting and deploy a 250-lb lead ball, the trail-ball, winching it out to about 15-feet below the *Trieste*, and Bob [Nevin] starts to drop ballast. When the ball hits bottom, the bathyscaph instantly becomes 250-lbs lighter and, when done properly, we come to a dead stop without ever touching bottom. Bob does it perfectly

> and we transition into horizontal propulsion. I report to the surface via the underwater telephone, "*White Sands*, this is *Trieste*; we're on the bottom, operating normally. No contacts, commencing search." Communications were muffled and garbled when the response was received about 50 seconds later.

Approaching the bottom, DOT-1 and -8 provided useable data part of the time but only at close ranges, 3,000 feet or less. DOT-7 provided good data out to 5,000 feet. The *Trieste* was often in acoustic shadow zones. Byrnes continued:

> Bob activated the CTFM [Continuous Wave, Frequency Modulated] sonar, which transmits a wavering sound ahead and to both sides of the *Trieste*. The *Trieste*'s top speed is only about two knots, so if the sonar indicates an obstacle, we can reverse motors and stop within about 500 yards. Since our visual limit was about 50 *feet*, we placed total reliance on the sonar for our safety during bottom searches.
>
> We drove for an uneventful hour, until suddenly the sonar beeped a warning. Bob reversed the motors and soon an object coalesced into visibility. We coasted to a stop, only to find an open beach umbrella resting on the bottom![20]
>
> About that time, I once again queried the DOTs and this time got a reply. Plotting the ranges on his manual plotting board, Bob determined that the *Scorpion*'s wreckage is northwest of our position. Inching along two hours later I obtained another DOT response, which plots the debris field to the northeast. Bob and Harry take turns lying on their stomach or kneeling to peer out of the forward window. Suddenly Bob exclaimed, "There's a billfold! There's money scattered on the bottom."

Nevin considered retrieving the billfold with the mechanical arm, but second thoughts concerning the impact on the owner's family made him decide to pass on the idea. The three men in the *Trieste* saw clothes, books, and plastic mattress covers scattered on the ocean floor. And there was an aluminum pantry cabinet filled with *unbroken* plates, saucers, glassware, and coffee cups. Jackson wanted to retrieve items from the debris field that could be analyzed for clues to the *Scorpion*'s demise. He identified some red pieces of plastic on the bottom as parts from a main battery cell.

A hovering bathyscaph is subject to the same laws of action and reaction that act on a spacecraft in orbit, in weightlessness. Reaching out the manipulating arm had pushed the *Trieste* backward. Grabbing an item and retracting the arm to place the object in a collection basket had pulled them forward. Nevin had vented a little gasoline to make the bathyscaph heavier and so give himself a steadier platform from which to make his recoveries. And, the *Trieste*'s recovery arm was able to pick up a ship's sextant that was observed on the ocean floor.

The time spent recovering items allowed the bathyscaph's skegs to sink into the bottom ooze. The *Trieste* may have become stuck. Some ballast was dropped. No response. The motors were backed. Still no change. Battery power was approaching the red zone. More shot was dropped, and the engines were reversed. Still no change.

There was no sudden lurch, no visual reference of movement. But after minutes that seemed to them like hours, the crew sensed that the bathyscaph's stern was rising. The craft slowly tilted, and within the sphere unsecured items began to slide forward. The *Trieste* was pulling herself out of the mud. They could see the motion of the silt as they backed out, just as the external lights dimmed and flickered out with the last electrons from the *Trieste*'s main battery. Byrnes once again: "There was still battery power for the electronics, so I keyed the underwater telephone to report, '*White Sands, Trieste*: found *Scorpion*. Took photos. Retrieved Items. Battery zero. Conditions normal. We're on our way up.'"

Analysis of photography from Dive #9 and from the *Mizar* would confirm that the major sections of the *Scorpion* wreck lie in a depression. The recovered sextant would end as a prized accession at the Navy Museum in Washington, D.C. Like the expelled propeller and shaft, the presence on the bottom of the sextant was completely inexplicable—it was normally kept in a wooden box in a metal cabinet in the wardroom. The pieces of battery cover plate were to have a major impact on the investigation into the cause of the *Scorpion*'s loss. Dive #9 was a fine and fitting finale to the *Trieste*'s operations at the *Scorpion* site.

The *Trieste*'s photography was a valuable addition to the *Mizar*'s coverage, showing a wide distribution of battery components in the *Scorpion* debris field and thus supporting the view that the main battery had exploded. There were no indications of an explosion external to the pressure hull, caused by, say, a foreign attack or by a circular-running torpedo, theories offered by many "experts." All photography was consistent with the conclusion that an implosion at great depth had instantaneously destroyed the *Scorpion*'s hull. Both the acoustic information and the photography unequivocally established that the *Scorpion* was destroyed by a single implosive event—the collapse of the pressure hull.[21]

Probably the most significant evidence obtained from the *Trieste*'s nine-dive investigation was the seemingly innocuous piece of red plastic identified by Jackson and retrieved at his specific request on Dive #9. The Portsmouth (Maine) Naval Shipyard subjected the item to microscopic, spectrographic, and X-ray diffraction analysis and found that "the general battery damage is violent. The high velocity intrusion of pieces of the flash arrestor into both the inside and outside surfaces of the retrieved plastic cover attests to violence in the battery well." The battery cover plates were found to be impregnated with alumina from the flash arresters, a condition consistent with a battery explosion that occurred *before* the ship imploded.[22] That explosion—actually two battery explosions at essentially the same moment—was the "event" that led to the death of the *Scorpion*.

Homeward Bound

The *Apache* fueled from the *White Sands* a second time on the afternoon of 1 August. After the Dive #9 debrief on 2 August, Commodore Gautier climbed up to the *White Sands*' bridge and ordered, "Wrap it up, we're going home."

It would take almost 15 hours to de-gas and secure the *Trieste* for transit. Meanwhile, the *Apache* rigged for towing the *White Sands*. Gautier released the *Ruchamkin* from the task unit on 2 August. Dr. Craven and a few others departed on her for the one-day express transit to Praia.

The *Apache*, towing the *White Sands* with the *Trieste* embarked, took four days to reach Praia, which she did on the afternoon of 6 August. Commodore Gautier, Captain Jackson's technical advisory group, and the civilian technical representatives quickly disembarked. Most returned directly to the United States. A couple of the bachelor techreps who had spent little of their pay for the last two or three months bummed a flight to Spain and bought new sports cars to drive around Europe.[23]

With Task Unit 42.2.1 dissolved, Bob Nevin, as officer-in-charge of the *White Sands* and the *Trieste,* once again was in control of the Integral Operating Unit. He granted liberty to their crews in Praia on 6–9 August. With the generous contribution of Craven's poker winnings and others from the visiting officers and techreps, the *White Sands*' Welfare and Recreation fund hosted a two-day, drunken blowout at the Lajes airfield.[24]

The *Apache* towing the *White Sands* carrying the *Trieste* sailed for the Panama Canal and San Diego on 10 August at 0750. It took the group almost a month to reach the Canal Zone, chased the last week by a hurricane. They arrived at the Atlantic entrance of the canal on 9 September.[25] From the Canal Zone, six *Trieste* officers flew back to San Diego, leaving only Nevin, as officer-in-charge, with the unit.[26] The homebound tow departed from the canal area on 12 September and arrived home on 7 October, to the cheers and band music of an official welcome.

The *Trieste* unit had departed San Diego on 3 February 1969, and returned on 7 October—a deployment of 246 days: the Integral Operating Unit—the *Trieste*, *White Sands*, and *Apache*—had spent 48 days in port, 61 days on station at the *Scorpion* site, and 137 days in transit. During that time there had been no deaths, no serious injuries; no one had "gone UA" (on unauthorized absence) or "missed movement" (sailing) and only one debilitating illness had been suffered (the *White Sands*' senior cook who suffered appendicitis).

The *Apache* had set the record for Navy's longest tow over a one-year period, without suffering incident or accident.[27]

The Navy recognized the superior performance of the *Trieste* unit with a Navy Unit Commendation for the period 3 February–7 October 1969, the third time that the *Trieste* program had been so honored. (The first two awards were from the 1963 and 1964 *Thresher* searches.)

Then-acting Secretary of the Navy John W. Warner wrote,

> As the only operational deep-submergence vehicle in the Navy, *Trieste II* [*sic*] was piloted to unprecedented working depths on nine occasions, thus ensuring the complete success of an operation of great significance to the United States. The officers and men of this Integral Operating Unit, through their superb teamwork, untiring efforts, and dedication,

merged their diverse elements into a unit, which provided a deep-ocean search capability for the United States. Their inspiring performance of duty reflected great credit upon themselves and was in keeping with the highest traditions of the United States Naval Service.

A Future for the *Trieste*?

Notwithstanding the celebration and pride of performance in what the *Trieste* unit had accomplished during the *Scorpion* operations, there were deeper forces at work that threatened to end the bathyscaph program. The raison d'être of the third version of the *Trieste* was to recover Soviet missile nose cones left on the floor of the Pacific after tests. However, the bathyscaph's participation in that role had been cancelled by the Chief of Naval Operations because of political and safety concerns.

Submarine detailers at the Bureau of Naval Personnel were advising young officers looking to supercharge their careers that the *Trieste* program now was just "marking time" and would not provide them any advantage.[28] Also, the high cost of operations and maintenance of the bathyscaph and its low operational availability rate made it uncompetitive with other Navy manned submersible programs—the *Alvin*, *Sea Cliff*, and *Turtle*.

Thus, with the termination of the *Trieste*'s participation in Winterwind the bathyscaph's value to the Navy was being reviewed. The current *Trieste* incarnation represented a capability to search and work at the ocean floor down to 20,000 feet. No other U.S. Navy deep submersible came close. The three *Alvin*-type deep submergence vehicles were rated to 6,500 feet and could not be upgraded to a 20,000-foot capability until at least 1984—if then.

The Stephan committee (i.e., the Deep Submergence Systems Review Group) of 1963 had levied upon the Navy a requirement to create and maintain the ability to search for and recover small objects down to 20,000 feet. Over the two years that followed the end of the *Scorpion* operation there would be a debate in the upper echelons of the Navy's deep submergence community as to whether or not to abandon that capability, at least with a manned submersible. Until a final determination was made the *Trieste* program would suffer under-funding, under-manning, and general neglect.

18 Neglect and Decline

With only two dives in the preceding year, Trieste's *20,000-foot capability was questionable. Budget constraints and the lack of trained personnel had left the* Trieste *in a degraded state. I was under orders from the new commodore to get the program resurrected.*

—Lieutenant Commander Malcolm Bartels,
Officer-in-Charge, *Trieste*

After their triumphant return to San Diego from the *Scorpion* operation, there was a rush to get the *Trieste* into overhaul. There could be no more dives until the shot containers were replaced and several other problems were corrected. The overhaul took place at the Ballast Point submarine complex in San Diego. The "teardown" would be overseen by the Naval Ship Systems Command and involve Mare Island shipyard workers, but the *Trieste*'s crew would be doing most of the work. Overhauls were planned for the *Apache* and *White Sands* as well.

The *White Sands* departed from San Diego on 9 February 1970 under tow by a fleet tug for Mare Island. There the *White Sands* was dry-docked, the underwater hull sandblasted to bare metal and repaired. The crew moved to berthing ashore, allowing an aggressive attack on a severe cockroach infestation in the dock. The three 500-horsepower azimuth thrusters that enabled the dry dock to keep in a precise position at sea were replaced with 750-horsepower units.

In mid-June, with the *White Sands* overhaul complete, a tug took her in tow, with all three thrusters activated for steering during the passage from Mare Island down the Napa River to San Francisco Bay. About halfway down the river the starboard thruster started to make noises indicating failure. A stopover at the Hunters Point shipyard in San Francisco Bay revealed no problems, and after a week's delay the *White Sands* was towed down to San Diego. The suspect thruster failed on the way; several months later the *White Sands* went into the Long Beach Naval Shipyard to have a replacement thruster installed.

The fleet tug *Apache* went into overhaul at San Diego from 15 December 1969 to mid-April 1970. Thereafter, until January 1971, she was assigned the ordinary duties of a fleet tug. The concept of a *Trieste* "integral" operating unit was coming unglued.

Meanwhile, work continued on the *Trieste.*

From early 1970 it had became apparent that the *Trieste* program was in decline. In 1966–1967 there had been up to 15 officers assigned, as the program was prepared for classified operations and trained pilots for the new-construction DSRV rescue vehicles and DSSV search vehicles. The number of officers on board declined markedly in 1968 with cancellation of the DSSV program, reduction in DSRVs from six to two, and the termination of *Trieste* participation in Winterwind.

A last-minute augmentation of officers in 1969—Lieutenant Ross Saxon and Lieutenant (Junior Grade) David Byrnes—brought the number to seven for the *Scorpion* dives. In 1970, Byrnes left to spend two years at the Naval Postgraduate School at Monterey, California, studying oceanography. Lieutenant Bill Leonard relieved Tony Dunn as assistant officer-in-charge in February 1970; Dunn had orders to the satellite navigation program, part of the Polaris/Poseidon submarine missile project. Lieutenant John Field was relieved in early 1970 by a newcomer, Lieutenant Albert L. Amaro, as the *Trieste*'s engineer.

While the *Trieste*'s officer count dropped to five, enlisted manning declined slightly, from 24 men, including eight chief petty officers, to 20 with five chiefs. The enlisted men who detached had experience in the *Trieste*'s operations and maintenance; they had limited opportunity to train their replacements, new to the unique bathyscaph world.

The situation was both inexplicable to and distressing for the officers and civilians in the *Trieste* program and the Submarine Development Group. The *Trieste* had proven capable of operational missions deeper than was possible for any other U.S. submarine or submersible. She had accomplished the *Scorpion* operation, which would have been impossible just 12 months earlier.

But, most simply: there was no longer a covert mission for the new *Trieste*. The Intelligence Community had dropped the surface-based recovery option from the Winterwind effort, and no one stepped forward to claim the orphaned bathyscaph. As early as mid-1968, the Office of the Chief of Naval Operations and the Office of Naval Research saw no missions in the offing.

As bathyscaph funding from sponsors expired, officer and enlisted billets continued to slip away, money for at-sea operating days vanished, and budgets for training and schools for the *Trieste* team evaporated. The program slumped into neglect and decline.

In the Pentagon, the Deep Submergence desk in OpNav, then occupied by Commander Brad Mooney, a former *Trieste* officer-in-charge, was undergoing a major reorganization. In late October 1969, coincident with the *Trieste*'s return to San Diego, Mooney's responsibilities were folded into the new Deep Submergence Systems Division, under Rear Admiral Mike Rindskopf.[1] The OpNav staff was being restructured to manage the new deep submergence capabilities coming on line, including the covert capabilities represented by the special projects submarine *Halibut* and the soon-to-be-completed, nuclear-propelled, deep submersible *NR-1*. And, the *Alvin*, *Turtle*, and *Sea Cliff* were being added to the list. Of course, the two DSRVs, when not engaged in training and rescue operations, were available for clandestine (submarine carried) "special ops."

At Ballast Point

On 18 February 1970, Jacques Piccard, his wife, and their 11-year-old son, Bertrand, returned to San Diego on a social visit and were welcomed at Ballast Point submarine complex. The *Trieste*'s officer-in-charge (Bob Nevin), Piccard, and his son were photographed for the press. Piccard's visit was handled cordially, and Bertrand would remember fondly the courtesy of a simulated dive in the trainer and the "diploma" presented to him as a "deep-submersible pilot."[2] The visit program did *not* include a detailed briefing on the extended and expanded operational capabilities of the "new" *Trieste*.[3]

The bathyscaph was at the time in overhaul and looked like a three-dimensional jigsaw puzzle on its cradle, components scattered around it. The most critical item of the work package was freeing up the shot pans and re-engineering to ensure that such a situation would never occur again. The electrical system, which had plagued most of the *Scorpion* dives, was given permanent fixes, and the external lighting was upgraded. Installation of a Kollmorgan binocular viewing system made it no longer necessary to view the bottom directly through the forward window. Instead, a person could see ahead with binocular vision without kneeling or lying on the sphere's deck, or two people could simultaneously look with monocular vision. The result was a 120-degree field of vision for search, 72 degrees with about 25 percent magnification. Four pan-and-tilt TV cameras were mounted outside the sphere, switchable to any of the three eight-inch monochrome TV monitors inside.[4] The problems with NAVNET during the *Scorpion* searches were corrected with faster electronics that recognized the first acoustic reply from each long-baseline transponder and ignored subsequent echoes, multiple-path signals, and reverberations.

In September 1970 and the Navy announced that the newly upgraded bathyscaph would carry the name *Trieste II* with the hull number DSV 1, a decision that has caused confusion ever since.

The DSV 1 had been activated as a covert operational vehicle in September 1967, the third U.S. bathyscaph of the *Trieste* program, and had frequently been referred to thereafter in both classified and unclassified documents as the "*Trieste III*." Now, three years later, this third bathyscaph was conflated with its predecessor, the *Trieste II*, despite the fact that there was hardly a nut or a bolt common to both vehicles. Now the Navy bestowed an official name and hull number on the previously covert bathyscaph *DSV-1,* as the "new" *Trieste II* (DSV 1), as well as hull numbers on the three already-named *Alvin*-type submersibles: the *Alvin* (DSV 2), the *Turtle* (DSV 3), and the *Sea Cliff* (DSV 4).

On 23 March 1970, Nevin entered the hospital with a possible ulcer. He was hospitalized for six weeks. Nevin's physician ordered six months of limited duties to follow, and Commodore Robert Gautier recommended that he remain at the Development Group, being assigned "light duties."[5] The decision to remove Nevin from day-to-day command responsibilities without relieving him as officer-in-charge was both a symptom and a cause of the continuing decline in the readiness and operational capabilities of the *Trieste* unit.

Nevin's command duties devolved upon Beau Myers for the *White Sands* and Tony Dunn and then Stretch Leonard for the *Trieste*. While in the hospital,

Nevin was joined for a few days by Commodore Gautier, who was admitted with kidney stones.

Dunn departed without relief, and Leonard assumed the duties of assistant officer-in-charge, and with it most of the weight of the command due to Nevin's limited-duty status. On 18 August, newly promoted Lieutenant Commander Leonard relieved Nevin as officer-in-charge of the *Trieste*. Simultaneously, Myers relieved Nevin as officer-in-charge of the *White Sands*. Among the noteworthy problems listed on the occasion of the change of command of the *White Sands* was that she was critically short of scuba divers, being allocated ten but having only three on board. Ross Saxon relieved Leonard as assistant officer-in-charge of the *Trieste* at the same time.[6]

The *Deep Quest* and *White Sands*

The *White Sands* was towed back to San Diego in late June 1970 and was dry-docked at Long Beach in September to replace the failed thruster. She then was involved in shakedown and training off San Diego from 4 through 7 October. On the evening of the 7th Lieutenant Commander Malcolm G. (Mal) Bartels, the prospective officer-in-charge of the *Trieste*, ordered the ARD to prepare for towing.

Bartels, the latest of four consecutive Naval Academy graduates to command the *Trieste* program, had completed a two-year postgraduate course in oceanography at the University of Washington in 1963–1964, had been one of the *Trieste* pilot-trainees in 1966, and had served on the staff of Submarine Development Group 1 in 1967–1968, during which time he had the opportunity to dive in the Japanese submersible *Yomiuri-Go*. Prior to returning to the *Trieste* as officer-in-charge, Bartels had been executive officer of the diesel-electric submarine *Catfish* (SS 339).

The submarine rescue ship *Chanticleer* (ASR 7) took the *White Sands* in tow and, in company with the Lockheed support ship *Transquest*, with Lockheed's deep submersible *Deep Quest* on board, proceeded to a position 12 miles off Point Loma where a World War II–era F6F Hellcat fighter had come down at sea in 1944. Former *Trieste* assistant officer-in-charge Larry Shumaker, now the chief pilot of Lockheed's *Deep Quest* program, had sighted the aircraft on one of his dives and had offered the assistance of Lockheed should the Navy wish to recover it.

An opportunity to prove the *White Sands'* ability to salvage as much as 20,000 pounds from the deep ocean floor was appealing to Commodore Gautier. At the site he boarded the *White Sands* with members of the San Diego press corps. The salvage plan was complex: Shumaker would take the *Deep Quest* down with a reel containing 4,800 feet of nylon line and would use both of the submersible's manipulating arms to thread a bridle through the aircraft's engine support frame, then pay out the line as he returned to the surface. There the line would be passed to the *White Sands* for the lift. Three critical actions were then to follow: breaking the 12,000-pound-plus aircraft out of the mud, transferring the load to the dry dock's 35-ton-capacity crane, and, finally, ballasting down to slide the aircraft—still underwater—into the docking well.[7]

On 9 October, Shumaker was successful in bridling the aircraft and lifting the three-inch line up to the *White Sands*. The lift began in the late evening, and the breakout from the mud was without incident. As the aircraft neared the surface, divers attached a six-inch nylon line and passed it under the ship to divide the load between the ARD's starboard winch and a pair of capstans on the port side. The aircraft was slowly pulled under the dock's keel.

The *White Sands* was ordered to stop her after thrusters for diver safety and was soon wallowing in the trough. This caused the aircraft to bob up and down and the six-inch nylon line to act like a rubber band—and dangerously strained. Chief Warrant Officer Bill Murphy on the dock quickly spliced a three-inch-thick lift line from the crane, and divers shackled that line to the aircraft just moments before the six-inch line parted.

Lieutenant Myers maneuvered the *White Sands* out of the trough and divers connected a new six-inch line from the crane to the aircraft.[8] It then was relatively simple to flood the dock, open the gates, and carefully swing the aircraft, inches below the surface, into the well deck. Then the gates were closed and the ship de-ballasted. The prize was high and soon dry.[9]

For this complex and somewhat harrowing recovery, Bartels, Myers, and Murphy received letters of commendations from the Commander, Submarine Force Pacific.[10]

Back in the Water

The *Trieste*'s first wetting after overhaul was on 27 October 1970, with Chief Warrant Officer Larry Hawks providing newly arrived Lieutenant Richard H.

Taylor and Richard L. Abbott a quick, tethered dip to 50 feet alongside the pier at Ballast Point. Taylor, who would relieve Ross Saxon, had been with the *Trieste* program as an enlisted electronics technician in 1964–1966. Abbott too had several years of enlisted experience prior to college and commissioning. He later would become the *Trieste*'s engineer officer.

On 4 November, Leonard and Saxon took Naval Ship Systems Command's project manager Don Johnson down to 2,132 feet off Point Loma as part of the *Trieste*'s post-overhaul trials. Johnson was required by regulations to be on board for the first deep dive after the overhaul. A month after this acceptance dive, on 6 December, Bartels piloted with Leonard and Hawks on Leonard's final dive. Bartels relates how he put the bathyscaph through its paces at about 3,000 feet, then, releasing more avgas, dived to 13,200 feet. After giving the bathyscaph time to settle out all the creaks and groans, Bartels turned to Leonard and stated, "I relieve you, sir."

Bartels formally relieved Leonard as officer-in-charge of the *Trieste* on the 17th. Bartels simultaneously relieved Myers as officer-in-charge of the *White Sands*. Leonard departed for his next duty station, as special projects officer on the nuclear submarine *Seawolf* (SSN 575), while Myers reverted to his permanent position as executive officer of the *White Sands*. In early January 1971, Taylor relieved newly promoted Lieutenant Commander Saxon as assistant officer-in-charge. Saxon would become the executive officer of the new submarine rescue ship *Ortolan* (ASR 22).[11]

Also in January, Captain Samuel H. Packer relieved Captain Gautier as Commander, Submarine Development Group 1. Gautier had orders to command an amphibious squadron in Norfolk. (Two years later he would return to San Diego as the commanding officer of the Naval Undersea Center.)

On 3 February 1971, Bartels exploited a slackening in the winter weather to make a 4,200-foot dive for his own and Taylor's training, under the watchful eyes of Hawks. This was the final dive of the winter season. There had been *exactly two dives* in the preceding operating year (October 1969–November 1970), one of them a 50-foot dip. Bartels was under orders from the new Development Group commodore to get the *Trieste* program "back on line."

Progress was slow.

In June 1971, Bartels arranged three tethered dips alongside the pier at Ballast Point to check out electrical systems and hull penetrations. These dips, to

50 feet, gave Taylor an opportunity to pilot, gave newcomer Lieutenant Commander Philip Stryker, the new assistant officer-in-charge, his baptismal bathyscaph dive, and took two enlisted crewmen (one on each dive) down to check out their equipment. The latter were Chief Interior Communications Technician Clifford E. Cline and Senior Chief Electronics Technician William D. Hood. Hood would eventually qualify as the *Trieste*'s first enlisted pilot.

Bartels was ready for an untethered dive on 8 July, and the venerable *Apache* towed the *White Sands* to sea off Point Loma. Taylor and Bartels were piloting, with Hawks on his final dive. The dive lasted only 30 minutes because of an electrical short and a low-battery alarm and was aborted at 800 feet. They returned to port on 15 July. Abbott relieved Hawks in August 1971. Upon the departure of Al Amaro, Dick Abbott became the *Trieste*'s engineer officer in December 1971, and in late 1972 he would relieve as assistant officer-in-charge.

While Bartels was scrambling to pull the *Trieste* program together, events were occurring that soon would demand the full attention of the Navy's bathyscaph "force."[12]

A Satellite Intervenes

The first of the third-generation U.S. spy satellites—Hexagon Mission 1201—was launched from Vandenberg Air Force Base, California, on 15 June 1971. Hexagon was the most complex unmanned electro-mechanical device yet placed in orbit by the United States. It had four film-carrying Re-entry Vehicles (RVs), each capable of independently returning to Earth with more than 50,000 feet of 6½-inch-wide film. Hexagon provided three-dimensional stereoscopic imagery that could reveal objects as small as two feet in diameter.

The four re-entry vehicles separately returned with a total of *33 miles* of film that were exposed over the Soviet Union and other areas of interest. RV-1 de-orbited on 20 June with 40,502 feet of film. Its parachute failed and it had to be recovered in the water north of Kauai, Hawaii, by divers from a range support ship. After its film was processed an official at the Photographic Interpretation Center in Washington exclaimed, "My God, we never dreamed there would be this much, this good! We'll have to revamp our entire operation."[13]

RV-2 with 53,194 feet of film re-entered on 26 June. It too experienced parachute problems but was successfully snagged in mid-air by an Air Force JC-130 Hercules aircraft. RV-3 de-orbited 10 July with 54,083 feet of film. This capsule also suffered a parachute failure and hit the water at high velocity, immediately sinking. RV-4 returned on 15 July with 25,797 feet of film, to be snagged in a standard aerial recovery by a Hercules.

This, the first KH-9 Hexagon mission, was an overwhelming success, with 119,473 feet of film recovered from three RVs. Still, almost one-third of the exposed film exposed lay on the ocean floor, some 300 nautical miles northwest of Pearl Harbor. Could RV-3's film vehicle be retrieved?[14]

Carl E. Duckett, the deputy director of the CIA for Science and Technology, quickly made contact with his counterpart in the Navy, Dr. Robert A. Frosch, the Assistant Secretary of the Navy for Research and Development. Frosch also managed the recently established National Underwater Reconnaissance Office (NURO).[15] He in turn had the staff director of the NURO, Rear Admiral Mike Rindskopf, "make some phone calls." Frosch was soon able to respond to Duckett with an encouraging, albeit unofficial, reply.

In 1971, several ex-*Trieste* officials were favorably positioned within the Navy establishment: Captain Don Walsh in Dr. Frosch's office; Dr. Andreas Rechnitzer, a program director in NURO, reporting to Frosch; "Buzz" Henifin, now a commander, worked directly for Rindskopf in his "white" role as Director, Deep Submergence Systems Division; and now-Commander Brad Mooney, the chief staff officer of Submarine Development Group 1 in San Diego.

At meetings in Washington on 27 July, Henifin stated that the *Trieste* was available and suggested a "hook and cable" retrieval of the film capsule. The precise location of the film capsule would need to be established by the Scripps Institution employing its Deep-Tow bottom-search system. On 4 August, the CIA tasked the firm of Perkin-Elmer to design a salvage device to be employed by the *Trieste* to recover the RV-3 capsule.[16]

Now John L. McLucas, director of the National Reconnaissance Office—responsible for satellite programs—formally requested a Navy deep-ocean salvage effort to recover the RV-3. Frosch formally responded on 18 August 1971, that the Navy would "be pleased to assist."[17] In an irony of fate, the *Trieste* was being tasked with a mission almost identical to the original Winterwind concept, although this re-entry vehicle was from a U.S. photo-reconnaissance satellite rather than a Soviet missile.

Scripps Institution had several classified contracts under way with the Office of Naval Research and with DSSP.[18] Fred Spiess of Scripps agreed to head up the search and recommended using the Military Sealift Command's oceanographic research ship *De Steiguer* (T-AGOR 12), which arrived in Honolulu on 2 October. On board the *De Steiguer* on 4 October, Rear Admiral Rindskopf and a CIA representative finally revealed to Spiess the true size and anticipated condition of the target, and then disclosed a rectangular search area, one and one-half by eight miles some 300 nautical miles northwest of Kauai.

Spiess' Deep-Tow search team would have to locate the film capsule precisely and mark it with acoustic transponders. The *De Steiguer* sailed for the search area on the evening of 6 October with the senior CIA representative, a McDonnell-Douglas observer, Spiess, and the Deep-Tow team on board.[19] From 8 to 17 October the *De Steiguer* conducted side-scan sonar surveys with the Deep-Tow fish. Only a few sonar contacts were plotted on a nearly flat ocean bottom that was almost completely devoid of rock outcroppings or biological formations.

Spiess felt more than a little personal satisfaction when the first photo of the target capsule was obtained on day ten. From 18 to 20 October, Spiess' team ordered 360-degree turns at one knot, by which it was possible to keep the fish's track within a small and predictable circle and to generate up to five photo opportunities per four-hour watch. This technique greatly improved the chance of obtaining useable photography and would become standard procedure for such a "hunt."[20]

At 0350 on 20 October, after more than 45 hours whirling above seafloor sonar contacts, the *De Steiguer*'s final camera run was completed. After developing the film the Scripps team commenced recovery of all buoys and bottom navigation transponders except for two DOTs and a pinger marking the target's location, one of these within 165 yards of RV-3.[21]

The target had been located on the deep abyssal plain less than ten miles south of the Mendelssohn Seamount—the remains of an underwater volcano whose ravines, gullies, and canyons would have likely swallowed the film capsule forever had RV-3 struck there.

In San Diego on 26 October, Spiess handed Brad Mooney the precise location of the target and the transponders' frequencies and codes—handwritten on a cocktail napkin. "This is the best I've got."[22] Spiess then briefed Mooney on heavy squalls in the area northwest of Kauai and the delays that such weather might cause the *Trieste*.

On 2 August 1971, 16 days *prior* to Dr. Frosch's official agreement to "assist" the National Reconnaissance Office, Commander Mooney had received a phone call from the Development Group's own senior, Submarine Flotilla 1, directing that the *Trieste* be prepared for a classified, deep-sea recovery operation. Initially scheduled for a six-week deployment to Hawaii waters, these mid-Pacific operations (i.e., "Mid-Pac Ops") would actually require almost ten months.

August found the *Trieste* officers and crew woefully under-trained after the neglect and decline of the previous 18 months. Mal Bartels had just two deep dives to his record, a 50-foot tethered immersion and a 30-minute, 800-foot dive that had been aborted. Dick Taylor had one deep dive in training (not counting an orientation dive made years before while an enlisted man in the *Trieste* crew), four 50-foot dips, and the same 30-minute 800-foot aborted dive that Bartels could claim. Stryker could cite only two 50-foot dives. The *Trieste* program had not a single certified deep submergence pilot attached. (Bartels had yet to complete his workbook and undergo his certification interviews). But operational exigencies meant that this critical operational deficiency would remain unacknowledged and have to be corrected on the fly.

Brad Mooney immediately ordered the *Trieste* operating unit to an area off San Clemente Island for a week of training, primarily to get bottom time and some experience with the long-baseline bottom navigation system. The *Trieste*'s logs show two dives to 4,800 feet, on 5 and 9 August, both with Taylor piloting, Bartels navigating, and Philip Stryker operating the sonar. In these two dives Bartels more than doubled his non-tethered dive time, Taylor multiplied his deep-diving experience, and Stryker made his first two deep dives.

The Perkin-Elmer "hay hook," designed by Leonard Molaskey and fabricated by the Naval Underwater Research and Development Center in San Diego, arrived on 17 September, accompanied by Molaskey as the company's technical representative. Called a "kludge" by the *Trieste* crew, the original configuration was designed to spring closed when the *Trieste*'s mechanical arm pulled a short lever.[23]

On 23 September, the Air Force trucked a 1,100-pound Mark 8 satellite training capsule to Ballast Point on a flatbed trailer. It was uncovered and open to prying eyes! Tests were conducted near the *White Sands*' berth to marry up Molaskey's kludge to the training device.[24]

Three civilian techreps were on board the *White Sands* on 27 September when the *Trieste* Integral Operating Unit departed San Diego for training with the precision bottom navigation equipment, the NAVNET computers, and the transponder interrogators. Three dives on 29 September, 6 October, and 11 October failed to prove that Molaskey's kludge could capture and retrieve the practice shape from the ocean floor. The NAVNET long-baseline bottom navigation computer failed, and casualties to other equipment frustrated the *Trieste*'s training.

The mission was nearly scuttled on 29 September, when a casualty to the *Trieste*'s bow winch at 4,200 feet cut the cable and dropped the claw. More than 45 minutes were required to relocate the kludge. Bartels then used the mechanical arm to recapture the claw and take it up to the surface.[25] On the 11 October dive Bartels successfully lowered Molaskey's claw onto the test shape, but could not close it to capture the training capsule because of problems with the manipulating arm. Nevertheless, Bartels advised Mooney that he was satisfied that the *Trieste* could accomplish the mission, given a modification to Molaskey's kludge—and that he was ready to deploy immediately and so evade the impending onset of winter weather in the target area.

Already two weeks behind schedule, the *Trieste* operating unit deployed without completely proving the kludge. The red-and-white training capsule remained on the ocean bottom off the Coronado Islands.

19 The Deepest Recovery

Almost no one was cleared for their operation, so everyone thought they were just playing games at sea.

—Lieutenant Commander Ronald J. Doyle,
Submarine Development Group 1

The *Trieste* Integral Operating Unit sailed toward the Hexagon recovery area northwest of Kauai from 12 October through 2 November 1971—three weeks at an average speed of 4.8 knots. Lieutenant Commander Luther Blevins, now the commanding officer of the tug *Apache*, was able to make such good time because of the *White Sands*' newly installed 750-horsepower stern thrusters. Lieutenant Commander "Mal" Bartels, officer-in-charge of the unit, was hoping to retrieve the lost satellite film capsule in a single dive, an operation that—ideally—could take as few as four days on site.[1]

Lieutenant Commander Brad Mooney, the chief staff officer of Submarine Development Group 1 in San Diego, flew into Honolulu on 29 October, and the next day saw the photos of the target taken by Fred Spiess ten days previously. The photos, which showed the parachute base of the re-entry vehicle, supported the evaluation that the capsule was damaged, but intact and largely buried in the bottom silt.

On 31 October, Mooney boarded the commercial submersible support ship *Maxine D* at Pearl Harbor, joining Len Molaskey of Perkin-Elmer, a techrep from McDonnell-Douglas, and a CIA representative, and sailed for the dive site. The following day they rendezvoused with the *White Sands*; Mooney came on board the *White Sands* and took on-scene command of the operation. The *Maxine D* then left, bound for San Diego, displaying flags signaling "Think Deep," the meaning of the *Trieste*'s Latin motto, *Pensate Profunde.*

On 2 November the *Apache* logged her 2000 position as 24°53.5'N / 161°48'W and cast off the *White Sands* tow. They had arrived at the dive site.[2] The first requirement was to gain contact with the transponders earlier dropped from the Navy research ship *De Steiguer*. Employing a portable SATNAV system on the *White Sands,* Mooney directed the dry dock to the coordinates written by Fred Spiess on a cocktail napkin in San Diego a week earlier. At that exact position, Mooney ordered, "Give me a ping. One ping only, please!"[3] The transducer on the *White Sands* rang out, and two deep-ocean transponders responded immediately.

The *Apache* dropped two BQN-8 transponders on 4 November, and the *Trieste* prepared for a dive to 16,400 feet. Loaded with 67,000 gallons of aviation gasoline, 32 tons of steel shot, and a special "kludge" to recover the film capsule, the *Trieste* submerged about 1815. Lieutenant Commander Phil Stryker piloted, Lieutenant Dick Taylor navigated, and "Mal" Bartels was on the sonar. They searched the bottom for six hours without success. The NAVNET computer was providing confused data, and the sonar contacts could not be visually acquired.

The next day the weather closed in so quickly that by 1630 the rough seas prevented the *Trieste* from being recovered into the *White Sands.* The *White Sands* was taken in tow by the *Apache* with the *Trieste* in tandem tow, 500 yards astern. Lieutenant Myers, the executive officer of the *White Sands,* wrote in a letter, "We managed to undock the *Trieste* and get off one dive. We didn't recover anything. Shortly after the dive, the weather closed in again and for almost a week now we have been thrashing about."[4]

Mooney decided to return the Integral Operating Unit to Pearl Harbor to await better weather. On 16 November, Rear Admiral Paul L. Lacy Jr., Commander, Submarine Force Pacific, met with Mooney, Bartels, and Stryker at Pearl Harbor to decide what course to follow. After a subsequent meeting with the senior CIA representative, Lacy ordered Mooney to remain ready to exploit any break in the weather.

When on 21 November a brief weather window opened the unit deployed again, with a sense of urgency. Analysis by the *Trieste* officers and separately by the Sperry representatives of NAVNET data from the 4 November dive indicated that the plot developed from Spiess' cocktail napkin was in error. The unit was back on site on 24 November, now under the tactical command of Captain Samuel Packer, commander of the Development Group. He tasked the *Apache* with resurveying the identities and relative positions of the deep-ocean transponders in the target field. The *Apache*'s survey determined not only that the locations of two of the DOTs had been transposed but also found that those two transponders were not 330 feet apart—as calculated from the original data—but *2,330 feet apart*! The *Trieste* had been searching almost 700 yards east of the target on 4 November, a significant error in the darkened depths. With a corrected plot—and luck—a second attempt could accomplish the recovery.

Delayed by encounters with large sharks and a killer whale, the 30 November dive did not start until the early evening, at 1745. The *Trieste*'s search at depth was impeded by an electrical surge, wiping part of the NAVNET computer's memory, and loss of the bathyscaph's Doppler sonar. The *Trieste* continued the search by dead reckoning. After almost eight hours submerged, the *Trieste* slowed to approach a sonar contact west of DOT-3 and sighted the errant capsule!

Momentum carried the *Trieste* beyond the target, and as Bartels maneuvered to return, a low-battery alarm went off. Bartels had no option but to immediately terminate the dive. An effort to mark the target with a mini-DOT carried by the *Trieste* was frustrated by a malfunction of the mechanical arm. The *Trieste* safely reached the surface and was taken in tow by the *Apache*.

Bad weather again closed in, ending recovery efforts for the year.

The 30 November–1 December dive revealed two new problems. First, during the 16,400-foot descent the *Trieste* had drifted so far from the target that it took two hours to get back to the search area; correcting for drift during descent in the next dive would mean more battery life for search and recovery.[5] Second, the malfunctioning of the mechanical arm at 16,000 feet required a redesign of the capture mechanism.

After a brief period in Pearl Harbor, the unit sortied to the dive site on 12 January, remaining there through 5 February 1972. Unrelenting foul weather prevented any dive attempt and finally, the *Trieste* unit headed back into Pearl Harbor for the season.[6]

On 20 January 1972, Lieutenant Commander Charles W. Hudiburgh relieved Bartels as officer-in-charge of the *White Sands*.

Mission Re-evaluation

Eastman Kodak had estimated that there was a good possibility of recovering useable film from the RV-3 canister even after immersion in salt water for more than three months. Now the film had been immersed more than six months.

CIA consultations on 3 February in Pearl Harbor with Rear Admiral Lacy, Captain Howard N. (Lou) Larcombe of the OpNav Deep Submergence Systems Division in the Pentagon, and with "higher authorities" from Washington (read CIA and NRO executives), ruled out unilateral cancellation of the operation by the Navy.[7] The question then became, did the National Reconnaissance Office (NRO) want to continue the recovery effort?

Within the NRO, there was more concern about security issues than about the value of the film. Colonel Frank S. Buzard, about to retire as Hexagon program director, argued that the unusual activity in international waters northwest of Oahu had been monitored by the Soviets, who could, if recovery efforts were halted, legally attempt a recovery of their own, potentially compromising one of the nation's most highly classified intelligence programs. This argument appeared to tip the scales, and the decision was made to continue the recovery effort in the spring of 1972.[8]

The Second Effort

After wintering over in Hawaii, the *Trieste* operating unit departed from Pearl Harbor on 8 April 1972, once again under the command of Brad Mooney—first for refresher training off Kauai, then on to the target site. To deal with the mechanical arm unreliability issue at depth, the *Trieste*'s engineer officer, Lieutenant Dick Abbott, and Chief Machinist's Mate Sam Goucher devised a work-around that would use a "cocked" trigger to close the claw. However, this device gave the *Trieste* only one claw release per dive.

On 12 April, "Beau" Myers was relieved by Lieutenant Antone (Tex) Texeira Jr., after 44 months as both executive officer and officer-in-charge of the *White Sands*.

The weather finally eased enough for a dive on 25 April. None of Spiess' transponders remained active, but the area was well charted, and DOTs emplaced in November and January still responded. With Bartels piloting, Stryker on sonar, and Taylor navigating, the *Trieste* descended for two hours and approached the ocean floor. There followed three hours of searching before they sighted a tangled mass of metal and wire northwest of DOT-3 and headed toward a large sonar contact. The bathyscaph's crew spotted two jagged, gold-foil capsule components and, 800 feet farther along, a segment of the parachute bridle. Finally, the *Trieste*'s large sonar contact became discernible as the round shape of the target capsule.

"Tally-ho, the fox!," the *Trieste* operators reported to Mooney by underwater telephone. What they had found, partially buried in the silt, was an internal sub-assembly of RV-3, festooned with thermal insulation pads on its periphery and with long filaments of film or plastic tape swaying in the deep-ocean current. The capsule had struck the ocean at about 300 miles per hour, an estimated 2,600-g deceleration—and had broken apart.

The men in the *Trieste* immediately recognized that this sub-assembly was *not* the component located and photographed by Spiess and the Deep-Tow team in October, but it *was* the all-important film-and-reel assembly.

Bartels made several "bounces" along the bottom to position the *Trieste* for the one-and-only claw-release possible for the dive. Each bounce stirred up the silt, requiring a long wait to see the target again. Once positioned and after taking photos with the *Trieste*'s external cameras, Taylor activated the winch to lower the kludge and envelope the target—and the *claw failed to close.*

Anxiety rose as five attempts to capture the target failed. On the sixth attempt, Abbott suggested—and Mooney relayed by underwater telephone to the *Trieste*—dropping the kludge onto the target and paying out additional cable. As the claw enclosed the target for the sixth time and the cable slackened, the release finally triggered, and the kludge closed around the film-and-reel assembly.

Taylor then winched the target off the bottom, and as the silt cleared he reported to the surface by underwater telephone, "Charlie Brown; Charlie Brown," the code phrase for a successful capture. As the capsule rose from the bottom, short segments of film could be seen falling from it. No matter, as it later proved—Kodak's later analysis of short strands of film revealed that the initial impact had shattered the brittle film into thousands of pieces.[9] During the ascent, Taylor reported that pieces of film "began breaking off in lengths varying from one inch to two or three feet. Almost halfway to the surface the film started breaking into two- to three-foot segments." By the time the *Trieste* neared the surface around 0230 on 26 April, more than nine hours after starting its dive, "pieces were hanging through the lines of the hook."[10] During the two-hour ascent Taylor alerted Mooney by underwater telephone that the divers needed to be "ready to recover the kludge and target as soon as possible because it's breaking up."

In the dark, early morning hours of the 26th, divers attached flotation devices to the claw about 100 feet below the surface, placed a black bag over the assemblage, and floated it up to the *White Sands.* With the kludge hanging about 35 feet below the surface and bobbing in the seaway, the weakened film stack suddenly began shedding internal components and almost all the remaining film was lost.

From the deck of the *White Sands*, Electrician's Mate 3rd Class Robin M. Salser saw "an object that was covered by a tarp . . . coming up to the surface and then what appeared to be small metal (glinting) fins/pieces float down and away from the package and divers."[11] The CIA representative would recall, "Everyone went from an emotional high to an emotional low in about one microsecond."[12]

The *White Sands*' crane lifted the remaining assembly over the "tailgate," and into the dock well. From there the crew transferred the tarp-covered "target" into a deck-mounted freezer, designed to prevent fungus or bacterial damage to recovered film. The freezer was locked and under armed guard until it was off-loaded in Pearl Harbor. The 30½ feet of film recovered by the *Trieste* had been exposed to salt water and divers' lights—and had no intelligence value.

The CIA representative, Mooney, and Taylor left the *White Sands* by helicopter the day following the dive. Taylor had orders to report as officer-in-charge of the rescue submersible *DSRV 2* (which later would be named *Avalon*).

After a slow plod homeward, the *Apache*, *White Sands*, and *Trieste* arrived in San Diego on 23 May. In a Navy press release that was printed in newspapers nationwide the next day, Bartels is quoted saying that the *Trieste* had recovered "a half-ton robot laboratory . . . from 16,400 feet . . . 400 miles north of Hawaii."[13] In the July 1972, issue of the journal *Undersea Technology*, a version of the press release was published under the title "*Trieste* Recovers Electronics Package from 16,400-ft. Depth." This article caused consternation in the NRO about the possibility of a security breach, but close reading revealed that nothing of a classified nature had been published.[14]

Retrospective

The *Trieste* team had located a small object in mid-ocean and recovered it from 16,400 feet—a significant milestone in the development of oceanographic technology and in U.S. Navy operational capabilities.[15] Despite the loss of film, the "customer" was satisfied with the outcome of the Hexagon effort. The partial recovery added to the National Reconnaissance Office's understanding of vehicle re-entry after catastrophic parachute failure and gave confidence that in such cases nothing of significance would remain on the ocean floor.

NRO director John McLucas praised the recovery, stating that the Navy had established and demonstrated "a unique capability vital to the security of the United States."[16] A CIA memo of 24 May 1972 cited lessons learned: that (1) reconnaissance assets and the recovery craft should be integrated into a single organization to eliminate delays and reduce errors communicating information; (2) methods should be developed to avoid weather-related delays in readying the recovery craft; (3) the "relatively poor" reliability of the equipment on board the *Trieste* should be addressed; and (4) a faster method of deploying the recovery craft should be developed.[17]

Lieutenant Commander Hudiburgh of the *White Sands* recorded in the ship's history for 1971 that the dry dock from "2 August through the end of the year was occupied by a major effort in preparing and conducting the deepest salvage-recovery operation yet attempted by man."[18] In addition to a highly unusual "Letter of Recognition," drafted by Commander Don Walsh for the signature of the Assistant Secretary of the Navy, the *Trieste* team received the Navy's Meritorious Unit Citation for the mid-Pacific operations from 1 September 1971 to 23 May 1972.[19] Bartels was awarded the Navy Commendation Medal as officer-in-charge and pilots Stryker and Taylor both received the Navy Achievement Medal. Individual letters of commendation from the Commander, Submarine Development Group 1 were presented to all *Trieste* crewmen, as well as to the three civilian technical representatives who participated.

Also, Walsh himself received his second Legion of Merit for his 3½ years in the Pentagon supporting both Assistant Secretary Robert Frosch and John Warner, the Under Secretary of the Navy from 1969 to 1972 (and Secretary of the Navy from 1972 to 1974).

Resurrection and Redemption

The Hexagon recovery resurrected the *Trieste* program and propelled it into high-visibility status within the Navy Department. The next several years would see plans and programs to build a more effective system around the *Trieste*, with a modern and capable support ship that could deploy the *Trieste* at a sustained speed of at least 12 knots.

On 13 July 1972, after a stand-down in San Diego for cleaning and repairs, the *Trieste*'s first post-Hexagon operation began: a search for one of two F-4 Phantom fighters that had crashed off of North Island, San Diego. The dives were opportunities for pilot training: Dive #2-72 on 15 July gave Abbott his second *Trieste* dive, with Stryker piloting and Bartels in the jump seat. Abbott's third dive came three days later, with Stryker and Senior Chief William Hood; Dives #5-72 and #6-72 on 7 and 9 August, respectively, saw Hood and Abbott under instruction and Stryker, then Bartels instructing.

In July 1972, Abbott was promoted to lieutenant commander, and with Stryker's departure in August he became assistant officer-in-charge of the *Trieste*. Stryker's next assignment was command of the Naval Facility Pacific Beach, Washington (a SOSUS sound surveillance facility). Lieutenant Raleigh D. Baker reported for *Trieste* duty in September 1972.

The *Trieste*'s next underway period generated three dives on the wreck of the submarine *Segundo* (SS 398), which had been sunk by U.S. submarine torpedoes two years earlier in a "Sinkex" training exercise.[20] The *Segundo* had sunk with towing cables still attached. On the bathyscaph's third dive to the wreck, on 26 September, Abbott recalled:

> On the dive, we became entangled in cables and Bartels, piloting, stated he was caught—he could not turn right or left or go ahead. Baker [on his first bathyscaph dive] suggested trying backing out, which worked. We immediately surfaced and the after part of the *Trieste* was still under water from a cable caught on the after skegs [feet]. A diver reported the cables went down over 75 feet and had things attached. Trying to cut the cable broke our cable-cutter. . . . The divers eventually managed to get the massive cable off and the *Trieste* righted itself. Later Bartels asked

> each of us if we would dive on the *Segundo* again and we all refused. My memory is very clear[:] . . . who could forget almost being trapped on the bottom.[21]

In November 1972, Lieutenant Frank G. Charlton joined the *Trieste* team; he was one of the very few officers ordered to the *Trieste* without having volunteered for deep submergence duty.[22] On 7 December, for Dive #10-72, the last of the year, Lieutenant Commander D. Patrick Raetzman from Development Group 1, a former officer-in-charge of the *DSRV 1,* rode with Bartels to certify Abbott.

The next four dives were for pilot training, off San Diego in January–February 1973, going down to 4,500 feet. On 24 May, with the *Trieste* back in the water after two months of maintenance, Abbott took Baker and prospective officer-in-charge Lieutenant Commander Roger B. Whitaker on a six-and-a-half-hour dive. Two days later, Abbott took Baker and Hood on a training dive in the San Diego operating area. On 28 June 1973, Whitaker relieved Bartels as officer-in-charge of the *Trieste.*[23]

DSV Manning Issues

Up to this time the submarine detailers in the Bureau of Naval Personnel had had a surplus of officers volunteering for deep submergence, allowing the bureau to establish very high criteria for *Trieste* assignments. Candidates were selected from among lieutenant commanders who were submarine qualified, had completed executive officer tours on submarines, had qualified for submarine command, and had post-graduate education in oceanography. After Don Keach, four successive OINCs had those qualifications, and all were Annapolis graduates.

However, after the *Scorpion* operations, the BuPers submarine detailers began to warn off "hard chargers" from deep submergence in general and the *Trieste* program in particular. Further, the CURV system's triumphant recovery of the H-bomb off of Palomares in 1966 demonstrated that the advantages of *un*manned deep submergence platforms over manned vehicles had become more and more salient. (As early as 1979, between one-third and one-half of all Navy deep submergence missions were being accomplished with unmanned systems.)[24]

Thus, manned deep submergence was no longer attracting career-minded, hot-running officers. Worse, the pool of volunteers was shrinking at a time when several new deep submersibles were entering the fleet. This gap opened billets on the *Trieste, Turtle, Sea Cliff,* and the two DSRV rescue vehicles to officers with extensive previous enlisted service. Also, the requirement that prospective officers-in-charge be former submarine executive officers was dropped, and soon the requirement for post-graduate education in oceanography also fell aside.

Roger Whitaker was one of the first products of the revised selection process. A graduate of the University of Kansas, Whitaker had received his commission in 1961 through the Navy Reserve Officer Training Corps program. Prior to reporting to the *Trieste* program he had had junior officer tours on two diesel-electric submarines and served as an aide to the Commander, Submarine Force Pacific. He had been on the staff of Submarine Development Group 1 since 1970 as the DSRV Project Officer when in 1973 he was ordered to the *Trieste* program as prospective officer-in-charge.

And Another Recovery

Since 1968 the Naval Oceanographic Office had been developing a deep-towed sensor system called Teleprobe. In many respects it replicated Scripps' Deep-Tow system and the towed system on the *Mizar.*[25] The Teleprobe team had in 1972 been embarked on the *De Steiguer* and sent into a submarine operating area west of San Francisco. The Teleprobe fish was being towed about 20 feet above the ocean floor when it struck the bottom; the towline kinked, snapped taut, and broke. Teleprobe #1 was lost. Attempts from the *De Steiguer* to grapple and recover the fish were unsuccessful.

On 26 May 1973, the fleet tug *Apache* got under way from San Diego with, once again, the *White Sands* in tow and the *Trieste* nestled in the docking well. En route to the Teleprobe dive site foul weather forced the unit to seek shelter in Drakes Bay, north of San Francisco. A week later, at the first break in the weather, the *Apache* towed the *White Sands* on to the dive site, where the *Trieste* was launched, fueled, and shotted for a dive. The weather turned again, and so quickly that the bathyscaph could neither be boarded for a dive nor recovered. Abbott described the situation: "too rough to dive, too rough to dock, nothing to do but keep the *Trieste* in tow and head downwind. Too much speed would possibly damage her."

The winds and seas posed a problem for the *Apache*. She could not safely handle the dry dock with the bathyscaph in tandem tow. The *Apache*'s skipper, now Lieutenant Michael D. Barker, called for help; a second tug was dispatched from San Francisco to tow the *Trieste* while the *Apache* towed the *White Sands*, both to Drakes Bay, where the waters were calm enough to de-gas and re-dock the *Trieste*. Subsequently the *Apache* towed the *White Sands*, with the bathyscaph safely in her docking well, to the Alameda naval air station in Oakland. During the weather-related maneuvering outside of Drakes Bay the *White Sands* suffered a casualty to one of her thrusters that would require a shipyard visit for repair.[26]

The *Apache* departed for San Diego to repair here own storm damage. The *White Sands* with the *Trieste* was towed across San Francisco Bay to the Hunters Point Naval Shipyard for dry-docking. Both remained there for almost a month.

On 18 July the *Apache*, which had departed San Diego on the 13th, arrived at Hunters Point.[27] Two days later the *Trieste* Integral Operating Unit sailed from Hunters Point for the dive site, but once again the weather turned foul and again the unit sought shelter in Drakes Bay. Seeing a possible break in the weather on 27 July, Whitaker took a gamble and ordered the *Apache* to hook up for a transit to the dive site, about 45 nautical miles from Drakes Bay.[28] Dive #5-73 put the *Trieste* over the target, the sunken Teleprobe. Abbott piloted, Baker assisted, and the new officer-in-charge, Roger Whitaker, in the jump seat. The dive went to 10,699 feet and lasted ten hours, 36 minutes submerged. Baker provided a description of this deep-ocean recovery:

> We found it by Braille—that is we bumped into it. Plan A was to use the manipulating arm to attach the bow winch hook onto a preselected lifting point on the sled. When that manipulator failed, Plan B was to drag the winch hook over the sled in the hope of engaging some element of the sled's frame.[29]
>
> I continued trying to snag the Teleprobe for a while without success. I backed off a couple feet and then drove up to the fish. I overshot my approach and sort of bulldozed the fish, which rolled it over about 90 degrees. Now exposed was one of the runners on the sled's bottom. After a few tries, I got the hook threaded through the gap between the runner and the sled's operating components. I

> swear I heard a "snick" when the hook snared the runner. After some thoroughly naval exclamations and profanity, we notified the surface "Charlie Brown, Charlie Brown" indicating we had captured the target and were on the way up.[30]

Abbot related the public's interest in the successful recovery: "We flew back to San Diego commercially, the officers and some of the *Trieste* crew. Someone on the aircraft told the pilot of our adventure, and he took the aircraft to 10,699 feet altitude and announced it on the intercom for the whole plane to hear, to see how deep we had been by looking down to the land below."[31]

Meanwhile, after a brief port call in San Francisco to offload the recovered "fish," the Teleprobe team, and several of the *Trieste* officers and crew, the *Apache* took the *White Sands* with the *Trieste* within her docking well under tow en route to San Diego, arriving on 8 August. A Navy public release identified the Teleprobe as "an unmanned deep submergence sled . . . carrying oceanographic equipment" that would cost the Navy over a quarter of a million dollars to replace.

The *Trieste* program was displaying growing sophistication and maturity as an operational deep-sea search and recovery system. It should be noted that there were nine dives at the *Scorpion* site in 1969 but that the Hexagon recovery in 1971–1972 required just three dives, and the Teleprobe recovery in 1973 only one. Still, the *Trieste* remained a system very sensitive to sea conditions and weather extremes.

On 1 August 1973, while en route San Diego under the care of the *Apache*, the *White Sands* (ARD 20), still a "yard craft," was reclassified as a deep submergence support ship and soon was placed in commission as a "ship"—the USS *White Sands*, hull number AGDS 1. The designation indicated that she was a general auxiliary ship (AG) for deep submergence (DS) support. Her first and only *commanding officer* as AGDS 1 was Lieutenant Commander Charles W. Hudiburgh.

Winding Down

Two *Trieste* dives were conducted in September 1973: on the 20th, to certify Baker and on the 22nd to certify Hood.[32] On both dives Abbott and

Lieutenant Commander John F. Cameron from the Development Group 1 staff were examining officers. Cameron was an ex-officer-in-charge of the Navy's deep submersible *Turtle* (DSV 3).

The *Trieste*'s last dive as a component of the Integral Operating Unit occurred on 14 October 1973 during a transit to the Mare Island shipyard. This was Abbott's last dive, and on it he took Lieutenant Clarence L. (Les) Parsons and Electrician's Mate 3rd Class Wesley R. Howe on their first deep descents.[33] The dive, to 12,311 feet, was made to test the manipulating arm at significant depths. The dive ended early, never reaching the bottom, owing to a main battery short. Several months later, while the *Trieste* was in overhaul at Mare Island, Lieutenant Commander Abbott was relieved as assistant officer-in-charge by Parsons and joined the staff of Submarine Force Pacific at San Diego.

The *Apache* made her last Navy tow on 31 January 1974. Two months later, on 23 March, the Navy decommissioned the "oldest tug in the navy," the 32-year-old *Apache*. Many of the *Apache*'s crew were transferred to the *Trieste*'s new mother ship, the future USS *Point Loma* (AGDS 2).[34] And, the *White Sands* was stricken from the Navy List on 1 April 1974, marking the end of a significant chapter of the *Trieste* story.

Meanwhile, in December 1973, the *Trieste* had entered the Mare Island yard for a 16-month overhaul.

Notwithstanding its operational capabilities, the *Trieste* remained a low-service-rate/high-cost system. The *Trieste* had made 31 dives during the years 1971–1975, *an average of only six dives per year*. This compared to 328 dives for the two DSRV rescue submersibles (an average of 33 dives each per year) and 426 for the *Turtle* and the *Sea Cliff* (42 dives per year on average).

The highest service rate for a deep submergence vehicle was that of the *Alvin*, Navy-owned and operated by the Woods Hole Oceanographic Institution. The *Alvin* made 295 dives during 1971–1975, an average of 59 dives per year—almost ten times the service rate of the *Trieste*.[35]

Still, the *Trieste* was the only choice for manned vehicle requirements deeper than 6,500 feet, the only true "deep diver" in the U.S. Navy. The *Trieste* program was about to enter its final era, as a seasoned and mature operational system for work deeper than any other manned system could go.

20 A New Support Ship

I watched the scene outside my viewport as metal slowly crumpled; then I saw shimmering bubbles. "Gasoline," I cried out, "in the water!" That was our buoyancy bleeding away—the only force that could bring us back up.

—Dr. Robert D. Ballard, Woods Hole Oceanographic Institution

There was a pause after the 1973 operations while the Navy conducted a major re-evaluation of the *Trieste* program and of the bathyscaph's future as a Navy asset. The Air Force and the CIA had acknowledged that the ability to recover small objects down to 20,000 feet was a necessity and should remain a national capability.

But there was an obvious requirement to deploy the bathyscaph in a more expeditious manner than had been possible in the past. The *Trieste* had arrived north of Kauai in the late autumn of 1971, having taken almost a month in transit at four to five knots. The most costly aspect of the decisions made to meet this "speed" requirement was to acquire more capable—*and faster*—support for the *Trieste.* Choices were few, as ships with docking facilities were either operational amphibious ships or were too small to handle the bathyscaph and its equipment, fuel requirements, and personnel.

At the time, Don Walsh was between duty stations. He recalled that "the deficiencies became very evident [during] the *Scorpion* ops at the Azores." Looking through *Jane's Fighting Ships,* he found the USNS *Point Barrow* (T-AKD 1).[1]

The *Point Barrow* was a one-of-a-kind cargo ship, built with a docking well and a strengthened hull to support the construction of the U.S.-Canadian Distant Early Warning radar installations in far northern areas. After modifications in 1965 the *Point Barrow* had been used to carry Saturn rocket boosters from California to a test site in Mississippi and then on to Cape Canaveral. She had also transported the massive SPS-32/33 radar antennas for mounting on the nuclear-propelled aircraft carrier *Enterprise* (CVAN 65) and cruiser *Long Beach* (CGN 9). In 1970 the *Point Barrow* had been taken out of service and laid up at Brooklyn, New York.

In 1973, Lieutenant Commander Millard S. Firebaugh, an engineering duty officer on the staff of Submarine Development Group 1, was ordered to put together a team to evaluate the *Point Barrow* as a potential bathyscaph mother ship.[2] The group, consisting of officers from the Development Group and from the *Trieste* program, included two former *Trieste* officers, John Howland and Dick Abbott, both now lieutenant commanders. Robert Meade, a civilian engineer with the Deep Submergence Systems Project, met the team in Brooklyn to assist.

The evaluation team concluded that the *Point Barrow* was suitable for the role but would require major modifications. Firebaugh obtained a financial commitment from OpNav and returned to the Development Group with a "done deal." Abbott was ordered back to Brooklyn with a ship-riding team of five officers from the Development Group, including Lieutenant Commander James F. Howick, a former commanding officer of the *Apache* and 23 enlisted men. They were to become acquainted with the *Point Barrow*'s strengths and quirks and help develop the modification plan. (Abbott later would become executive officer of the new bathyscaph mother ship.)

Firebaugh contacted the Navy's engineering program at the Massachusetts Institute of Technology (where he had been a student) to request assistance. Lieutenant Commander Michael Nickelsburg and Lieutenant Stephen A. Beckley soon arrived in San Diego from MIT to develop the specifications for the *Point Barrow* conversion.

Converting the *Point Barrow*

The last commanding officer of the *Point Barrow* and the first commanding officer of the USS *Point Loma* would be Commander Wilson R. Whitmire, a former commanding officer of the diesel-electric submarine *Volador* (SS 490). Whitmire was now in Washington; that assignment was interrupted after 18 months by the offer to command the future bathyscaph support ship.

He jumped at the opportunity and took command of the *Point Barrow* on 26 February 1974. Whitmire cobbled together a skeleton crew and on 29 April got the ship under way from San Diego and proceeded to a Long Beach shipyard for her conversion.

Because of her original role as a cargo ship, the *Point Barrow* had been constructed to commercial standards, not to Navy specifications. In addition to the modifications to support bathyscaph operations, the Navy required a multitude of changes to her commercial-level systems. As originally built, there was no redundancy in her steam propulsion system, nor was it possible to isolate the port from the starboard main steam piping. In the event of a casualty to the primary steam loop the ship would have to go "cold iron" until repairs could be effected. Also, the ship burned a high-viscosity black fuel oil known as "Bunker C"; conversion to the new Navy standard fuel meant steam-cleaning all the fuel tanks, fuel lines, valves, and fittings to remove the sludge of Bunker C.

To dock and recover the bathyscaph in the open sea the ship's docking well was fitted with a wave-energy break (like the "beach" on the *White Sands*) and a stage on which the *Trieste* would rest, so that wave energy would dissipate around and under the bathyscaph and not shove it around the docking well. Probably the most expensive and time-consuming item was making the ship able to store and handle aviation gasoline. In addition to dedicated tanks for 100,000-plus gallons of avgas, the yard installed pumps and monitors, systems to maintain a nitrogen or carbon dioxide "blanket" in partially filled avgas tanks, and special firefighting equipment.

The *Point Barrow* was designed to ballast the docking well's sill to a depth of ten feet. Docking the *Trieste* required 14 feet, thus 1,492 tons of lead was installed in the wing-wall ballast tanks to push the stern lower.

Less complex but time consuming were the habitability modifications: berthing and messing facilities to accommodate a Navy crew of more than 260 plus the *Trieste* crew of about 25 men. (The earlier T-AKD civilian crews had numbered from 45 to 65 men.) Also installed were maintenance shops for the *Trieste*'s mechanical, electrical, and electronic systems.

The *Point Barrow* conversion was first undertaken by the Fellows and Stewart Shipyard of San Pedro. However, in September 1975, the Navy was obliged to terminate the contract because the yard had run its coffers dry and was bankrupt. The Long Beach Naval Shipyard had to arrange an expeditious removal of the *Point Barrow* from the yard to avoid the ship becoming embroiled in legal entanglements that would further delay the conversion. The California Shipbuilding and Drydock Company received the contract to complete the work while the ship was berthed at the Long Beach shipyard.

On 26 February 1976, the ship emerged from the yard and on 3 July 1976, was placed in commission as the USS *Point Loma* (AGDS 2)—the first AGDS having been the *White Sands.*[3] During the AGDS 2 conversion, on 23 January 1975, Lieutenant Commander James T. Worthington II, a former *Trieste* officer, relieved Lieutenant Commander Jimmie Gray, formerly the executive officer of the *White Sands*, as the "new" ship's executive officer. This established a precedent that at least one of the two senior *Point Loma* officers would have previous experience in deep submergence.

Lieutenant Commander Abbott relieved Worthington as the ship's executive officer on 11 March 1977.[4] Commander Norman H. (Bud) Branchflower Jr. would relieve Whitmire shortly after the *Trieste*'s first operational dive from the *Point Loma* in April 1977.

The *Point Loma*'s workup between completion of her conversion 26 February 1976, and her first operational dive with the rehabilitated *Trieste* on 19 February 1977 involved shipyard tests, sea trials, and extensive tests with a "pig" simulating the *Trieste*. Training for handling avgas was especially taxing, as it involved securing all spark-producing equipment while exercising firefighting crews in how to respond to an avgas spill or fire.

Awaiting the *Point Loma*

Meanwhile, the *Trieste* herself was in overhaul from December 1973 to March 1975, formally a major teardown, inspection, rebuild, renewal, upgrade, and reassembly. Deep submergence vehicles had been in existence long enough for the Navy bureaucracy in Washington to establish standards, requirements, procedures, and documentation for their maintenance of every type. After five years of operations the *Trieste* required an overhaul at Mare Island that would take almost 15 months. Abbott departed early in the overhaul, but Lieutenant Commander Whitaker, Lieutenants Clarence L. (Les) Parsons, and Raleigh Baker suffered through the entire exhausting process.

In addition to a complete teardown, inspection, and rebuild of the float, the Hahn & Clay sphere was removed and a new sphere machined to replace it. This Mare Island Naval Shipyard–produced sphere was a three-port affair called the "MINSY sphere."[5] (The MINSY sphere was the twin to the Hahn & Clay sphere, spin-formed by Lukens Steel in 1965. Originally intended for a new Navy bathyscaph, it had been stored in an unmachined condition at Mare Islands for eight years. The proposed bathyscaph never appeared, apparently a victim of budget cuts.)[6]

With the MINSY sphere came a major revamping of electronic, communications, navigation, and lighting systems. When it emerged from overhaul, the *Trieste* had 962 amp-hours of 120-volt main-propulsion battery power, plus 5,000 amp-hours of 24-volt power for lighting, sonar, communications, and external equipment. This was in addition to the emergency 12-volt and 24-volt batteries in the sphere that powered the shot valves and the electromagnets holding jettisonable materials in place. External lighting was upgraded to 12 lamps with a combined power of 4,600 watts, with four of the lights steerable. The monochrome TV system had three cameras on pan-and-tilt mounts, feeding to three monitors and a video recorder inside the sphere. One of the two forward TV cameras had a zoom lens and a highly sensitive Vidicon tube. Also, three 70-mm film cameras were mounted externally, each with an individual strobe and film for 400 exposures. The manipulating arm was fitted to grasp, cut, or lift items from the ocean floor.[7]

Following the overhaul, the still strangely named *Trieste II* (DSV 1)—it was actually, of course, the third bathyscaph to carry the name *Trieste*—was towed to the Alameda naval air station at Oakland for tests, trials, and certification dives. Baker piloted three 50-foot tethered dips pierside from 25 April to 14 May, with a mix of shipyard inspectors and *Trieste* crewmen on board.[8] After these preliminary tests and a "FAST cruise"—a simulated dive while "made fast" to the pier—the *Trieste* was towed down to San Diego. Shortly thereafter, Lieutenant Les Parsons transferred to the diesel-electric submarine *Salmon* (SS 573).[9] Two years later, as his tour on board the *Salmon* was nearing an end, he submitted his "dream sheet" for future assignments—asking to return to the *Trieste*.

On 16 August 1975, Whitaker arranged for the civilian-manned Navy tug *Ute* (T-ATF 76) to tow the *Trieste* to a position off San Diego for a

4,000-foot dive, which was to be Baker's last dive (prior to his return to the program in 1983 as officer-in-charge) and Master Chief Electrician George G. Ellis' first deep dive. (Later, as a civilian, Ellis would join the *Alvin* program at Woods Hole.)

The next deep dive, on 4 February 1976, was the post-overhaul acceptance dive to certify the bathyscaph to 10,000 feet. Towed into position and supported there by the *Florikan* (ASR 9), the *Trieste* was taken down by Whitaker to 10,000 feet with Chief Ellis and Don Johnson, the Naval Sea Systems Command's certification authority; this was Johnson's second bathyscaph dive. Upon surfacing the *Trieste* again was an operational deep submergence vehicle.

Back to Work

The first operational employment for the bathyscaph after this overhaul was a series of five dives from 20 February to 26 May 1976, locating and salvaging parts of a lost SH-3 Sea King helicopter off San Diego. Two of the dives produced pilot certifications. On 30 March, Whitaker took Chief Ellis and Lieutenant Commander Richard D. Waer on a dive. Waer was a former *Trieste* engineering officer and pilot, thereafter on the Development Group staff, later the officer-in-charge of the *Turtle* (DSV 3), and still later the commanding officer of the *Point Loma*. Waer was the examining officer for both Whitaker's certification as a deep submergence pilot and Ellis' certification as enlisted hydronaut #16.

On 26 May, Lieutenant Commander James K. Newell piloted, with Whitaker and Commander Ronald Doyle on board. This was Newell's third deep dive since February, and Doyle was certifying him as a deep submergence pilot. Upon surfacing, Newell relieved Whitaker as officer-in-charge of the *Trieste*. With bachelor's and master's degrees in electrical engineering from Auburn University, Newell had served on the diesel-electric submarines *Trout* (SS 566) and *Trigger* (SS 564), as well as on the research submarine *Albacore* (AGSS 569). Newell would be officer-in-charge of the *Trieste* until 1978.

The crash of an F-14 Tomcat fighter aircraft off the coast of Ensenada, Mexico, attracted the attention of the *Trieste* for two dives in August 1976. The Development Group's deep-submersible support ship *Maxine D* provided towing and security services in the operating area. Newell piloted both the 22 and the 28 August dives, with Aviation Machinist's Mate Chief John R. Turdevich and the *Trieste*'s George Ellis. Turdevich was on the dives to assess the debris visually. The F-14 wreckage lay at almost 4,000 feet, and each of the dives took in excess of ten hours. On 11 January 1977 the *Maxine D* returned to the F-14 site with the *Trieste* in tow for Dive #1-77. Lieutenant Kenneth W. Hanson, now the assistant officer-in-charge, piloted, with Newell and Engineman 1st Class H. W. (Pete) Smith with him, going down to recover the transponders marking the debris field.

Newell would use the *Maxine D* or the *Florikan* for six additional dives on the 1975 SH-3 site off San Diego: three dives returned to the crash site to recover the main rotor and tail rotor; the other three were primarily for *Trieste* pilot training, between 18 November 1976 and 27 January 1977. On the 27th dive Hanson piloted with Newell and Captain Charles R. (Chuck) Larson, now Commander, Submarine Development Group 1, on board. Larson and Newell were examining officers to certify Hanson as a deep submergence pilot, while Larson received firsthand experience in bathyscaph diving.[10]

All missions during this period were to less than 6,500 feet and could have been more economically handled by the *Turtle* or *Sea Cliff*, but these dives were needed to train *Trieste* pilots. These dives were the last without a specialized mother ship.[11] Operating without a mother ship was backbreaking, wearing work for the *Trieste* crew.

Working with the New Team

Both Whitmire and Worthington remained on board the *Point Loma* long enough to conduct the first operational mother ship–supported dive of the *Trieste*: Dive #4-77 on 19 February 1977, to 4,000 feet, lasting ten hours, 15 minutes to the west of Santa Catalina Island. A newcomer, Lieutenant Thomas G. Vetter, piloted his first bathyscaph dive, with Hanson as instructor and Ellis as co-pilot. On this dive the *Trieste* discovered the first-ever abyssal "whale fall," the descent of a dead but intact whale to the deep ocean floor. Ellis used the manipulating arm to recover a section of the whale's jawbone and other bone segments.[12] Worthington departed on 11 March, and Whitmire was relieved a few weeks later.

On 20 April the *Point Loma*'s new commanding officer, Commander Bud Branchflower, and new executive officer, Lieutenant Commander Dick Abbott, took the ship to a point 800 nautical miles northwest of San Diego for the *Trieste* to dive to 17,246 feet for the official 18,000-foot certification. Newell piloted, with Hanson and Ellis along, spending nine hours, 46 minutes submerged. With that certification and the demonstration by the *Point Loma* of her ability to launch and recover the bathyscaph in mid-ocean, the *Trieste* was ready for the Sea Floor Geophysical Research Program, sponsored by the Office of Naval Research. Of the officers assigned to the *Trieste,* only Newell and Hanson were certified officer pilots; Master Chief Ellis was a certified enlisted pilot.[13]

On 31 May 1977, the *Point Loma,* with the *Trieste* tucked in, left San Diego for the Caribbean. She arrived at Rodman in the Panama Canal Zone on 13 June for two days of crew liberty and the embarkation of National Geographic Society and Woods Hole teams. The *Point Loma* transited the canal on 15 June.[14]

The *Trieste* was now participating in an ONR/Woods Hole program that had two primary purposes. The first was to use the *Alvin*, *Trieste*, and the towed-camera system ANGUS (Acoustically Navigated Geological Underwater Survey) to investigate a small tectonic "spreading center" situated in the Cayman Trough, between the North American Plate and the Caribbean Plate.[15] This was one of deepest spreading centers in the world, its central volcanic axis occurring at a depth of 20,000 feet.[16] The second reason for the Cayman Trough program was to maintain the momentum created by the success of the French-American Mid-Ocean Undersea Study—that is, Project Famous.

Dr. Robert D. Ballard was Woods Hole's chief investigator for this portion of the program, and his team included William M. (Skip) Marquet, who would assist in a later *Thresher*-site re-visit. Ballard had extensive experience with bathyscaph diving, having spent three years (1972–1974) training with the French and diving in their bathyscaph *Archimède*.[17] The National Geographic team was led by Gilbert M. Grosvenor and included noted photographer Emory K. Kristoff. Kristoff's photography was to record vividly the Cayman Trough dives by the *Trieste*.

The original plan was for a four-dive program. Tom Vetter explained the *Trieste*'s role in this way:

> I would design dive plans that would enumerate what each dive was to accomplish, the right suite of cameras, tools, sample baskets, and other

> items necessary to accomplish the mission. I would then vet it with the guest team, with the *Trieste* assistant officer-in-charge [Ken Hanson] and with the *Point Loma* executive officer [Dick Abbott] to work out any issues. At that point, it became the official dive plan.
>
> In Ballard's case, the dive plans called for diving to the top of newborn undersea volcanoes to collect lava samples—the freshest deep lava on the planet. National Geographic was there for another reason, they wanted to document the first-of-a-kind exploration: of oceanographers working at 20,000 feet depth.[18]

However, before the first Cayman Trough dive Branchflower required hospitalization for acute intestinal distress and was taken by small boat to Georgetown, Grand Cayman Island; he then was flown to the U.S. naval base at Guantanamo, Cuba, for treatment. Abbott took command until a temporary replacement could arrive. That replacement was the chief staff officer of Submarine Squadron 4, in Charleston, South Carolina—former *Point Loma* skipper Commander Wilson Whitmire. He remained on board for the Cayman Trough dives, leaving the ship in Guantanamo when Branchflower returned after a brief hospitalization.

The *Trieste* conducted three dives into the Cayman Trough, south and west of Cuba, with Ballard in the sphere for all three. During 17–18 June the *Point Loma* re-embarked the bathyscaph, dropped two deep-ocean transponders to mark the planned dive site, and then retired to protected waters off Georgetown for repairs to the *Trieste*'s batteries.[19] Upon completion the bathyscaph was launched, fueled, and shotted, and towed back to the dive site.

On 23 June, on Dive #6-77, Ballard went to 20,236 feet for ten hours submerged with Newell piloting and Ellis assisting. They bottomed on pillow lava, which had glass-like edges that cut the rubber on the shock absorber of one of the skegs, requiring its later replacement. Newell and Ellis on this dive were able to collect lava samples for Ballard, including a softball-sized piece of lava, "still crackling," from near the tectonic plate boundary. This was the deepest object recovered from the ocean floor to that date.[20] Back on the surface the *Point Loma* towed the *Trieste* back to protected waters off Georgetown.

On 26 June the *Trieste* was brought alongside the *Point Loma*'s port quarter to replace a failing battery pack. The *Point Loma* was riding high, and wave action forced the *Trieste* close to its hull. As the ship's stern sank in a swell, a protruding pipe flange punched into the starboard side of the float's center tank causing a serious avgas leak. Electronic Technician 2nd Class Daniel DeVoe was bathed in avgas when he went over the side to drive a wooden plug into the hole with a bronze hammer. (All *Trieste* hand tools were bronze, to preclude sparks.) Still, the hole necessitated de-gassing, de-ballasting, and re-docking the bathyscaph for repairs by a submarine-qualified welder flown down from the Charleston Naval Shipyard in South Carolina. Meanwhile, the crew enjoyed three days of liberty in Georgetown.[21] There was no anchorage near Georgetown, because the shelf dropped off abruptly from the island, and so instead of anchoring and the *Point Loma* continuously steamed at slow speeds in the lee of the island.

News of the death by heart attack of the noted oceanographer Dr. Bruce C. Heezen on 21 June 1977 while on board the nuclear-propelled deep submersible *NR-1* caused sorrow on the *Point Loma*. Heezen had been programmed for several upcoming *Trieste* dives in October. Ballard recalled: "Bruce Heezen was a good friend of mine and scheduled to use the *Trieste* after me in the Puerto Rican Trench. When he died in the *NR-1,* I had the crew on the *Point Loma* fashion a bronze plaque commemorating his great contribution to marine geology, which we took to the bottom of the Cayman Trough and placed on a fresh lava flow."[22] Heezen's planned dives were passed to Ray Freeman-Lynde and Mike Rawson of the Lamont-Doherty Geological Observatory at Columbia University.

On 2 July, with repairs completed, the *Trieste* was re-launched, fueled, shotted, and towed back to the site. Dive #7-77 the next day took Ballard to 19,310 feet for 16 hours, 35 minutes submerged, with Hanson piloting and Ellis assisting. The third dive, on 14 July with the same team, was aborted when the bathyscaph's forward pan-and-tilt camera struck a cliff face as it reached the bottom. Total time submerged was three hours, 16 minutes—the vertical transit time to descend to 15,750 feet and ascend back to the surface.

The *Trieste*'s fathometer could not be expected to detect a cliff-face, and Newell had been unable to see the cliff in time to reduce his speed of descent, leading to destruction of the camera at the bathyscaph's "chin." Newell emptied the ballast tubs for an immediate surfacing but dropped no emergency weights. The *Trieste* was designed to return safely to the surface with any two of the 12 buoyancy tanks empty of avgas, but there was no compromise to any of the buoyancy tanks during this dive.

This event ended the *Trieste*'s participation in the Cayman Trough program. After re-docking the *Trieste*, the *Point Loma* proceeded to Georgetown to offload the Woods Hole and National Geographic teams. A one-day transit took the ship to Guantanamo Bay for minor upkeep work, three days of on-base liberty for the crews, and a chance to meet the returning Branchflower after his hospitalization.

The *Point Loma* left Guantanamo on 20 July and arrived in Charleston on 23 July to embark a Naval Ocean Systems Center team led by chief scientist Sherman L. Williams. Also in Charleston, the crews took on shot for the bathyscaph, fuel, and stores before departing on 26 July for secret operations.[23]

21 An Epiphany

It was like a light bulb snapping on inside my head[,] a real epiphany. We had much wrong during those years, attempting to provide the Trieste *with navigational autonomy.*

—John VanVoorhis, Sperry Technical Representative

The Naval Reactors directorate—headed by Admiral H. G. Rickover—was the moving force behind most of the *Trieste*'s dives in 1977 and 1979. His purpose was to respond to the Marine Protection, Research and Sanctuaries Act of 1972, which caused major re-evaluations of radioactive dumping sites off both the U.S. East and West Coasts.

In 1976, Glen L. Sjoblom at Naval Reactors contacted the OpNav Deep Submergence Systems Division in the Pentagon to request assistance in locating a sunken barge containing the reactor from the *Seawolf* (SSN 575), the second U.S. nuclear-propelled submarine. The *Seawolf*'s original liquid-metal-cooled reactor had been replaced by a standard, pressurized-water reactor after her initial operations. A towed barge had carried the original *Seawolf* reactor off the Atlantic coast to a point some 120 nautical miles due east of the Delaware–Maryland line, where it was scuttled in 9,600 feet of water. Sjoblom later expanded his request, adding investigations of both the *Thresher* and the *Scorpion* wreckages to determine the levels of radioactivity at those sites.

Sjoblom arranged for Dr. Sherman Williams, of the Knolls Atomic Power Laboratory in Ballston Spa, New York, to participate in the *Trieste* dives, and the Naval Ocean Systems Center in San Diego assigned Jon R. Losee to install and monitor special radiological instrumentation on the *Trieste*.[1]

The *Seawolf* 's reactor had been dumped in April 1959. Because the towing and scuttling of the barge had been undertaken by a commercial firm, and in view of the limits of the navigational accuracy of that era, there was some uncertainty as to the location of the scuttling, Sjoblom contacted Woods Hole to help in the search and it provided a towed, side-scanning sonar system to look for the barge.[2]

At the request of the Environmental Protection Agency, the *Alvin* made several dives on nearby low-level radioactive-waste dump sites in 1974–1976.[3] However, the *Alvin*, certified only to 6,500 feet, could not search the 9,000-foot *Seawolf* location. The Navy arranged for the *Trieste* to dive on the *Thresher* wreckage off the New England coast and on the nuclear waste site, both during the month of August 1977, the best-weather month for Atlantic operations.

Losee reached the *Trieste* in Charleston on 26 July, as did Dr. Williams, to be the chief scientist for the operation. Losee's instruments were to be lowered by the *Trieste*'s bow winch directly into the sediment or onto a section of the wreckage to capture samples and measure radioactivity.

On 4 August the *Point Loma* reached the point where the *Thresher* had been lost and began to survey the area and mark the dive site. *Trieste* Dive #9-77 on the *Thresher* site began in the evening of 7 August, and 18 hours, 37 minutes later the bathyscaph surfaced, during the afternoon of the 8th. Lieutenant Commander James Newell piloted with Lieutenant Kenneth Hanson as co-pilot and Losee working the sensors. The wreckage of the nuclear submarine could not be located. The Development Group staff directed an expanding-square search, but the plan concerned the *Trieste*'s Lieutenant Commander Thomas Vetter: "I did not think that the *Trieste*'s limited battery power would provide a high probability of success using a blind search method. Therefore, as the *Trieste* futilely plowed an expanding square in the barren mud, I took notice when they reported finding a stoetzeroonie and read off the grid code. I knew where the wreck was in the stoetzeroonie grid."[4]

On 12–13 August, Dive #10-77, with Hanson piloting and Master Chief Electrician George Ellis assisting, once again took Losee to the *Thresher* wreckage on a 20-hour, 18-minute dive. The work went expeditiously. The new videotape recordings of the *Thresher* remains were amazingly clear and distinct.[5]

Losee observed that the site "looked like a junkyard." One hull section was relatively intact; there they took readings and found a low level of Cobalt-60 from reactor piping, but no isotopes from the reactor core. The Cobalt-60 signal was close to ocean background radiation levels.[6]

Vetter perceived a problem that would require the *Trieste* to return to the *Thresher* site: "Nowhere in the debris field could we find the stern section. The *Mizar* and the *Trieste* had seen it in 1963–1964, but we had no navigational clues to find it again."[7] The *Thresher* dives had found no significant radiological contamination at the wreckage that was visited, but they had failed to locate the section containing the reactor. This meant, as Vetter had foreseen, there would have to be another visit to the *Thresher* in the future.

On 13 August the *Point Loma*, with the *Trieste* embarked, transited south to a point 120 miles due east of the Delaware–Maryland state line. The previous towed-sensor search had pinpointed several barge-sized targets to be investigated by the *Trieste*. Dive #11-77, on 19–20 August, was made by Newell, piloting, Chief Electrician's Mate Larry J. Porter, assisting, and Dr. Williams, helping identify the barge among those objects. The 14-hour, 29-minute (submerged) search was unable to locate any radioactive waste; the several large sonar contacts proved to be "rafted boulders," large rocks that had been carried to sea by icebergs and deposited at random onto the ocean floor. Williams realized that further dives by the *Trieste* would be fruitless and terminated the operation. In 1980 the Navy publicly declared that it did "not plan to retrieve the *Seawolf* reactor vessel."[8]

Lieutenant Commander Dick Abbott, executive officer of the *Point Loma,* later recalled a major upgrade to the *Trieste*'s navigational suite following these *Thresher* dives: "Originally, locating the *Trieste* on the bottom was by a navigational system on the bathyscaph [i.e., the NAVNET with the Mark 15 computer]. It never worked. It always gave out a lot of unreadable garbage. Sperry's techrep John Van Voorhis designed a computer system, with the *Point Loma*'s skipper Bud Branchflower, to automatically plot the positions of both the *Trieste* and the *Point Loma* in real time. It could pinpoint the *Trieste* on the bottom within a few feet. It was great!"[9]

The new navigation system was beyond "great"—it was revolutionary. The combination of an accurately surveyed transponder field, electronic positioning on the surface, precise location of the *Trieste* within the transponder grid, and accurate plotting enabled the plotting team to *know* where the *Trieste* was, to map her movements and geolocate the targets she found in latitude/longitude coordinates. Vetter explained, "From the beginning, the bathyscaph's navigation had relied on data only the *Trieste* could see—provided the NAVNET worked, which it never did. Often, we were just wandering in the dark, like a drunk at night looking for his car keys on the lawn with a flashlight."[10]

VanVoorhis and Vetter began experimenting with surface-based navigation in late 1976, tracking the *Trieste* within the transponder field and plotting its movements by hand.[11] They learned to analyze the reported clues and to vector the *Trieste* toward the target. The learning curve was slow and steep. They came to realize that navigation was the key to success in this mission.[12]

Transitioning from evolutionary growth and training in plotting to real-time situational awareness required automation. It required that the precisely timed acoustic replies from transponders and from the *Trieste* be converted into X-Y coordinates and plotted in the control van on the *Point Loma* at set intervals, as often as every minute.

VanVoorhis eventually selected on Westinghouse ATNAV deep-ocean transponders that were buoyed and tethered to float 300 to 400 feet above the ocean floor. However, they were only part of the solution.[13] The first rough system soon was in place and the team in the *Point Loma* control van was attempting to get useable plotting from it during later *Thresher* dives.[14]

The first iteration did not achieve a perfect application, but VanVoorhis, Branchflower, and Vetter adjusted, tested, and revised the system until it could be tested with a real dive. The system provided a new capability: the *Trieste* could be vectored toward a target not only while at depth but also during its descent. This reduced bottom search time and allowed more time at the target.[15]

After the 1977 deployment the Navy continued to employ Sperry techrep VanVoorhis in deep submergence navigation and refined the new tracking methodology, which eventually became standard procedure for all of the Navy's deep submergence vehicles. The new methodology was employed by a special deep submersible support team of ship riders from Submarine Development Group 1 beginning in the 1990s.[16] (VanVoorhis, as a Sperry engineer/techrep, had worked on Navy surveying ships as well as on the nuclear-propelled submarine *Seadragon* [SSN 584] during her voyage to the North Pole in 1962; subsequently he was the lead Sperry techrep on the *Trieste* program and for the modification of the *Seawolf* to a special projects submarine.[17])

The Blake-Bahamas Plateau

The *Point Loma* and *Trieste* were in Charleston for upkeep and refit from 24 August though 15 September. Dr. Williams, Jon Losee, and the Ocean Center team left at that point, and a new group of scientists boarded. These were Roger D. Flood, Georges L. Weatherly, and Mark Wimbush, all from Woods Hole. In Charleston, the prospective officer-in-charge, Lieutenant Commander Les Parsons, rejoined the *Trieste* program.

On 16 September the *Point Loma* with the *Trieste* embarked set off for the Blake-Bahamas Plateau, a relatively flat area of the continental shelf east of Jacksonville, Florida, and extending southeastward through the Antilles. The *Trieste* made five dives down to 16,000 feet from 20 September to 2 October, with a key participant Dr. Flood, a marine biologist. The Flood dives were coordinated with experiments of Weatherly and Wimbush that involved dropping two instrument assemblies from the *Point Loma*. The first instrument was the "profiler," which measured the speed of deep-ocean currents at ten different altitudes above the bottom, from one inch to 100 feet; this instrument also precisely measured water temperature at four depths.

The profiler was launched from the *Point Loma* near midnight 18/19 September, and *Trieste* Dive #12-77 located and photographed it on the bottom. The profiler's anchor was released by acoustic signal on 29 September and the buoyant device was recovered by the *Point Loma*.

A second instrument, called the "tripod," would take measurements of water current speed and direction close to the bottom and shoot time-sequence photographs of bottom sand furrows. The tripod was launched from the *Point Loma* 90 minutes before the profiler. It was located by the *Trieste,* Dive #13-77, just 900 yards east of the profiler. On the instructions of Flood, Hanson "bulldozed" the tripod into a position to photograph automatically a mud-furrow rim. Later, when its recovery signal was sent, the tripod failed to release its anchor. On 1 October, Dive #16-77, the *Trieste,* with Newell piloting, relocated the tripod and recovered the device, surfacing with it in the grasp of the mechanical arm.[18]

Chief Ellis piloted Dives #14-77 and #15-77. The former was aborted early—after four hours, 45 minutes submerged—because of electrical problems. The next dive, #15–77, was the first and only *Trieste* dive without an officer on board: Ellis piloted and Porter assisted. Once at depth and "skiing" along the bottom, the bathyscaph bounced on a ridge and dragged its stern section down through the mud. Ellis did not recognize this as a problem until he tried to drop shot to ascend at the end of the dive.

The bathyscaph began to rise, but a marked "up bubble" indicated that though the bow was rising the stern was heavy. Ellis realized that the after tub's shot release valve was clogged. The two discussed the situation by underwater telephone with the *Point Loma.* The controller on the surface authorized Ellis to drop the after shot container if necessary. Flood felt certain that doing so would mean cancellation of the remaining dives in his series.[19]

They surfaced without further problems. Divers were able to clear the after shot release valve, and the next dive was unaffected. When the Development Group discovered that there had been an "all-enlisted" dive, it mandated that future *Trieste* dives have at least one officer on board.

Dive #16-77, with Newell, Hanson, and Flood, over the night of 1/2 October, was the last of the series. Flood summarized his experience of the *Trieste* as a scientific work vehicle: "The mechanical arm didn't work on some dives, and there was a problem with an external camera, but enough things worked enough of the time that we all thought the dive series was very successful and got us what we wanted."

Subsequently the *Point Loma*, carrying the *Trieste*, steamed to Port Everglades, Florida, for a three-day liberty call on 6–9 October and for an exchange of scientists.

The Puerto Rico Trench

Three dives into the Puerto Rico Trench were scheduled for Raymond P. Freeman-Lynde and one for Michael Rawson of the Lamont-Doherty Geological Observatory at Columbia University. On 15–16 October Newell and

Parsons took Freeman-Lynde into the Mona Canyon, down to 19,924 feet, spending 16 hours, 41 minutes submerged.

With relatively rapid turnaround times, on 20–21 October, Hanson and Ellis carried Rawson to 16,597 feet to study the northern face of the Navidad Bank.[20] On 23–24 October, Newell and Chief Porter took Freeman-Lynde to 16,434 feet to study the eastern face of the same bank.[21] This was Porter's qualification dive.[22] Finally, on 28–29 October, Hanson and Ellis dived with Freeman-Lynde to 20,000 feet to study the southern face of the Navidad Bank.[23] Freeman-Lynde found the *Trieste* unable to chip away rocks from the cliff faces at specified levels as he had wanted, because the manipulator was under the bathyscaph's "chin," positioned specifically for bottom recoveries.

Homeward Bound

Upon completion of the Navidad Bank dives, the *Point Loma* re-docked the *Trieste* and proceeded to St. Thomas in the U.S. Virgin Islands. There the scientists offloaded their samples and equipment and departed, while the Navy crew enjoyed high-quality liberty. Newell and Hanson flew back to San Diego; Parsons and Vetter remained on the *Point Loma* for the voyage home. The *Point Loma* passed through the Panama Canal westbound on 6–7 November, turned north, and ran into the worst weather that most of the men on board had ever experienced.

Called the Tehuantepecer, or Tehuano, these gale-force winds of the tropical eastern Pacific had belabored the *Trieste* once before, back in the *Apache* and *White Sands* days.[24] It normally blows from the north or northeast at 20 to 45 knots, with occasional gusts to 100 knots. The *Point Loma* encountered the Tehuano south of Acapulco, winds in excess of 80 knots on her starboard bow. She could not maintain her course; she was blown to seaward and on occasion rolled more than 35 degrees. Because of the *Point Loma*'s high sail area forward, there was concern that she might capsize. Commander Branchflower ordered partial ballasting down to reduce the rolling. Unable to maintain his heading with the rudder alone, Branchflower maneuvered with engines as well, holding the rudder almost hard right. The *Point Loma* clawed her way northward until she cleared the Tehuano system.

It was a storm that few who were on board would ever forget.

Having completed 15 dives at five different dive sites in the Atlantic, the *Trieste* reached San Diego on 17 November 1977. The Atlantic dives demonstrated operations of just over 20,000 feet after the bathyscaph had qualified to 18,000 feet off San Francisco. Further, she had carried out the deepest manned recoveries from the sea floor: volcanic samples for Ballard from 20,236 feet in the Cayman Trough, and cliff rock samples for Freeman-Lynde at 20,242 feet off the Navidad Bank. The *Trieste* and the *Point Loma* were awarded the Meritorious Unit Commendation for the Atlantic deployment.

22 Maturity

I still have dreams of being down there—not just occasionally but frequently. I felt a presence. We knew we were entering sacred ground, and in some way we were being guided and helped on this dive.

—Lieutenant Commander C. Lester Parsons,
Officer-in-Charge, *Trieste*

In the three months of the *Trieste*'s stand-down from mid-November 1977, to mid-February 1978, Lieutenant Commander Kirk Newell and Lieutenant Ken Hanson were relieved as officer-in-charge and assistant officer-in-charge by Lieutenant Commander Les Parsons and Lieutenant Tom Vetter, respectively. Master Chief Petty Officer George Ellis also departed, leaving the program with only Parsons as a certified *Trieste* pilot.

Parsons had served on the submarines *Menhaden* (SS 377) and *Harder* (SS 568) before his first assignment to the *Trieste* program in 1973–1975, during which the bathyscaph was mostly in the shipyard; he made exactly one dive during that first tour. From the *Trieste* Parsons went to the diesel-electric submarine *Salmon* as operations officer and navigator, returning to the *Trieste* program as prospective officer-in-charge in 1977.

The first *Trieste* dive under this "new management" was on 26–27 February 1978: west of San Diego, Parsons piloted to 11,945, with Chief Petty Officer Larry Porter and Lieutenant Thomas Vetter assisting. Porter had been qualified as a pilot during the 1977 Atlantic operations but was not yet certified. In March 1978, Parsons was again piloting, with Porter assisting and newcomer Lieutenant William J. Beres on his first dive; the *Trieste* went to 13,297 feet to retrieve a deep-ocean transponder off San Clemente.

There was then a four-month interruption of *Trieste* dives because of budgetary constraints. On 31 July, Porter took his certification dive, with Parsons examining and Vetter along for instruction. Porter then was detached to report to his next assignment, having been certified as enlisted hydronaut #20.

Back to Work

The Navy had emplaced three deep-ocean transponders at depths of over 16,000 feet near Midway Island in 1970. They were powered by radioisotope generators, each with more than a pound of Stronium-90 to provide electrical power for five years of operation. The system generated timed acoustic pulses for the precise positioning of hydrophones in an underwater array, part of the missile-impact location system to determine U.S. missile accuracy.[1] Now the transponders were to be retrieved by the *Trieste*.

The *Point Loma* got under way from San Diego with the *Trieste* on board in late July 1978, and arrived in Pearl Harbor ten days later to take on board a team from the Naval Ocean Systems Center. On 15 August, in Pearl Harbor, Commander Richard Waer relieved Commander Bud Branchflower as commanding officer of the *Point Loma*. Waer had been a *Trieste* pilot in 1965–1966 and subsequently the officer-in-charge of the submersible *Turtle*.[2]

The *Point Loma* was delayed in Pearl Harbor for two weeks due to a Navy-wide diesel fuel shortage. The *Point Loma* finally was fueled and on her way from Pearl Harbor to Midway on 17 August; she arrived off Midway four days later.

The nuclear transponders had been precisely placed near Midway with their locations carefully plotted in reference to the Sound Surveillance System (SOSUS) array of acoustic sensors. Vetter recalled, "We dropped our DOTs [from the *Point Loma*] in a triangular pattern, and then navigated amid the DOTs while we dropped small explosive charges overboard to detonate and obtain concurrent positions by SOSUS. This allowed us to translate the SOSUS coordinates of the target to our transponder field coordinates."[3]

But before the first *Trieste* dive the weather and sea conditions forced the *Point Loma* to seek shelter at Midway, allowing the crew an unusual atoll liberty. Subsequently, Dive #4-78 submerged at 1000 on the 29th with Vetter piloting, Parsons in command, and Beres navigating. This was Vetter's certification dive; he would surface as hydronaut #69. The surface tracking party on the *Point Loma* was able to provide a vector to the target transponders immediately on the *Trieste*'s making the bottom, and within only 30 minutes they sighted the first target.

The target transponder was fully embedded in bottom mud, with only the cabling for the hydrophone and float protruding. Vetter made repeated attempts to attach the *Trieste*'s winch hook to the cables, but these were badly corroded and broke when Vetter tried to hoist the device from the mud. He then attempted to push and pull the generator with the manipulating arm, abandoning the attempt when hydraulic fluid oozed out of the arm. At that point he tried to bulldoze the generator free, again to no avail. Parsons decided to desist at 1845, and the *Trieste* broke the surface at 2029.

A number of alarms during the mission had indicated trouble with external electrical systems. On the surface it became clear that the electrical system had serious problems that would require help from a shipyard. The Midway operation was terminated on 30 August.[4]

The *Point Loma* with the *Trieste* on board returned to Pearl Harbor and then sailed on to San Diego. About halfway between Honolulu and San Diego the *Point Loma* experienced an evaporator casualty that caused contamination of the boiler feed water. The *Point Loma* was dead in the water for some ten hours while repairs were made. She ultimately arrived at San Diego on 29 September. With both the *Trieste* and *Point Loma* needing repairs, there was no diving activity between October 1978 and April 1979. (In December 1978, Beres relieved Vetter as assistant officer-in-charge; Vetter soon left the Navy.[5])

After this lengthy hiatus, April and May 1979 saw four *Trieste* dives in the San Diego area, primarily for pilot training. On Dive #1-79, on 25–26 April, Parsons took a new arrival Lieutenant James R. Denzien on his first deep submergence and *Trieste* crewman Chief Electronics Technician James Scales to 2,640 feet. (Scales later would qualify as enlisted hydronaut #23.) Four days later Beres piloted, with Parsons assisting, to take Captain Charles R. MacVean on a dive to 3,830 feet. This continued the tradition of new Development Group commanders gaining such exposure to deep submergence diving.

The next dive, on 12 May, was Beres' certification dive; he piloted with Parsons as his examination officer and Chief Warrant Officer Burton P. Tharp on his maiden bathyscaph dive. Like Porter and Vetter before him, Beres left the program soon after his certification dive.

Also, Lieutenant Commander Abbott was relieved as executive officer of the *Point Loma* by Lieutenant Commander Bruce Hartman in July 1979, shortly before the *Trieste* deployed to the Atlantic.[6]

Back into the Atlantic

In early July 1979, the *Point Loma*, with the *Trieste* embarked, again sailed for the Atlantic. After clearing the Panama Canal, the *Point Loma* made her first port call at Norfolk to take on board scientists from Woods Hole and technicians from the Naval Ocean Systems Center with their radioactivity monitoring equipment. This year, as in 1977, Jon Losee led the radioactivity monitoring team.

During the 1979 Atlantic deployment the *Trieste* would once again become involved with "black" operations, and all but one of seven dives that summer and fall would be classified either as to the identity of the target or the precise location of the dive.[7] Dive #5-79, on 1 August to 12,000 feet, was the only unclassified dive of the deployment. Lieutenant Denzien piloted with Chief Petty Officer James Scales assisting and Electrician's Mate 2nd Class Christopher Lampman the third man in the sphere. That dive was made to visit one of the two unmanned Woods Hole "deep ocean stations" south of Cape Cod, Massachusetts. The first had been established in 1971 at various depths between 5,700 and 6,200 feet on the continental slope, approximately 110 nautical miles almost due south of Cape Cod; the second, laid in 1975 was at 12,000 feet, 190 nautical miles southeast of the cape. Both stations monitored growth of marine life. The *Trieste* retrieved biological samples from that station.

The first classified dive of the season was Dive #6-79 on 6 August, with Tharp piloting, under the instruction of Lieutenant Commander Parsons with Losee monitoring radiation-measuring instruments. This dive achieved the belated radiation examination of the submarine *Thresher*'s stern section. The previous (1977) investigation had located most of the *Thresher* wreckage, but not the stern (see chapter 21). Losee explained,

> The aft end of the *Thresher* was at the upstream end of the debris field and was mostly buried into the sediment—the open end above the sediment was bare inside. The Cobalt-60 signal was at a hull plate break on the top of that hull section, just above the point at which the hull was buried. The fact that that hull section was buried deep into the sediment and was at the up-stream end of the debris field indicates that it was heavy. Couple that with the Cobalt-60 signal and one

> can be fairly certain that it contained the reactor, and thus was the aft end. The stern section was visited in 1977 but not positively identified until 1979.[8]

The *Point Loma,* with the *Trieste* snugly in her docking well, then proceeded to Halifax, Nova Scotia, for a two-day liberty call, reportedly the first international port call for a U.S. Navy ship with a mixed-gender crew, three female officers having been assigned to the mother ship. The port visit was extended for seven days because of the delayed arrival of an electrical cable for the *Trieste* that had been manufactured by Mare Island and air-freighted to Halifax.

And Back to the *Scorpion*

The *Point Loma* carrying the *Trieste* next sailed to a position 370 nautical miles southwest of Praia da Vitória in the Azores for a series of dives on the *Scorpion* wreckage. Dive #7-79 submerged on 27 August and surfaced 24 hours, 44 minutes later. This was not only the longest dive of the entire *Trieste* program, but the most productive—and most nearly flawless—dive of the bathyscaph's history. Tharp piloted, with Parsons in the co-pilot seat and Jon Losee as the third man in the sphere. Again, the purpose of those dives on the *Scorpion* wreckage was to monitor possible radiological contamination.[9]

Commander Clark D. Sachse of the Naval Sea Systems Command had been an advisor to Captain Harry Jackson's on-site technical advisory group during the 1969 dives on the *Scorpion* by the *Trieste*. He had arranged for Parsons and Tharp to visit Mare Island when the nuclear submarine *Snook* (SSN 592) was in the yard. They went through the *Snook* to familiarize themselves with what they would be seeing at the *Scorpion* site. Sachse also produced all the photos from the *Mizar* and *Trieste* investigations in 1968–1969, a model of the debris field, and a map of the wreck's sections.[10] As a result, Parsons felt, "We were very well briefed, well prepared, and knew exactly what we would be seeing. We had a check sheet of work to accomplish, where we had to visit, and what we had to do at each section of the hull."

Sasche and Parsons did not want to land in the middle of the debris field and unnecessarily disturb the site. They decided instead to touch down some distance away and move cautiously into the area. Chief Electronics Technician Charles Copeland, the surface controller in the tracking van on the *Point*

Loma, described dive preparations: "We vectored the *Point Loma* to the transponder drop locations as suggested by Sperry techrep John VanVoorhis and then directed the *Trieste* to the submergence point. I had an automatic plotter that positioned both the *Point Loma* and the *Trieste* in real time on a plot where we had drawn the locations of the *Scorpion*'s major hull sections, sail, and propeller shaft."

Special preparations for what they knew would be a very long dive included taking two lunches each, fruit, bottles of drinking water, *and* empty bottles for urination. Parsons had instituted a new method of submerging more quickly to reduce vertical transit time. He explained:

> A submariner is naturally paranoid about weight and buoyancy. We calculate every change of equipment, analyze water temperature, salinity, and developed precise measurements for water displacement and compensation. However, to speed the process of submerging, I decided to ignore compensation because we were wasting critical time to achieve an exact negative buoyancy to submerge. Instead, we calculated compensation but did not drop shot if we were anywhere near what we expected. Instead, when we flooded our ballast tanks we were heavy, and we submerged immediately. Once we were free of the surface effect, we would start dropping shot and by 1,000 feet we would be under control and descending at about 150 feet per minute. From submergence to bottoming at 11,200 feet for the *Scorpion* ops was about 85 minutes.

Tharp also had developed a method to eliminate dead time waiting for the cloud of silt to disperse that engulfed the bathyscaph when it landed on the ocean floor. The new procedure involved getting a vector to the target from the tracking van when about 100 feet above the bottom. As the silt cloud boiled up, Tharp would turn on the motors and drive away on that heading. The *Trieste* would touch down a couple hundred yards away in a silt-free area, saving as much as an hour of bottom time.

On this occasion, having received the vector from the tracking van, Tharp headed in the direction indicated. Suddenly, what looked like a telephone pole appeared in front of the bathyscaph, its top in line with the viewport. Tharp readily admitted, "There are times when panic is the correct response to a situation and this was one of them. I put all the motors in reverse and watched the *Trieste* come to a stop a few feet from the 'telephone pole,' later identified as one of the *Scorpion*'s [retractable] masts that had broken loose and fallen out."

After several seconds in reverse, Tharp was satisfied that he had backed the *Trieste* sufficiently clear. After explaining the situation to Parsons and Losee, he turned back to the viewport "just in time to see the top of the pole slowly drift up to the viewport, not more than inches away." In backing the *Trieste* away the trail ball had not moved as much, and the *Trieste* was dragged forward by the trail ball, again into a hazardous proximity. Tharp backed away a second time, then proceeded on an oblique course to the bow section.

There Losee commenced his investigation. The *Scorpion*'s outer torpedo tube doors were partially open, and Losee made gamma measurements to determine if any radiation had leaked from the forward compartment, which contained two Mark 45 nuclear torpedoes; no radiation above background was detected. The *Trieste* then moved to the broken after end of the bow section. Losee used gamma sensors attached to the bow winch, lowering them onto the sediment for measurements. Again, only background radioactivity was detected.

The submarine's stern section was next. The tracking van supplied a vector, but Parsons did not know the distance. After about five minutes Parsons ordered, "Stop!" The stern section of the *Scorpion*, lying in a depression, was directly in front of the *Trieste.*

Losee took gamma measurements at several locations as far into the crater as he considered safe. Cobalt-60 was detected at all of them, indicating that the reactor compartment was present. No reactor fuel readings were detected, meaning that the reactor remained intact. They also took about 500 photos of the stern section, some looking into the open end, and obtained some videotape of a telescoped portion of the hull.

Then Losee "parked" the bathyscaph alongside each section, waiting there for an hour for the radiation instruments to take their readings. Then Losee directed Tharp to the next location.

Topside, in the control van on the *Point Loma,* John Van Voorhis, the Sperry techrep who had developed the precision acoustic navigation system being used, related:

> We had by this time, become very confident in our ability to track the *Trieste*, and in our ability to provide accurate vectors and distance estimates to the next target section. The system had become so accurate, that I was able to detect the sway of the reference DOTs as they circled 300 feet above their anchors. I estimate that we were able to track the bathyscaph and plot the target sections to within a 25-foot radius. When they asked for vectors to each successive section, our recommendations were precise. We were so confident that we were even applying corrections for bottom currents measuring less than a half-knot.

Parsons took samples of the bottom silt at the locations selected by Losee, thrusting clear bottles carried externally into the mud. Parsons also retrieved some piping from the ocean floor with the manipulator arm. Once in position to look into the open end of the stern section, Parsons tried to deploy an improved Tortuga remotely operated vehicle but could not get it to work. The surface controller then gave Tharp a vector to the *Scorpion*'s propeller and shaft, where they took several photographs. Another vector from the surface took them to the sail structure, where they took hundreds.

During the record dive the *Trieste* was in continual communications with the surface via the UQC acoustic underwater telephone. Parsons concluded his dive description: "Once we had finished with the photography of the sail, I asked 'What now?' We had spent so much time, just sitting and waiting for the radiological instruments to get their readings that we still had power, even after 23 hours submerged." Losee in the sphere and all in the control van all agreed—it was time to end the dive. The *Trieste* surfaced with 24 hours, 44 minutes submerged—a bathyscaph record.

The sensors had worked well, and all of the needed data had been retrieved. In the control van Copeland was struck by how simple it had been to vector the *Trieste* between hull sections. Everything was on the plot, and once Parsons could see the bow section, the surface controller could drive the bathyscaph around as in a computer game. Jack Brandt, one of the *Trieste*'s divers, recalled that "the dive was so long that the batteries for the restraining magnets were nearing their end, and we got to them just before they started dropping equipment into the ocean."

The next dive on the *Scorpion* site took Commander Sachse to the wreck for 17 hours, five minutes, surfacing on 1 September. He was trying to determine the physics of the loss, the physical condition of the hull sections, and what the location of the wreck elements on the ocean floor revealed about the mechanics of the events that had destroyed the submarine. Denzien piloted with Senior Chief Scales assisting. A third dive, on 5 September, was aborted because of electrical problems before reaching the bottom.

Parsons and Tharp took Dr. Williams to the ocean floor on 11 September, spending 20 hours, 18 minutes submerged. This was fourth dive on the *Scorpion* site and the last of the series—not that it was planned that way. When the *Trieste* was at 11,300 feet and taking biological samples, the *Point Loma* suffered a major casualty to one of her two main feed pumps, critical elements of the propulsion plant; the remaining pump ran hot, eventually overheating and cracking its casing. This shut down both engines and both steam turbo-generators. The ship went dead in the water with nothing more than an emergency gas-turbine generator for electrical power.

While the *Point Loma*'s engineers wrestled with the problem, the commanding officer, Dick Waer, transmitted the Navy equivalent of an "SOS." With no propulsion and the *Trieste* at depth, a number of major emergencies were possible. First, the *Point Loma* might drift away and be unable to render assistance to the *Trieste* on resurfacing; second, the *Point Loma* was the *Trieste*'s only "lifeguard," and there was no longer any way to keep commercial shipping clear; and, third, when the *Trieste* surfaced she might, though the tracking van remained in acoustic communications with her throughout, come up directly into the bottom of the *Point Loma*, with fatal consequences.

The Commander-in-Chief Atlantic Fleet responded to the situation decisively. The ship that could arrive soonest was the nuclear-propelled missile cruiser *Texas* (CGN 39), then bound for the Mediterranean. She was ordered to make all possible speed to the dive site.

Meanwhile, the *Trieste* was obliged to surface "independently" as best it could. Parsons and Tharp later related their emotions on this final surfacing at the *Scorpion* site. In Tharp's words, "With our dive goals completed, we headed to the surface as cautiously as we could. Once surfaced, we were taken in tow by the *Point Loma*'s motor whaleboat and towed over to the ship. Without power, she could have easily drifted over the horizon leaving the *Trieste* alone in the middle of the Atlantic Ocean. Instead, the *Point Loma* had drifted away a couple of miles, then turned around, and drifted back to us. When we came up, there she was a short distance away. It still gives me the chills."[11]

The arrival of the *Texas* was dramatic. She came over the horizon with a bone in her teeth—nautical jargon for a large bow wave—making close to 35 knots. Without slackening speed she passed down one side of the *Point Loma*, crossed her stern, and raced back up the opposite side, blasting away on her "topside" speakers, "The Eyes of Texas Are Upon You." The *Point Loma* was without propulsion for three days, during which time the *Trieste* was unmanned and tended by the ship's motor whaleboat. Repair parts were dropped from an Azores-based P-3 Orion aircraft within a day of the emergency casualty and the *Texas* supplied other needed parts to repair the electrical motors.

Meanwhile, a fleet tug had been dispatched in case towing was required. Before the tug could make an appearance, however, the *Point Loma*'s engineers had one of the two main feed pumps on the line so that the ship could re-dock the *Trieste* and limp into Ponta Delgada for further repairs.

However, Admiral Harry D. Train II, commanding the Atlantic Fleet, was livid. He had diverted a nuclear-missile cruiser from an aircraft carrier battle group to "babysit" the re-examination of a wreck. He demanded, "Get that ship [the *Point Loma*] out of my fleet." The commander of Submarine Force Atlantic, Vice Admiral Kenneth M. Carr, realized that the *Trieste* was committed for one more dive and so stated to his boss. Train relented, and Wilson Whitmire, a captain, once again was designated to oversee the *Trieste*'s operations in the Atlantic, for one last dive.[12]

Tongue of the Ocean

Following two days at Ponta Delgada, the *Point Loma* proceeded to Norfolk for an extended maintenance period. There Parsons was asked to brief the *Scorpion* dives to Vice Admiral Carr. Parsons, who had been told to remain flexible with regard to schedule, was paged at the beach, and without time to return to the ship to change clothes, he made his briefing to Carr and about 30 members of his staff in jeans and a T-shirt.

When he was finished, Carr shook Parson's hand, said, "Well done," and departed without another word. Parsons later recalled, "I thought a well done from a vice admiral was a good outcome, and forgot all about it, until in San Diego when I was advised that Carr had gotten me a Meritorious Service Metal. It is not often that a vice admiral looks out for a lowly lieutenant commander."[13]

On 16 October, the *Point Loma* with the *Trieste* left Norfolk for the Tongue of the Ocean, a deep region between two islands in the Bahamas where the Navy's Atlantic Undersea Test and Evaluation Center (AUTEC) had been established. Captain Whitmire was in tactical command. The *Trieste*'s final dive in the Atlantic, #11-79 on 19 October, saw Denzien and Parsons piloting, along with Chief Electronics Technician Craig Wilken, down to 4,943 feet for 19 hours, ten minutes. The purpose of the lengthy dive was to recover a Mark 70 mobile submarine simulator.[14]

The Mark 70 was about eight feet long and 12 inches in diameter. To recover it the *Trieste* crew constructed a special lifting device to attach to the bow winch. Tharp, in the control van on the *Point Loma*, recalled that the AUTEC range controllers had the *Trieste* on their tracking system and were automatically feeding data to the tracking van. The van's plotter had a nice presentation of the *Trieste* slowly corkscrewing downward toward the bottom. When the *Trieste* reached the bottom and visibility cleared, there was a long, skinny green device in front of it. AUTEC informed the *Trieste* crew that the green device was nothing important and advised them to search to the north.

Denzien drove northward, then east, then looped back southwest under Tharp's direction, returning to their starting point. After hours of transit on the bottom there outside the viewport was once again, not surprisingly, the long skinny, green thing. Whitmire was in the tracking van, and when the *Trieste* wound up back in the original position just as the plot indicated, Tharp got a slap on the back from him for vectoring the *Trieste* so precisely.

Parsons again asked AUTEC, "Could that green-painted device be our target?" He moved the *Trieste* in close and read off a serial number. After several moments AUTEC confirmed that it was indeed the target, after all. Parsons unlimbered the mechanical arm and brought the "green thing" up to the surface, 19 hours into what should have been a five-hour task. On completion of the dive, the *Point Loma* entered Eleuthera, and Captain Whitmore departed.

1979 Retrospective

Despite the *Point Loma*'s engineering problems, the four-dive *Scorpion* investigation and the Tongue of the Ocean dive demonstrated real situational awareness in the tracking van and a mature and responsive deep submergence system able to react expeditiously and work effectively to depths of 20,000 feet. The *Trieste* program had come of age and could be confidently employed almost anywhere required on the deep-sea floor, albeit with observable—that is, to suspicious eyes—surface support.

In 1978, Congress had permitted the Navy to assign women to sea-duty billets on support and noncombatant ships, in addition to ships already authorized for women, i.e. hospital ships and civilian-crewed auxiliary ships. In the second half of 1979 the *Point Loma* became one of the first additional ship types to accommodate women. She had three women officers on board during her 1979 deployment in the Atlantic. En route home to San Diego, at the Rodman naval station in the Canal Zone, a fourth female officer reported on board for duty.

Several additional women officers as well as enlisted women arrived shortly after the *Point Loma* reached San Diego. During her 1981 overhaul the *Point Loma* was given habitability modifications to isolate a 42-berth compartment for female enlisted quarters with its own shower and toilet facilities. The two-, four-, and six-person rooms assigned to chief petty officers and officers were relatively easy to modify for use by either gender. It would not be until the ship emerged from her 1981–1982 overhaul that a full allocation of 42 enlisted women and eight women officers would report on board, and by that time the *Point Loma*'s service to the *Trieste* program had come to an end.

Senescence and Retirement

The budget cycle got worse each year, trying to convince [the Department of Defense] to retain the manned deep submersibles, which were generally considered to be Navy toys.

—Commander George H.Verd, Deep Submergence Systems Division, Office of the Chief of Naval Operations

The *Trieste* program entered 1980 with confidence and a reputation to guide the bathyscaph's future employment. Submarine Development Group 1 planned a series of scientific dives by the (third) *Trieste* in the eastern Pacific, off southern Mexico.

In preparation, the *Point Loma* took the *Trieste* into the southern California operating areas on 4 February for a training dive to 4,821 feet, with assistant officer-in-charge Chief Warrant Officer Burt Tharp piloting, Lieutenant Commander Les Parsons in command, and Lieutenant Jim Denzien under instruction. The next dive, on 14 March, was Tharp's certification dive as a deep submergence pilot (#78), with Parsons examining, and newcomer Lieutenant Deloss M. (Bud) Lester on his first bathyscaph dive. Also, this was Commander Richard F. Grant's first operation as the new commanding officer of the *Point Loma*.

The upcoming expedition was affiliated with the Geological Exploration of the Mid-America Trench to investigate plate tectonic dynamics, under the auspices of the Scripps Institution of Oceanography. The effort sought to identify a permanent disposal site for high-grade nuclear waste—radioactive waste with half-lives in hundreds or thousands of years.[1]

The first dive into the trench off Acapulco was on 14 April, to 16,141 feet. Denzien was piloting on his certification dive, Parsons was in command, and Scripps oceanographer Dr. Peter F. Lonsdale was with them. Parsons had a bet with Lonsdale that they would not be able to locate or recognize the subduction plate boundary. To the surprise of Parsons and to the satisfaction of Lonsdale, they found a sharp and distinct line of cliffs on the ocean floor, exactly where Lonsdale's sources indicated. Denzien followed the boundary line for several hours—and there was no question that Parsons owed Lonsdale a fifth of the most expensive Scotch on the planet.

Five days later Lester piloted the *Trieste* under the instruction of Tharp, taking Lonsdale into the same area and bottoming at 15,540 feet. The three men had almost 24 hours on the ocean floor. Afterward the *Point Loma* and *Trieste* crews received liberty in Acapulco and then sailed back to San Diego.

Reaching San Diego on 4 June, the *Trieste* made a local dive to 2,600 feet for vehicle check and pilot training; Lester piloted with Denzien in command and Electronics Technician 2nd Class Michael Yermakov on his first (and only) deep dive. This dive was in preparation for deployment to Hawaii to recover an unmanned, cable-controlled, deep submergence vehicle lost west of Kona.

Hawaii Area Dives

The *Point Loma* left San Diego with the *Trieste* embarked about 10 July 1980, arriving in Pearl Harbor ten days later. Following briefings by the Submarine Force Pacific staff and a few days of liberty, the *Point Loma* with the *Trieste* nested in her docking well transited to a point 15 miles west of Kona for Dive #6-80, to 15,500 feet.

The target for the first dive was a locally produced remote underwater work vehicle lost in January 1980 in about 15,200 feet of water when its control cable parted during training. Press reports reported the value of the device as $20 million in two articles and $750,000 in a third. Whatever its dollar value, the Navy wanted it recovered.[2]

Hawaii Undersea Research Laboratory personnel boarded the *Point Loma* at Pearl Harbor and assured the *Trieste* team that they knew exactly where the vehicle was lost and that the ocean floor there was flat and sandy. On the "first" dive Lester piloted, Parsons was in command, and Machinist's Mate 3rd Class Richard D'Aniello rode along, having drawn the lucky number in the dive lottery. At depth, the dive revealed not a flat sandy surface but a bottom broken by ravines, gigantic boulders, lava flows, rock outcroppings, and a general jumble so extreme as to stymie the most intensive of searches.

Sonar was useless in that environment. Depth changes of hundreds of feet marked cliffs and ridges, any one of which could have hidden the vehicle only a few dozen feet from the bathyscaph. After a single dive Parsons declared the search ineffective, and the effort was abandoned.[3]

The next two dives, both to depths beyond 13,000 feet, were on a classified target to the east of the Hawaiian Islands. The target location was provided to the *Trieste* team in the form of an exact latitude and longitude, and the bottom was described as flat, sandy, and unobstructed. The dives of this sequence are recorded in the official dive log, but errors have been introduced by hand transcription and security imposed at the time to obscure the data. The dives on 9 August and 12 August 1980 are the only two-man dives known to have been made by the *Trieste II* (DSV 1).[4] On the first two-man dive, Tharp and Lester dropped onto what was this time as advertised, a relatively flat ocean floor. Tharp "motored" for several miles in search of the target but had to quit on encountering the approaches to a volcano. The *Trieste* lacked sufficient power to "fly" safely up the side of the volcano's rim wall. On this extensive transit, however, the *Trieste* passed through a field of manganese nodules. To Tharp "they looked like baked potatoes. We did not have a basket attached so we had to leave our newfound wealth behind."[5]

The second two-man dive, on 12 August, took Parson and Tharp into the volcano's caldera, a rather hazardous undertaking. Tharp recorded:

> Everywhere you looked there were sharp outcroppings that could have cut the *Trieste* open, leaving us there forever. On top of everything else, both Les and I were exhausted. We had had some problems getting the *Trieste* ready to dive. By the time we got things right it was late at night and both Les and I had been up for 18 hours. Our OTC [officer in tactical command], who was not a deep submergence guy, insisted that we start the dive immediately. That put Les and me in the most dangerous place we have ever been, having already been up more than 24 hours.[6]

The volcano into which the bathyscaph descended did not appear on the charts on board the *Point Loma*; the *Trieste*'s crew named it the Laura Seamount, after Tharp's daughter. Tharp had been the first to exclaim at the unexpected tracing appearing on the precision depth recorder on board the *Point Loma* as they approached the dive site. The unclassified logs for the Hawaii dives of 6–15 August are not comprehensive and, again, reflect deliberate misdirection.[7] Following the third dive in the Hawaii area the *Point Loma* returned to San Diego with the *Trieste* nested in her docking well.

Warning Voices

The *Trieste* program was getting long in the tooth; it was an aged and expensive system, suitable only for operations that other, less expensive systems could not perform. Commander George H. Verd, then the OpNav Deep Submergence Plans and Programs Officer in the Pentagon, summarized:

> One of my jobs was to sell the use of our [deep submergence] vehicles to the oceanographic community; these included the *Trieste*, *Turtle*, *Sea Cliff*, and *NR-1*. However, the most frustrating and time consuming efforts were to justify and defend our budgets annually (the current year including the two subsequent years) up through the chain of command and in formal budget hearings, while trying to include [Research and Development] for new potential systems. Any tweak of the budget at any level meant reprioritization of what was most important as to which deep submergence capabilities we could maintain.[8]

Every year it was becoming more difficult to convince the Department of Defense to sustain funding for the *Trieste* program. Verd related, however, that "even in the offices of the Chief of Naval Operations, there was an estimation that the funds needed to keep the *Trieste* program active was [*sic*] not worth the few dives it could accomplish each year, nor were issues of safety acting in its benefit."[9]

The *Trieste* program was kept active and operational solely to accomplish missions beyond the depth capabilities of the Navy's other manned submersibles—the *Alvin*, *Turtle*, *Sea Cliff*, and the nuclear-propelled *NR-1*. For certain specific operations at lesser depths there were also the two DSRV rescue vehicles, which, being transported and supported by submarines, provided a clandestine capability.[10]

Most of the *Trieste* dives during the period 1979–1984 could more easily and more economically have been accomplished with the other deep-diving vehicles and were assigned to the *Trieste* primarily for training and proficiency. Those missions that the *Trieste* did undertake involved only a few dives per year, with the maintenance and operating costs for a crew of three officers and 12 to 16 enlisted men, plus the *Point Loma*, with her crew of 260 officers and enlisted personnel.

An ability to locate and recover small objects down to 20,000 feet was not solely a Navy concern, but received sponsorship as well from the Air Force and the Central Intelligence Agency. Looking forward, two lines of development offered opportunities to retire the aging, expensive, and increasingly hazardous *Trieste* system: first eventually extending the depth range an *Alvin*-type deep submergence vehicle to 20,000 feet, and, second, developing unmanned systems to recover small objects down to 20,000 feet. The Navy pursued both.

The deep submersible *Alvin,* launched in 1964, was certified to 6,000 feet in 1965; in 1973, the *Alvin* was fitted with a titanium sphere and certified to 12,000 feet. In 1976, the *Alvin* further closed the depth gap by certifying for 13,124 feet with the same sphere. Meanwhile, plans to modify the *Turtle* and *Sea Cliff* for depths greater than 6,500 feet were being developed. When the *Alvin* shifted to a new titanium sphere in 1976 her old sphere was retested and installed on the *Turtle*, which then was certified for 10,000 feet. Finally, in 1983, the *Sea Cliff* was fitted with a titanium sphere and the following year it was certified to 20,000 feet. This last event allowed the *Trieste* to be retired.

The Navy had planned to produce—as follow-on to the *Trieste* program—two manned submersibles for operations to 20,000 feet. Those two Deep Submergence Search Vehicles (DSSV) were being sponsored by the Deep Submergence Systems Project. The DSSVs were to be carried into operating areas by submarines, which could recover and support the vehicles, enabling clandestine missions that were not possible with the *Trieste* and the other manned DSVs. But the two DSSVs were cancelled because of budget constraints (as were four of the planned six DSRV rescue submersibles). These cancellations also were in part the result of the development of unmanned recovery systems for deployment from submarines.

Powering Down

Meanwhile, in San Diego, Lieutenant Lester detached from the *Trieste* without relief in September 1980; Chief Warrant Officer Tharp assumed the duties of assistant officer-in-charge. In mid-1981 the *Trieste* was placed in a reduced operating status while awaiting the outcome of the *Sea Cliff* conversion. Upon the *Sea Cliff*'s certification to 20,000 feet the *Trieste* was to be retired.

The reduced operating status of the *Trieste* program led Lieutenant Commander Parsons to take on a second assignment on the staff of Submarine Development Group 1. He also became the officer-in-charge of the Unmanned Vehicles Detachment, a group of "ship riders" that deployed a towed and remotely controlled recovery system down to 20,000 feet.

As another result of the *Trieste*'s reduced operating status, the *Point Loma* was sent to the Mare Island for modifications related to an additional mission—support ship for the Trident submarine-launched ballistic missile tests in the Pacific. After the work was completed in April 1982, the *Point Loma* was never again involved with the *Trieste* program. (The *Point Loma*'s Trident mission continued until 1993; she was decommissioned on 27 September 1993 and sold for scrap.)

Meanwhile, with only two officers assigned—Parsons and Tharp—workup commenced to qualify two chief petty officers as *Trieste* pilots. On 5 February 1981, Parsons took the *Trieste* to 2,200 feet off San Diego with Tharp and Chief Electronics Technician Delbert D. Hart. On 9 July and 15 December, the same trio took the bathyscaph to moderate depths (5,300 and 3,600 feet), with the *Pigeon* (ASR 21) in July and the *Florikan* in December supporting and towing.

With only three dives in 1981 and no support ship expected to be available in 1982, Parsons arranged for a series of five *Trieste* dives from a barge operating from Wilson's Cove of San Clemente Island for the period May through June 1982. Preparations were extensive and laborious. A barge was acquired from San Diego port operations and extensively modified by the *Trieste*'s crew: adding a generator, freshwater tanks, two air compressors, and four support vans. Everything loaded was weighed, so Tharp could calculate the loaded barge's stability, balance the load, and so ensure that the barge was seaworthy for the 40-mile tow to Wilson's Cove.

Senior Chief Petty Officer Daniel Rean, future assistant officer-in-charge of the *Trieste*, recalled that "dive logistics were a nightmare. Each dive consumed

30 tons of iron shot as ballast and none of that could be reused for the next dive. Four additional loads of shot (120 tons) had to be carried on the barge along with a forklift and barrel clamps that were needed to lift the shot and pour it into the shot hopper. Hundreds of gallons of fuel for the generator and several barrels of compensation fluids were loaded along with about 500 gallons of bottled water."[11]

The bathyscaph was towed to and from the dive site by an LCM-8 landing craft and, with the exception of messing and berthing the first few days, the *Trieste* was able to operate independently. The first dive was with Parsons and his prospective relief and former *Trieste* pilot, now–Lieutenant Commander R. D. Baker, and Chief Hart. The remainder of the dive series was with Tharp piloting and two enlisted *Trieste* crewmen in the sphere.

On 25 June 1982, at the conclusion of this series, Baker relieved Parsons as officer-in-charge. On 10 December—the last dive of the year—Tharp and Baker took Hart for his certification dive. Hart emerged from the *Trieste* sphere as enlisted hydronaut #28.

Like Parsons before him, Baker was returning to the *Trieste* program for a second tour. Baker had enlisted in 1961 and qualified in submarines a year later on the diesel submarine *Carp* (SS 338). In 1969 he had become a warrant officer on the USS *Greenfish* (SS 351); in 1972 he was commissioned as a lieutenant (junior grade). From 1972 to 1975 he was with the *Trieste* program, qualifying as deep submergence pilot #48. After earning a college degree and a tour on the submarine tender *Holland* (AS 32), he moved to the staff of Submarine Development Group 1 as logistics officer, 1979–1983. When in 1983 he relieved Parsons as officer-in-charge of the *Trieste*, it was as an additional duty. After 15 months as officer-in-charge, Baker recommended to the Commander Submarine Development Group 1 that Chief Warrant Officer Tharp, his assistant officer-in-charge, become his relief.

From the *Trieste* program Baker became the executive officer of the submarine tender *Simon Lake* (AS 33) in 1985–1987 and then served again on the staff of Submarine Development Group 1 until 1989, that time as new developments officer, with additional duty as officer-in-charge of a group of ship riders responsible for deep submergence search and recovery.

Parsons later became commanding officer of naval facility (i.e., SOSUS station) at Pacific Beach, Washington. (This was the same "NavFac" that former

Trieste assistant officer-in-charge Lieutenant Commander Philip C. Stryker Jr. had commanded in 1973–1975). In 1985, Parsons became executive officer of the Trident Refit Facility at Bangor, Washington, then retired after 30 years of naval service.

The 1982 series of *Trieste* dives working from a barge out of Wilson's Cove was a remarkable success. In August Tharp was called on to repeat the operation and arranged a quick test dive and pilot checkout on 12 August with himself, Chief Hart, his certified enlisted pilot, and soon-to-be-promoted-to Senior Chief Rean. On 2 September 1983, Tharp relieved Baker, "fleeting up" to become the last officer-in-charge of the *Trieste*. Simultaneously, Rean relieved Tharp as assistant officer-in-charge.

The End Cometh

The last 18 months of the *Trieste*'s active operations gave members of the bathyscaph crew chances to descend into the depths. Rean recalls that the lottery top choice was not well received by all: "Some winners would swap the ticket with someone who really wanted to make a dive. Our leading electrician won the hat trick several times but never went on a dive. He told me that he was afraid that he would not survive. The funny thing was that none of the pilots ever thought like that. We knew that at the depths that we operated at, if something went seriously wrong we would not have time even to realize that we were dead."[12]

Working out of Wilson's Cove during October 1983, Rean was certified as enlisted hydronaut #33 on the 10th, searching for a malfunctioning San Clemente test-range hydrophone. On 17 October Tharp succeeded in locating and recovering a faulty hydrophone, from a depth of 4,100 feet—thus ending the *Trieste*'s career.

The 17 October 1983 Dive, #3-83, about the 132nd dive of the third variant of the *Trieste* bathyscaph, was the final dive of the program.

In Retirement

In June 1983, the *Sea Cliff* began a seven-month conversion and overhaul at the Mare Island shipyard, receiving the titanium sphere that would make it

Table 23-1. *Trieste* Dive Record

Vehicle	Dives
Trieste I	120 dives (including one unlogged classified dive)
Trieste II	56 dives
Trieste II (DSV 1) (*Trieste III*)	132 dives* (including classified dives)
Total	317 program dives, including dips and classified dives

* The unclassified dive logs for the "*Trieste III*" (DSV 1) show 129 dives. There were six or seven dives in 1967 during its classified workup for Winterwind that were not recorded in the unclassified logs. Two dive entries were erroneously duplicated in 1980. Thus the total number of dives for the "*Trieste III*" (DSV 1) was 127 plus six or seven, and minus two, giving 131 or 132 dives total.

capable of diving to 20,000 feet. Once operational, the *Sea Cliff* relieved the *Trieste* as the nation's deepest-diving manned search-and-recovery vehicle.

The *Trieste* was deactivated on 18 May 1984, a milestone marked formally on 15 June at a ceremony held at the Ballast Point submarine facility in San Diego. *Trieste* alumnus Rear Admiral J. Bradford Mooney, as the Chief of Naval Research, gave the keynote address: "The *Trieste* made possible a whole family of submersibles—some operated by remote control. The people who will man them will not read history; they will continue to make it."

Already recognized as a hot-running lieutenant commander in 1963, he lit a fire under his career trajectory in his tour as officer-in-charge of the *Trieste*, leading all the way to flag rank. After holding key OpNav positions related to deep submergence activities, Mooney served as Oceanographer of the Navy and then Chief of Naval Research. Retiring in 1987, Mooney was elected to the National Academy of Engineering in 1988 and subsequently held several important oceanography-related positions.

Captain Don Walsh, deep submergence pilot #1 with a PhD in oceanography, also participated in the *Trieste* retirement ceremonies. The last officer-in-charge of the *Trieste*, Chief Warrant Officer Tharp, capped off the retirement proceedings with a sailor's prayer: "Our thanks to God for giving us the talent, drive, and good fortune to successfully pull off some outrageous stunts and live to tell about it."[13] Tharp received orders to the staff of Submarine Development Group 1. Electrician's Mate 1st Class Ramon Garcia was the last crewman assigned to the program, and among his final official acts was to emboss late-arriving philatelic requests from the stamp-collecting public.

The original *Trieste* has been for many years at the Navy Museum at the Washington Navy Yard, Washington, D.C. The third variant—known as the *Trieste II* (DSV 1)—is on display at the Naval Undersea Museum, Keyport, Washington.

Thus, the deepest pioneers are preserved.

24 Postscript

There now exists a fully operational deep (hadal) ocean exploration system (mother ship, unmanned landers and submersible) that is capable of repetitive dives anywhere in the World Ocean.

—Don Walsh

In the more than three decades since the retirement of the third *Trieste* bathyscaph, there have been extensive and impressive developments in deep submergence technology. Most noticeably there has been a major shift away from human-occupied deep submersibles to remotely operated and autonomous vehicles.

The manned *Sea Cliff* and *Turtle* both were retired in 1998. The rescue submersible *Avalon* was retired in 2000 and the *Mystic* in 2008, the same year as the nuclear-propelled submersible *NR-1*. This left the *Alvin* as the only manned deep submersible financially supported by the U.S. Navy, operated by Woods Hole. The *Alvin* remained in service into the third decade of the 21st Century.

The manned submersible programs in France and Japan have reflected the same reality as unmanned underwater vehicles are cheaper and safer methods of exploring and working in the ocean depths. This was true for military and commercial/civilian activities.

In the United States, Howard R. Talkington of the Naval Research Laboratory has been recognized as the "father" of the Navy's remote-controlled submersibles, especially of the Cable-Controlled Underwater Research Vehicle (CURV), which lifted the hydrogen bomb lost off Palomares, Spain, in 1966. A decade later, in 1976, he prophesied that "remotely manned" vehicles could do practically anything a manned vehicle could do—better, faster, cheaper, and more safely.[1] By 2020 there were hundreds of unmanned submersibles being operated throughout the world, designed to meet military, oil industry, civilian, and scientific requirements. Also, when required these unmanned systems can reach the greatest depths of the oceans. The Japanese cable-connected vehicle *Kaiko* landed on the floor of the Challenger Deep—a depth of almost 36,000 feet—in a series of dives in 1995–1998. In 2009 the American-developed *Nereus* vehicle also reached the bottom of the Challenger Deep.[2]

The firm Ocean Infinity in Houston, Texas, has fielded a highly automated bottom-search system consisting of a mother ship supporting ten unmanned vehicles. These systems can search up to 400 square miles per day at depths to 19,685 feet. In July 2019, the firm located the French Navy's submarine *Minerve* in the Mediterranean after only 18 days of searching. The *Minerve* had been lost in January 1968, and despite repeated searches over more than 50 years, her wreckage had not been found.

In the field of academia, semi-autonomous "landers" have become de rigueur, especially in the study of biologics at different depths. A lander can be equipped with some combination of cameras, bait, and special sensors and then dropped into the water to free-fall to the bottom. There it deploys bait or takes samples; at the end of its intended stay at the bottom a timer or a coded acoustic signal orders the lander to jettison ballast and return to the surface for recovery.

Several countries have continued limited programs in the manned submersible field; China and Russia have continued to develop advanced, deep-depth, manned submersibles. Indeed, the Russian nuclear-propelled *submarine* Project 20120/*Sarov*—known by the nickname "Losharik"—consists of an inner pressure hull of several titanium spheres; she has been publicly credited with an operating depth of between 8,000 and possibly *20,000 feet.*[3] Other Russian-built special-mission "submarines" have lesser depth capabilities, but still are rated as deep-diving craft intended for both oceanographic and military roles.

China also is reported to continue the development of human-occupied submersibles specifically intended to study the Mariana Trench.

How Deep Is Deep?

During the seven decades from the first report of extreme depths near Guam from HMS *Challenger* in 1875 (26,850 feet), to their measurement by HMS *Challenger II* in 1952 (35,760 feet), there were differing opinions as to the location of the deepest depth in the world oceans. The principal candidates were the Challenger Deep in the Mariana Trench, the Mindanao Trench, Japan Trench, Kuril Trench, and Tonga Trench.

The Challenger Deep is a long, narrow canyon on the ocean bottom where the Philippine Plate rides up over the dense Pacific Plate and drives it deep into the Earth's mantle. Even as late as 2019—60 years after the Fisher Expedition of 1959 proved that the Challenger Deep was the deepest point in the Northern Hemisphere—there existed the possibility that the Horizon Deep in the Tonga Trench in the Southern Hemisphere might be even deeper.

The Mariana Trench "bottom" reached by Jacques Piccard and Lieutenant Don Walsh on 23 January 1960, in the original bathyscaph *Trieste* was at 35,814 feet, in the western basin. The two men stayed there 20 minutes. More than a half-century later, in 2012, James Cameron tested the central basin with an automated "lander" and then correctly chose the eastern basin for his manned solo dive. On 25 March, Cameron dived into the eastern basin of the Mariana Trench, alone in his purpose-built *Deepsea Challenger*, about 25 miles east of the *Trieste*'s 1960 dive. On this dive to 35,787 feet (plus or minus 31 feet), Cameron spent two hours, 34 minutes exploring the bottom and navigating northward to investigate the nearby boundary slope.

The Five Deeps Expedition carried out in 2019 by Commander Victor Vescovo, U.S. Navy Reserve (Ret.), erased the last remaining doubts as to the location of the deepest spot. Before the Vescovo expedition, the Challenger Deep had been mapped as a single canyon about 30 miles in length and reaching depths 1,500 feet below that of the surrounding trench. However, multibeam scanning with a 3rd-generation Kongsberg EM124 sonar on Vescovo's expedition ship *Pressure Drop* in 2019 identified three separate basins arranged in a southwest-to-northeast line, each exceeding 35,600 feet in depth.[4]

On 28 April 2019, Vescovo solo-dived into the now-confirmed deepest area of the Trench—the eastern basin—bottoming his submersible *Limiting Factor* at 35,843 feet (plus or minus 10 feet), a new world depth record. He then toured the alluvial bottom for three hours, four minutes. On 1 May Vescovo made a second solo dive into the same area of the Eastern Pool, achieving a depth only a hair less (i.e., within the range of uncertainty) than his 28 April dive.

On 5 May the *Limiting Factor* was piloted by Patrick Lahey of Triton Submarines, which had designed and built the craft, with Jonathan Struwe of the commercial classification and certification agency. They bottomed in the eastern basin at 35,843 feet (plus or minus 23 feet). This dive saw history's first certification of a commercial submersible to "unlimited depth" by a classification agency.

Vescovo's *Limiting Factor* undertook five deep-ocean dives into the Mariana Trench in ten days, displaying technical maturity and operational proficiency never seen in earlier deep submergence operations, especially not the venerable *Trieste* bathyscaphs. Commercial manned deep submergence had come of age. Also, no longer reliant on the financial largess of governments, the deep submergence community had demonstrated the ability to take a manned submersible anywhere in the world's oceans, conduct multiple dives, and do productive work on the ocean floor.

As a final demonstration of proficiency and flexibility, Vescovo interrupted his dive schedule to ascertain whether the Horizon Deep in the Tonga Trench was deeper than the Challenger Deep. The detour to the Horizon Deep was capped by a solo dive by Vescovo on 5 June 2019, in the *Limiting Factor* to 35,509 feet. The Horizon Deep proved to be just 335 feet shy of the Challenger Deep's record depths and so the second-deepest point on the ocean floor.

Vescovo's Five Deeps Expedition of 2018–2019 into the Puerto Rico Trench, the South Sandwich Trench, the Java Trench, the Mariana Trench, the Tonga Trench, and the Molloy Deep in the Arctic Ocean concluded in August 2019. It was a tour de force, netting Vescovo a number of world firsts and conclusively proving that the Challenger Deep in the Mariana Trench is the deepest point in the world's oceans.

Vescovo returned to the Challenger Deep for six dives, 8–27 June 2020. Among those dives, Vescovo escorted the first woman into the world's deepest waters, astronaut Kathryn D. Sullivan, a former head of the National Oceanic and Atmospheric Administration. Upon surfacing she made a telephone call to the International Space Station to announce her dive to the first two American astronauts to have boarded the station using commercial transportation (provided by the SpaceX firm).

As an acknowledgement to his forerunners, Vescovo also escorted Kelly Walsh to the bottom of the western pool, the site that his father, Don Walsh (along with Jacques Piccard), made a world record dive into the Challenger Deep 60 years earlier. This not only celebrated the 60th anniversary of Walsh's dive but was the first manned revisit of the western pool since 1960 and the first father/son assault on the Challenger depths, albeit 60 years apart.

The *Trieste*'s legacy lives on.[5]

But the *First* First . . .

The *first* vehicle to reach the bottom of the deepest point on Earth, therefore, was the U.S. Navy bathyscaph *Trieste*. From 1953 to 1984, the *Trieste*—actually three generally similar vehicles carrying the same name—undertook an estimated 317 dives, including shallow dips and classified operations. These dives accomplished oceanographic and military tasks that were not possible with any other platform or device at the time.

Today, after the retirement of the vast majority of its human-occupied deep submersibles, the U.S. Navy's continues deep-ocean projects and programs with a flotilla of unmanned systems for both operational and intelligence purposes. These operations carry on the work first demonstrated by the bathyscaph *Trieste*—that Opened the Great Depths.

Appendix A
Trieste Leadership

Year	Officer-in-Charge	Assistant OINC	Leading Chief Petty Officer
TRIESTE			
1958	LT R. B. Davey (medical absence) (Dec. 1958)		MRC E. John Michel (Dec. 1958)
1959	LT R. B. Davey (medical absence)	LT Don Walsh (Jan. 1959)	MRC Erhard John Michel
	LT Don Walsh (July 1959)	LT L. A. Shumaker (29 July 1959)	ENCS Harlan DeGood (May 1959)
1960	LT Don Walsh	LT Lawrence A. Shumaker	ENCS Harlan DeGood
1961	LT Don Walsh	LT Lawrence A. Shumaker	MRCS E. John Michel
1962	LCDR D. L. Keach (13 July 1962)	LT George W. Martin (22 May 1962)	PHC Carl W. Lewis
1963	LCDR Donald L. Keach	LT George W. Martin	PHC Carl W. Lewis
TRIESTE II			
1964	LCDR J. B. Mooney (Feb. 1964)	LT John H. Howland (7 June 1964)	MRCS E. John Michel (Dec. 1964)
1965	LCDR J. Bradford Mooney	LT John H. Howland	MRCM E. John Michel
1966	LCDR E. E. Henifin (20 May 1966)	LT John H. Howland	MRCM E. John Michel
1967	LCDR Edward E. Henifin	LT Frederick S. Forst (Jan. 1967)	ETCM M.J. Vogel Jr. (mid-1967)
	LCDR Edward E. Henifin	LT Robert R. Denis (Sep. 1967)	ETCM Merle J. Vogel Jr.
TRIESTE II* (DSV 1) *("TRIESTE III")			
1968	LCDR R. F. Nevin (May 1968)	LT Anthony T. Dunn (May 1968)	ETCM Merle J. Vogel Jr.
1969	LCDR Robert F. Nevin	LT Anthony T. Dunn	ETC W. T. Burrage
1970	LCDR Robert F. Nevin	LCDR Wm J. Leonard (Feb. 1970)	ETC W. T. Burrage
	LCDR Wm J. Leonard (18 Aug. 1970)	LCDR R. E. Saxon (18 Aug. 1970)	ETC W. T. Burrage
	LCDR M. G. Bartels (17 Dec. 1970)	LCDR Ross E. Saxon	ETC W. T. Burrage

1971	LCDR Malcolm G. Bartels	LT Richard H. Taylor (Jan. 1971)	MMC Samuel Goucher
	LCDR Malcolm G. Bartels	LCDR Philip C. Stryker Jr. (Jun 1971)	MMC Samuel Goucher
1972	LCDR Malcolm G. Bartels	LCDR R. L. Abbott (Aug. 1972)	MMC Samuel Goucher
1973	LCDR R. B. Whitaker (28 June 1973)	LCDR Richard L. Abbott	ETCS William D. Hood
1974	LCDR Roger B. Whitaker	LCDR Richard L. Abbott	ETCS William D. Hood
1975	LCDR Roger B. Whitaker	LT C. Lester Parsons (Mar. 1975)	ETCM George G. Ellis (June 1975)
1976	LCDR J. K. Newell (26 May 1976)	LT K. W. Hanson (Feb. 1976)	ETCM George G. Ellis
1977	LCDR James K. Newell	LT Kenneth W. Hanson	ETCM George G. Ellis
1978	LCDR C. L. Parsons (Jan. 1978)	LT Thomas G. Vetter (Jan. 1978)	ETCS Norman J. Brandt (Jan. 1978)
	LCDR C. Lester Parsons	LT William J. Beres (Dec. 1978)	ETCS Norman J. Brandt
1979	LCDR C. Lester Parsons	LT Deloss M. Lester (Apr. 1979)	ENCM John W. Rofinot (July 1979)
1980	LCDR C. Lester Parsons	CWO4 B. P. Tharp (Sep. 1980)	MMC Clifford Forrester (June 1980)
1981	LCDR C. Lester Parsons	CWO4 Burton P. Tharp	EMCS Angel Magdaluyo
1982	LCDR R. D. Baker (25 June 1982)	CWO4 Burton P. Tharp	EMCS Angel Magdaluyo
1983	CWO4 B. P. Tharp (2 Sep. 1983)	MMCS Daniel T. Rean (Sep. 1983)	MMCS Daniel T. Rean
1984	CWO4 Burton P. Tharp	MMCS Daniel T. Rean	MMCS Daniel T. Rean

Rank/Rate Abbreviations

CWO4 Chief Warrant Officer (W-4)
EMCS Senior Chief Electrician's Mate (E-7)
ENCS Senior Chief Engineman (E-7)
ENCM Master Chief Engineman (E-8)
ETC Chief Electronics Technician (E-6)
ETCS Senior Chief Electronics Technician (E-7)
ETCM Master Chief Electronics Technician (E-8)
LCDR Lieutenant Commander (O-4)
LT Lieutenant (O-3)
MMC Chief Machinist's Mate (E-6)
MMCS Senior Chief Machinist's Mate (E-7)
MRC Chief Machinery Repairman (E-6)
MRCS Senior Chief Machinery Repairman (E-7)
MRCM Master Chief Machinery Repairman (E-8)
PHC Chief Photographer's Mate (E-6)

Appendix B

Trieste Deep Submergence Pilots

All officers and enlisted men who were certified as deep submergence pilots on the *Trieste* were submarine qualified prior to reporting to the *Trieste* program. Both officer and enlisted pilots of deep submersibles are entitled to the sobriquet "hydronaut."

Certificate #	Name	Certification Date
1	LT Don Walsh	15 June 1960
2	LT Lawrence A. Shumaker	15 June 1960
3	LCDR Donald L. Keach	6 June 1962
4	LT George W. Martin	28 June 1962
5	LCDR John Bradford Mooney Jr.	10 March 1964
6	LT John H. Howland	13 May 1965
7	LT Donald E. Saner	5 January 1966
8	LT Richard D. Waer	27 September 1966
9	LT David F. Helms	3 February 1967
10	LCDR Edward E. Henifin	23 November 1967
11	LCDR Robert F. Nevin	13 April 1968
12	LT Charles H. Staehle	20 April 1968
13	LT John B. Field	12 December 1968
14	LT Anthony T. Dunn	13 December 1968
15	CWO2 Lawrence E. Hawks	18 July 1969
16	LT William J. Leonard	18 July 1969
17	LT Ross E. Saxon	21 July 1969
18	LTJG David T. Byrnes	31 July 1969
24	LCDR Malcolm G. Bartels	8 July 1971
25	LT Richard H. Taylor	9 August 1971
35	LCDR Philip C. Stryker Jr.	4 November 1971
45	LCDR Richard L. Abbott	7 December 1972
48	LT Raleigh D. Baker	30 September 1973
59	LCDR R. B. Whitaker	30 March 1976
60	LCDR J. Kirk Newell	26 May 1976
62	LT Kenneth W. Hanson	27 January 1977
65	LT C. Lester Parsons	16 October 1977
69	LT Thomas G. Vetter	29 August 1978
75	LT William J. Beres	12 May 1979
78	CWO4 Burton P. Tharp	14 March 1980
80	LT James R. Denzien	14 April 1980
84	LT Deloss M. Lester	9 August 1980

The first 18 U.S. Navy officer deep submergence pilots were qualified on the *Trieste*. Thereafter, pilots also qualified in the submersibles DSRV 1, DSRV 2, *Turtle*, *Sea Cliff*, and *NR-1*.

The certification date is the date of the certification dive.

Six *Trieste* chief petty officers qualified as deep submergence pilots and were assigned numerical certification numbers in a separate list of enlisted pilots.

Certificate #	Name	Certification Date
11	ETC William D. Hood	22 September 1973
16	ETCM George G. Ellis	30 March 1976
20	EMC Larry J. Porter	31 January 1978
23	ETCS James R. Scales	3 December 1976
28	ETCS Delbert D. Hart	10 December 1982
33	MMC Daniel T. Rean	10 October 1983

Rank/Rate Abbreviations

CWO2	Chief Warrant Officer 2 (W-2)
CWO4	Chief Warrant Officer 4 (W-4)
EMC	Chief Electrician's Mate (E-6)
ETC	Chief Electronics Technician (E-6)
ETCM	Master Chief Electronics Technician (E-8)
ETCS	Senior Chief Electronics Technician (E-7)
LCDR	Lieutenant Commander (O-4)
LT	Lieutenant (O-3)
MMC	Chief Machinist's Mate (E-6)
MMCS	Senior Chief Machinist's Mate (E-7)

Appendix C
Trieste-Related Awards

Individual Awards

Military Award	**Year**	**Individual**	**Operation**
Legion of Merit	1960	Don Walsh	Nekton
Meritorious Service Medal	1969	Robert F. Nevin	*Scorpion*
	1979	Clarence L. Parsons	*Scorpion*
Navy Commendation Medal	1960	Lawrence A. Shumaker	Nekton
	1963	Frank A. Andrews	*Thresher*
	1963	Donald L. Keach	*Thresher*
	1964	John B. Mooney	*Thresher*
	1970	Ross E. Saxon	*Scorpion*
	1972	Malcolm G. Bartels	Hexagon
	1978	Wilson R. Whitmire	USS *Point Loma*
	1978	Richard L. Abbott	USS *Point Loma*
Navy Achievement Medal	1967	E. John Michel	Winterwind
	1969	David T. Byrnes	*Scorpion*
	1969	Anthony T. Dunn	*Scorpion*
	1969	John B. Field	*Scorpion*
	1969	Lawrence E. Hawks	*Scorpion*
	1969	William J. Leonard	*Scorpion*
	1969	Ross E. Saxon	*Scorpion*
	1972	Philip C. Stryker Jr.	Hexagon
	1972	Richard H. Taylor	Hexagon
	1979	Burton P. Tharp	*Scorpion*

Unit Awards

Award	**Year**	**Organization**	**Operation**
Navy Unit Commendation	1963	*Trieste*	*Thresher*
	1964	*Trieste II*	*Thresher*
	1969	*Trieste II* (DSV 1)	*Scorpion*
	1969	USS *Apache*	*Scorpion*
	1969	ARD *White Sands*	*Scorpion*
Meritorious Unit Commendation			
	1972	*Trieste/Apache/ White Sands*	Hexagon
	1977	*Trieste II* (DSV 1)	Atlantic Ops
	1977	USS *Point Loma*	Atlantic Ops

Civilian Awards

Awards are listed in order of precedence. All data are from interviewees and from public records. Authors did not have access to official Navy records, thus awards to individuals who were deceased prior to 2017 may not appear in this listing.

Award	Year	Organization	Operation
Navy Distinguished Civilian Service Citation	1960	A. B. Rechnitzer	Nekton
Navy Superior Civilian Service Citation	1960	Robert S. Dietz	*Trieste*
	1963	Arthur E. Maxwell	*Thresher*
Navy Meritorious Civilian Service Citation Navigation	1978	John B. Van Voorhis	Bottom
Navy Distinguished Public Service Award	1960	Jacques Piccard	Nekton
Navy Merit Award for Group Achievement	1964	Naval Research Laboratory, Deep Sea Research Branch (Buchanan/*Mizar*)	*Thresher*

Appendix D

Trieste Support Ships

Two ships were "mother ships" assigned to support the *Trieste* bathyscaphs.

White Sands (ARD 20/AGDS 1)

Built in 1943–1944 as an auxiliary repair dock, the *ARD* 20 served in the Pacific until she was placed in reserve ("mothballed") in 1947. She was reactivated and converted into a mother ship for the *Trieste* in 1966–1967. Upon conversion she initially was redesignated as the *ARD(BS) 20,* the *BS* indicating bathyscaph support. In August 1967, the Navy named the dock the *White Sands* (ARD 20).

For the period September 1966 to July 1973 she was not a commissioned ship; rather her status was "in service," and she was commanded by an officer-in-charge. On 25 July 1973 she was commissioned as the USS *White Sands* (AGDS 1), her new designation indicating deep submergence support ship, and was assigned a commanding officer. She was decommissioned on 1 April 1974.

The dock's officers-in-charge and single commanding officer during her association with the *Trieste* were:

14 September 1966–25 January 1968	OINC—LCDR James W. Watts
25 January 1968–5 February 1969	OINC—LCDR Charles E. Poarch
5 February 1969–11 March 1970	OINC—LCDR Robert F. Nevin
11 March–18 August 1970	(Acting) OINC—LT B. E. Myers
18 August–17 December 1970	OINC—LT Beauford E. Myers
17 December 1970–20 January 1972	OINC—LCDR M. G. Bartels
20 January 1972–25 July 1973	OINC—LCDR C.W. Hudiburgh
25 July 1973–1 Apr 1974	CO—LCDR C.W. Hudiburgh

Point Loma (AGDS 2)

The Arctic supply ship *Point Barrow* (T-AKD 1) was completed in 1958 for operation by the Navy's Military Sea Transportation Service (later Military Sealift Command). She was constructed along the general lines of a dock landing ship (LSD), but with strengthened hull and bow and special insulation for Arctic operations. After extensive modifications in 1965, the *Point Barrow* carried Saturn rocket boosters from California to Mississippi for NASA's manned space flight program. She also transported large shipboard radar antennas, and she carried landing craft to Southeast Asia in 1968–1969. The *Point Barrow* was deactivated in 1974.

The ship was reactivated on 28 February 1974, for conversion into a mother ship for the *Trieste*. She emerged from shipyard conversion in February 1975 and was placed in commission on 3 July 1976 as the USS *Point Loma* (AGDS 2). The *Trieste* made her first dive from the ship in February 1976.

When the *Trieste* was placed in limited availability status in 1983, the *Point Loma* was converted into a missile launch support ship and was never again associated with *Trieste* operations.

The commanding officers of the *Point Loma* during her support of the bathyscaph were:

(26 February 1974)–7 April 1977	CDR Wilson R. Whitmire*
7 April 1977–11 August 1978	CDR N. H. Branchflower Jr.
11 August 1978–December 1979	CDR Richard D. Waer
December 1979–May 1982	CDR Richard F. Grant
May 1982–4 April 1985	CDR Donald J. O'Shea

Rank abbreviations as in appendices A and B, plus CDR = Commander.

*Whitmire assumed command of the *Point Barrow* on 26 February 1974; he assumed command of the USS *Point Loma* at her commissioning on 3 July 1976.

Notes

Chapter 1. Columbus of the Stratosphere

Epigraph: https://bertrandpiccard.com/family-tradition-auguste-piccard (accessed August 2020).

1. Roy Chapman Andrews is reputed to have been the model for the fictional character Indiana Jones.
2. "Mrs. Putnam Honored by the Belgian King; He Pins Order of Leopold on Her," *The New York Times* (14 June 1932).
3. "Piccard at Dinner Greets Lindbergh: 'I Am a Mere Balloonist,' He Tells Flier at Home of Amelia Earhart. Reveals Submarine Plan," *The New York Times* (24 January 1933). The *Times* published about 150 articles concerning Auguste Piccard from 1930 to 1960.
4. "Auguste Piccard Is Dead at 78; Stratosphere and Sea Explorer," *The New York Times* (25 March 1962).
5. "Piccard Has Lunch with Scientists Here," *The New York Times* (23 January 1933).
6. Robert S. Dietz, "Dives of the Bathyscaph *Trieste*, 1958–1963," transcriptions of 61 Dictabelt recordings, 87–88, Robert Sinclair Dietz Papers 1905–1994, Manuscript Collection 28, Archives of the Scripps Institution of Oceanography, San Diego, Calif. [hereafter Dictabelts].
7. "Piccard Tells His Story of Trip to Stratosphere: Escaped Death Narrowly," *The New York Times* (30 May 1931).
8. *"Le Discours de Seraing, Albert 1er, Seraing le 1er octobre 1927 à l'occasion du 110e anniversaire des usines Cockerill,"* http://www.reflexions.uliege.be/cms/c_10605/le-discours-de-seraing, accessed 16 March 2018.
9. On 4 November 1927, Capt. Hawthorne C. Gray, U.S. Army Air Corps, reached an altitude of 43,380 feet in an open-gondola helium balloon. Gray died of asphyxia prior to returning to Earth, and thus his record could not be officially accredited. Lt. (later vice admiral) Apollo Soucek, U.S. Navy, flew a Wright-Apache fixed-wing aircraft to an altitude of 43,168 feet on 4 April 1930, then a world record for aircraft.
10. "To Seek Atom Clue 50,000 Feet in Air," *The New York Times* (9 September 1930). Safety measures for the flight included not only the spare oxygen generator but personal parachutes for Piccard and Kipfer and a separate parachute for the gondola.
11. Piccard's telegram to his wife was succinct: *"Gut gelandet"* (good landing). Gurgl telegram 28 May 1931, 1635, no. 1860.
12. Both medals are in silver. The smaller is 36 millimeters in diameter; the larger weighs 82.63 grams and measures 60 millimeters.
13. "Piccard on Flight into Stratosphere," *The New York Times* (18 August 1932).
14. "Professor Piccard's Own Log Book of His Second Stratosphere Ascent," *The New York Times* (20 August 1932).

15. Piccard and Cosyns continued to make balloon flights for several years. The most headline-worthy attempt, in 1937, was made notable by a hydrogen fire while on the ground that destroyed the balloon and threatened the lives of both the aeronauts. "Piccard Escapes Fire of His Huge Balloon; Cosyns, Associate in Studies of Stratosphere, Also Safe in Accident near Brussels," *The New York Times* (26 May 1937).
16. Auguste Piccard, *"Le Professeur A. Piccard parle de la stratosphère et de l'avenir de l'aviation"* [Professor Piccard Talks of the Stratosphere and of the Future of Aviation], recorded lecture, *Polydor* disque 78 tours no. 522.529 (1933), time 5:46–6:13. Partially available in French at https://www.youtube.com/watch?v=3k3IfHkPKXU.

    The first transatlantic commercial passenger flight (Berlin to Brooklyn) landed on 11 August 1938, after a 25-hour crossing. It was not until the first transatlantic commercial flights of the supersonic Concorde aircraft on 24 May 1976, a 3½-hour crossing at 50,000 feet, that Piccard's prediction became a reality. The Concorde era ended in 2003; today's airliners fly London to New York in about six hours.
17. The 1933 stratospheric balloon and gondola *Century of Progress*, constructed by the Dow Chemical Corporation for the Chicago World's Fair, was designed by Auguste and his twin brother Jean Felix Piccard. See http://lcweb2.loc.gov/service/mss/eadxmlmss/eadpdfmss/1998/ms998008.pdf, 4.
18. *New National Biography of Belgium* (Brussels: Emile Bruylant, 1979), s.v. "Cosyns, Max." The morphing into a caveman of Cosyns' bust by Demanet signified Cosyns' international reputation as a spelæologist.

Chapter 2. A Balloon into the Depths

Epigraph: Piccard, *Earth, Sky, and Sea*, 26.

1. "Piccard Now Ready for Deep-Sea Dive," *The New York Times* (11 September 1948).
2. "Piccard Planning Ocean Descent: Expects to Go Down 5 to 7 Miles," *The New York Times* (30 July 1939).
3. Carl Chun, *Aus den Tiefen des Weltmeeres* [From the Depths of the Ocean] (Jena, Ger.: Gustav Fischer, 1903).
4. Jacques Piccard and Robert S. Dietz, *Seven Miles Down: The Story of the Bathyscaphe* Trieste (New York: G. P. Putnam's Sons, 1961), 41.
5. "Piccard Plans a Sub-Sea 'Balloon' for Wider Study of Ocean's Bed: Would Reach New Depths in a Free-Moving Metallic Sphere without Cables," *The New York Times* (28 November 1937).
6. "Piccard Planning Ocean Descent."
7. Jean Marchand, *"Au Fond de Oceans avec de Professeur Piccard"* [To the Depths of the Oceans with Professor Piccard] *La Science de la Vie* (January 1939), 2–12.
8. Ibid., 12.
9. Auguste Piccard, *"Le projet d'une exploration sous-marine belge"* [The Belgian Project of Underwater Exploration], *Actes de la Société Helvétique des Sciences Naturelles* 120 (1940), 33–45.
10. Auguste Piccard, *Earth, Sky, and Sea* (Oxford: Oxford University Press, 1956), dedication page: "To my son Jacques Piccard, Croix de Guerre."
11. Piccard and Dietz, *Seven Miles Down,* 46.

12. "Piccard Test Dive Takes Him Down 82 Feet: Bathyscaphe Overhaul Mapped for New Bid," *The New York Times* (28 October 1948).
13. Piccard, *Earth, Sky, and Sea,* 63.
14. "Piccard Dives Called Off after Hitch in Heavy Sea," *The New York Times* (5 November 1948).

Chapter 3. Admirals of the Abyss

Epigraph: Piccard, *Earth, Sky, and Sea,* introduction.

1. Titoist forces occupied the area in and around the city of Trieste on 1 May 1945, and British and New Zealand forces arrived the following day. It was not until January 1947 that a United Nations resolution formed the Free Territory of Trieste, in two zones: Zone A occupied by U.S. and British troops and Zone B (twice the size of Zone A) occupied by Yugoslavian forces. Zone A included the city of Trieste, its shipyards, and industrial facilities and consisted mostly of Italian speakers. On 5 October 1954, the London Memorandum, signed by the United States, the United Kingdom, Italy, and Yugoslavia, ceded Zone A to Italian administration; Zone B had been incorporated into Yugoslavia since 1947. This de facto recognition was solemnized by the Treaty of Osimo in 1975, which was ratified by both Italy and Yugoslavia two years later. The Free Territory of Trieste ceased to exist in October 1954.
2. Yolanda's married name was Yolanda Versich Agoral.
3. Pietro Spirito, *Trieste a stelle e strisce* [Trieste Stars and Stripes] (Trieste, It.: MGS, 1994).
4. Piccard and Dietz, *Seven Miles Down,* 52–54.
5. Aldo Chirini, ed., *L'Esplorazione delle Profondita Marine e I Batiscafi* 'Trieste,' 'Archimède' & 'Alvin' [Deep Ocean Explorations by the Bathyscaphs "*Trieste*," "Archimède" & "Alvin,"'], AMA 82 (Trieste, It.: Associazione Marinara Aldebaran, January 1999), 4, http://www.cherini.eu/pdf/Batisfera.pdf.
6. Pietro Spirito, "La Madrina del Batiscafo *'Trieste'*" [The Godmother of the Bathyscaph *Trieste*], Historical Diving Society Italy, 17, no. 51 (April 2012), 6–8.
7. Dietz, Dictabelts, 78.
8. Chirini, *L'Esplorazione delle Profondita Marine*, 7.
9. The 60-foot dip on 25 August was not logged as an official dive.
10. Piccard and Dietz, *Seven Miles Down*, 69.
11. Auguste Piccard, *Earth, Sky, and Sea*, 133–134.
12. Alberto Romero, "Victor Aldo De Sanctis," Historical Diving Society Italy *HDS Notizie,* no. 22 (February 2002), 25–29. De Sanctis (1909–1996) was a founding father of Italian underwater cinematography. In 1946 he founded, with Franco Cristaldi, the studio Vi.de. S. He was the director of the underwater photography in the 1953 film *I sette dell'Orsa maggiore.* He was responsible for the design and development of the *Trieste*'s underwater lighting systems in 1952 and had one dive in the bathyscaph. See http://www.hdsitalia.org/files/documenti/HDSN22.pdf (accessed 3 August 2017).
13. Auguste Piccard, *Earth, Sky, and Sea*, 135–136.

Chapter 4. White Knight

Epigraph: Lill, "Oceanographic Activities in the Geophysics Branch," 253.

1. Office of Naval Research, "The Bathyscaph Conference (Morning Session)," 20 January 1958, Washington, D.C., 4–6. Iselin was director of Woods Hole 1940–1950 and again 1956–1958.

2. Gary Weir, *An Ocean in Common: American Naval Officers, Scientists, and the Ocean Environment* (College Station: Texas A&M University Press, 2001), 318.
3. Dietz, Dictabelts, 99.
4. "*Trieste* Wives Share Interest, Enthusiasm," *San Diego Union* (13 February 1960). Also Piccard and Dietz, *Seven Miles Down*, 72.
5. Piccard and Dietz, *Seven Miles Down*, 13.
6. Dietz, Dictabelts, 79–80.
7. Harvey M. Sapolsky, *Science and the Navy: The History of the Office of Naval Research* (Princeton, N.J.: Princeton University Press, 1990). Also see Marc Rothenberg, ed., *The History of Science in the United States: An Encyclopedia* (New York: Garland, 2001), 411–412, https://web.archive.org/web/20140716213818/http://www.csupomona.edu/~zywang/onr.pdf.
8. Gordon G. Lill, "An Interview with Dr. Gordon Lill," by David K. Van Keuren, Naval Research Laboratory (Washington, D.C., 20 March 1995), 2–8.
9. Gordon G. Lill, "Gordon G. Lill Oral History Interview," American Heritage Center (6 February 1978), accession #8469, box 1, folder 14, University of Wyoming, Laramie.
10. Dietz, Dictabelts, 79–80.
11. Robert S. Dietz, "Bathyscaph *Trieste*," Office of Naval Research London, Technical Report #71-55, 10 August 1955. Also Piccard and Dietz, *Seven Miles Down*, 15–18.
12. Dietz's prescience was astounding. The *Trieste*'s dive into the Challenger Deep occurred just prior to the fourth anniversary of this presentation.
13. W. S. von Arx, ed. "Proceedings of the Symposium on Aspects of Deep Sea Research, 29 February–1 March 1, 1956," Pub. No. 473 (Washington, D.C.: Committee on Undersea Warfare, National Academy of Sciences, 1957), 121–136.
14. Ibid., 135.
15. Ibid., 172–176. The one word changed was "enormous," which became "far-reaching."
16. Ibid., iii.
17. Dietz, Dictabelts, 26.
18. Ibid., 101.
19. Piccard and Dietz, *Seven Miles Down,* 75–80.
20. Gary Weir, *An Ocean in Common*, 317. Also "Dr. Piccard Arrives for Scripps Talks," *San Diego Union* (7 March 1956), and "Piccard Repeats Lecture to Accommodate Big Crowd," *San Diego Union* (20 March 1956).
21. Charles B. Bishop, "U.S. Navy Involvement with DSV *Trieste*," *Marine Technology Society Journal* (Winter 2009), 15. Two secondary references, both by Will Forman, make identical statements: "Interested parties in the National Academy of Science got the Secretary of the Navy to lean on the head of ONR to 'do something about the *Trieste*'"; "From Beebe and Barton to Piccard and *Trieste,*" *Marine Technology Society Journal* (Winter 2009), 30; and *The History of American Deep Submersible Operations, 1775–1995* (Falstaff, Ariz.: Best, 1999), 27. The Bishop mission to Castellammare in late 1956 may have reflected such an initiative.
22. Piccard and Dietz, *Seven Miles Down*, 131.
23. Office of Naval Research, ONR Contract N62558-1342. NEL participation was funded under IO 15401, NE 120221-847.13 (NEL L4-1). See Dietz, Dictabelts, 27.
24. Capt. William G. Stearns II, USN, was plans officer for anti-submarine warfare for the Commander, Fleet Air Forces Eastern Atlantic and Mediterranean, and was interested in the

potential of passive acoustic detection of Soviet submarines. At Naples he met Andy Rechnitzer, who extended an invitation for a dive. Stearns II, "Three Generations of Piccards: Three World Records" (unpublished paper, 20 April 1999), available at: http://bertrandpiccard.com/files/miki_document/27/traditionfamiliale_enbref_usnavy-5190e26d0fd69.pdf.

25. Weir, *Ocean in Common*, 318.
26. Published papers from the 1957 *Trieste* dives include: Robert S. Dietz, "Deep-Sea Research in the Bathyscaph *Trieste*," *New Scientist* (April 1958); Dietz, "1000-Meter Dive in the Bathyscaph *Trieste*," *Limnology and Oceanography* (January 1959), 94–101; Arthur E. Maxwell, "The Bathyscaph: A Deep-Water Oceanographic Vessel, Part 1—A Report on the 1957 Scientific Investigations with the Bathyscaph, *Trieste*," *U.S. Navy Journal of Underwater Acoustics* (April 1958), 149–154; Jacques Piccard and Robert S. Dietz, "Oceanographic Observations by the Bathyscaph *Trieste* (1953–1956)," *Deep-Sea Research* (1957), 221–229.
27. Gordon G. Lill, "Oceanographic Activities in the Geophysics Branch, Office of Naval Research, 1950–1959," *Proceedings of the Royal Society of Edinburgh, Section B: Biology* (January 1972), 253.
28. Piccard and Dietz, *Seven Miles Down*, 132.
29. Andreas B. Rechnitzer, "The 1957 Diving Program of the Bathyscaph *Trieste*," NEL Report 941 (San Diego, Calif.: Navy Electronics Laboratory, 28 December 1959).
30. Office of Naval Research, "Bathyscaph Conference," 47.
31. Ibid., 55–56.
32. Switzerland, being landlocked, never had nor likely ever will have a navy. Mentions of the "Swiss Navy" in maritime societies are ironic. Office of Naval Research, "Bathyscaph Conference," 61.
33. "Navy Plans Studies with Bathyscaph," *San Diego Union* (8 February 1958).
34. Arthur E. Maxwell, quoted in Weir, *Ocean in Common*, 319–320.
35. Piccard and Dietz, *Seven Miles Down*, 115.
36. "Lill Oral History Interview," 20–21.
37. Piccard and Dietz, *Seven Miles Down*, 133; also Dietz, Dictabelts, 103.
38. "Bathyscaph Due for Sea Studies," *San Diego Union* (13 August 1958). Also see Don Walsh, "The Bathyscaph *Trieste*: Technological and Operational Aspects, 1958–1961," NEL Report 1096 (San Diego, Calif: Navy Electronics Laboratory, 27 July 1962), 57.
39. Dietz, Dictabelts, 74.
40. Dr. Heinfield Voss, Historisches Archiv Krupp, e-mail to Lee Mathers, 28 August 2013.
41. Piccard and Dietz, *Seven Miles Down*, 135.
42. Dietz, Dictabelts, 74–75.
43. Voss to Mathers.
44. Ibid.
45. Dietz, Dictabelts, 74–75.
46. "Lill Oral History Interview," 20–21.
47. Rear Adm. J. Edward Snyder, USN, Oceanographer of the Navy, foreword to R. Frank Busby, *Manned Submersibles* (Washington, D.C.: Office of the Oceanographer of the Navy, 1976), iii.
48. "Lill Oral History Interview," 20–21.
49. Arthur E. Maxwell, quoted in Weir, *Ocean in Common*, 320. If not an error in memory, the $1 million mentioned in Maxwell's oral history interview may include grants to the Navy Electronics Laboratory for management of the program and the first year of operation, and modification expenses in preparation for Project Nekton.

Chapter 5. The New World

Epigraph: Piccard and Dietz, *Seven Miles Down*, 124.

1. In 1959–1960 the *Antares* was converted to a stores-issue replenishment ship and redesignated AKS 32.
2. "Navy's Bathyscaphe *Trieste* Arrives for Sea Studies," *San Diego Union* (24 August 1958). Don Walsh, "Review of Bathyscaph *Trieste* Operations 1958–1961" (San Diego, Calif.: Navy Electronics Laboratory, December 1961), 1 and 57, incorrectly records that the *Trieste* arrived in San Diego on 2 September. The *San Diego Union* used the spelling "bathyscaphe" until 4 October 1958, when it shifted to "bathyscaph," which appears in all official Navy documentation. *The New York Times* and other national news organization used the French spelling for years thereafter.
3. "Navy's Bathyscaphe *Trieste* Arrives for Sea Studies." This was a poor estimate. Piccard would not arrive until late October, after which two months of preparations would be required before the first dive.
4. Ibid.
5. "Pacific Sub Chief Hails Bathyscaph," *San Diego Union* (4 October 1958). Grenfell subsequently served a Commander, Submarine Force Atlantic, with the rank of vice admiral.
6. Walsh, "Review of Bathyscaph *Trieste* Operations," 2.
7. "Jacques Piccard Arrives in S.D.," *San Diego Union* (24 October 1958).
8. Dietz, Dictabelts, 114.
9. Although having the rank of captain, Styles as a flotilla commander was called by the honorific "Commodore."
10. Don Walsh, "Going the Last Seven Miles," Naval Academy Alumni Association *Shipmate* (April 2000), 18.
11. Dietz, Dictabelts, 114.
12. Ibid.
13. John F. Light achieved notoriety as a U.S. Navy diver, as a civilian salvage diver, and as an underwater cameraman. After his dive on the *Trieste* he lobbied for a second *Trieste* dive but without success. He then turned his skills to making a film of the *Lusitania* and made 42 dives on the wreck. In 1967, Light purchased the wreck from Liverpool and the London War Risks Insurance Association for £1,000. See Piccard and Dietz, *Seven Miles Down*, 120.
14. Dietz, Dictabelts, 115. The YFU proved not to be well suited for towing the *Trieste* and its use was not repeated.
15. Dietz, Dictabelts, 75.
16. Dr. Heinfield Voss, Historisches Archiv Krupp, e-mail to Lee Mathers, 28 August 2013; also "*Pressnotiz* for the Indian Industries Fair, New Delhi," Krupp AG, 14 November 1961, 2–3.
17. Jacques Piccard, introduction to Krupp's *Pressnotiz* for the Indian Industries Fair, New Delhi, 2–3.
18. Voss to Mathers.
19. Dietz, Dictabelts, 115–117.
20. Voss to Mathers.
21. "Italian Prince to Visit City," *San Diego Union* (15 April 1959); also "Bathyscaph Crew Aided by Prince," *San Diego Union* (17 April 1959).
22. The new paint scheme consisted of Amercoat 85 (white) and Amercoat 33 (blue for stripes on the float), both selected for their anti-fouling characteristics. See Andreas B. Rechnitzer "Summary of Bathyscaph *Trieste* Research Program (1958–1960)," NEL Report 1095 (San

Diego, Calif.: Navy Electronics Laboratory, 2 April 1962), 51–55; also Walsh, "Review of Bathyscaph *Trieste* Operations," 4.

23. "Test Dives Set for Bathyscaph," *San Diego Union* (23 April 1959).
24. Busby, *Manned Submersibles*, 39; "331 Million Paid Ship Lines . . . ," *San Diego Union* (29 March 1959); "Harbor Notes," *San Diego Union* (17 April 1959); and "Danish Ship Unloads Cargo at B Street," *San Diego Union* (28 April 1959).
25. Busby, *Manned Submersibles,* Ibid.,136.
26. Davey had an impressive career, capped off by assignment as deputy program director of Program D (Advanced Technology) of the National Underwater Reconnaissance Office, 1971–1973, under the direction of Dr. Rechnitzer. As a captain, he returned to the program as the deputy program manager of the Special Naval Collection Program under Naval Intelligence, 1978–1980.
27. "Eggs Pass Stiff Sea Dive Test," *San Diego Union* (21 May 1959).
28. Walsh, "Review of Bathyscaph *Trieste* Operations," 7.
29. Rechnitzer, "Summary of the Bathyscaph *Trieste* Research Program Results," 8.
30. "Bathyscaph Dives to Near Record," *San Diego Union* (22 May 1959).
31. Rechnitzer, "Summary of the Bathyscaph *Trieste* Research Program Results," 8.
32. "Navy's Bathyscaph Goes to Work," *San Diego Union* (30 May 1959).
33. Piccard and Dietz, *Seven Miles Down,* 124.
34. Walsh, "Technological and Operational Aspects," 67.
35. Walsh, "Review of Bathyscaph *Trieste* Operations," 9–10.
36. Ibid., 10; Charles B. Bishop, "U.S. Navy Involvement with DSV *Trieste*," 15.
37. Walsh, "Going the Last Seven Miles," 18–19.
38. Walsh, "In the Beginning . . . A Personal View," *Marine Technology Society Journal* (Winter 2009), 10.
39. Walsh, "Going the Last Seven Miles," 18–19.
40. Rechnitzer, "Summary of the Bathyscaph *Trieste* Research Program Results," 51. During the spring of the 1959 diving season, the Amercoat system proved inadequate in preventing severe biologic fouling of the underwater portions.
41. Walsh, "Review of Bathyscaph *Trieste* Operations," 13–14.
42. Ibid., 14.
43. Ibid., 15.
44. Ibid.
45. Ibid., 16–17.
46. "Some Facts behind the Push to Explore 'Inner Space,'" *San Diego Union* (28 June 1959).

Chapter 6. The Ocean's Deepest Hole

Epigraph: Piccard, telegram to Yolanda Versich, 24 January 1960, quoted in Pietro Spirito, "Yolanda: The Woman Who Named the Piccard Bathyscaph *Trieste*," *Il Piccolo* (29 March 2009), 21.

1. Elizabeth Noble Shor, *Scripps Institution of Oceanography: Probing the Oceans 1936 to 1976* (La Jolla, Calif.: Scripps Institution of Oceanography, 1977), 291, http://scilib.ucsd.edu/sio/hist/gc29s5.pdf (accessed 29 July 2017).
2. The *Sumner*'s 2010 survey located the 36,070-foot hole (plus or minus 130 feet) at precisely 11.326344°N / 142.187248°E. Surveys from 1992 to 2008 had indicated that the deepest hole was at about latitude 11.22°N. GPS-derived data used by Gardner et. al., in 2010 placed

it 6.87 statute miles north of earlier calculations. See James V. Gardner, Andrew A. Armstrong, Brian R Calder, and Jonathan Beaudoins, "So, How Deep Is the Mariana Trench?" *Marine Geodesy* 37 (2014), 1–13; also Victor L. Vescovo e-mails to Lee Mathers, July–August 2019.

3. Piccard and Dietz, *Seven Miles Down*, 130.
4. Handwritten contemporaneous log maintained by Robert J. Hynosky, leading radioman on the USS *Lewis* [hereafter Hynosky archives].
5. Piccard and Dietz, *Seven Miles Down*, 147.
6. Walsh, "Review of Bathyscaph *Trieste* Operations," 21. See also Georges S. Houot, "Deep Diving by Bathyscaphe off Japan," *National Geographic Magazine* (January 1960). A footnote on page 150 reads, "As this issue went to press, word came of *Trieste*'s record-breaking descent to 18,600 feet—more than 3½ miles—in the Mariana Trench off Guam. The U.S. Navy dive by Dr. Andreas B. Rechnitzer and Swiss Jacques Piccard bettered the previous record by a full mile."
7. E. John Michel, "In the Trenches . . . Topside Remembrances by the Chief of the Boat, DSV *Trieste*," *Marine Technology Society Journal* (Winter 2009), 20–21.
8. Ibid., 51.
9. John D. Cawley, "Review of Marine Navigation Systems and Techniques," Project Trident Report 1510165 (Washington, D.C.: Bureau of Ships, 1965).
10. Dietz, Dictabelts, 110.
11. Ibid.
12. Dennis C. Jensen, a Philco technical representative stationed on Guam, volunteered his spare time to assist in readying the *Trieste* for diving. His initiative resulted in a position at the Navy Electronics Laboratory as Rechnitzer's assistant for the *Trieste* program.
13. Walsh, "Technological and Operational Aspects," 63–68. Piccard's dive log reprinted in *Seven Miles Down* does not include these harbor dives as he was not the pilot.
14. Kurie continued as civilian head of the Navy Electronics Laboratory until he was debilitated by a serious stroke in late 1960.
15. "Navy Bathyscaph to Attempt 7-Mile Plunge in Pacific," *San Diego Union* (29 December 1959).
16. Walsh, "Review of Bathyscaph *Trieste* Operations," 23.
17. Walsh, "Conquering Inner Space," U.S. Naval Institute *Proceedings* (January 1985), 103.
18. Piccard and Dietz, *Seven Miles Down,* 149–156.
19. Walsh, "Review of Bathyscaph *Trieste* Operations," 24.
20. Rear Adm. William L. Erdmann, message COMNAVMARIANAS 080514Z JAN 1960 and Adm. Herbert G. Hopwood, message CINCPACTFLT 080745Z JAN 1960, Hynosky archives. Also "Navy Deep Sea Diving Craft Breaks Its Own World Record," Department of Navy press release, 8 January 1960, Guam, Marianas.
21. Log entry for 6 January1960: "Underway for operations with *Trieste* at 0930L. Reporters aboard from *Life, London Daily Telegram,* and *National Geographic*," Hynosky archives.
22. Piccard and Dietz, *Seven Miles Down,* 149–156.
23. Walsh, "Review of Bathyscaph *Trieste* Operations," 25.
24. Ibid., 18.
25. Piccard and Dietz, *Seven Miles Down*, 156–157 [emphasis added].

26. Walsh, "Review of Bathyscaph *Trieste* Operations," 24.
27. Piccard and Dietz, *Seven Miles Down*, 157–158.
28. The *Life* article went to Don Walsh, who produced "Our 7-Mile Dive to Bottom," *Life* (15 February 1960), 113–120.
29. Dietz, Dictabelts, 24.
30. Piccard and Dietz, *Seven Miles Down*, 157.
31. Walsh, "In the Beginning," 10.
32. Dietz, Dictabelts, 24–27.
33. Ibid., 26.
34. "Scientist Tells of 7-Mile Dive," *San Diego Union* (3 February 1960).
35. Walsh, "Review of Bathyscaph *Trieste* Operations," 26–27.
36. Log entry for 19–23 January 1960: [19th] "Departed Guam (1430L [local time]) for ops with *Trieste*." [21st] "Sked [scheduled] dive cancelled, *Wandank* (ATA 204) not on station." [22nd] "Returning to Guam to pickup explosives. Departing dive area 0830L. 1800L rdvu [rendezvous] with YTB 408 approx 60 miles off Guam. Load ammo (hand grenades) and mail. Departed for dive area ETA [estimated time of arrival] 230030L." [23rd] "0100L dropping grenades for soundings. 0831L *Trieste* went down for record dive. 1430L rcvd [received] word from *Trieste* that she had reached 37,800 feet (6300 fathoms). 1647L *Trieste* came up. Had picture taken with Dr. Rechnitzer. 1800L enroute Guam." [24th] "In port Guam 0830L [berth] Sierra 2." Hynosky archives.
37. Piccard and Dietz, *Seven Miles Down*, 161.
38. Walsh, "Conquering Inner Space," 106; Jacques Piccard and Thomas J. Abercrombie, "Man's Deepest Dive," *National Geographic Magazine* (August 1960), 224–226. Available at http://www.deepseachallenge.com/the-expedition/1960-dive/.
39. Piccard and Abercrombie, "Man's Deepest Dive," 224–226.
40. Piccard and Dietz, *Seven Miles Down*, 162.
41. Walsh, "Our 7-Mile Dive to Bottom," 116.
42. Michel, "In the Trenches," 21.
43. Piccard and Dietz, *Seven Miles Down*, 163.
44. For Rolex history, http://www.rolexmagazine.com/2010/01/deepest-deep-sea-dive-in-history-us.html. Also, Walsh, e-mail to Lee Mathers, 15 September 2015. Walsh was entirely unaware of the Rolex "conspiracy" until the company's advertising displayed the watch and its history. Knowing the prohibition on using Navy property for commercial purposes, he would have stopped the plan had he known of its existence.
45. Michel, "In the Trenches," 21–22. Another source quotes the depth of *Trieste* as 10,000 feet in the response message.

Chapter 7. The Deepest Dive

Epigraph: Voss to Mathers; "*Pressnotiz* for the Indian Industries Fair, New Delhi," 5.

1. Piccard and Abercrombie, "Man's Deepest Dive," 224–239.
2. Piccard and Dietz, *Seven Miles Down*, 165. Slightly different depths for the multiple thermoclines are given in Piccard's *National Geographic* article and in Walsh's *Life* magazine article.
3. Forman, "From Beebe and Barton to Piccard and *Trieste*," 31; also Voss to Mathers; and Piccard, "Introduction to Krupp AG's *Pressnotiz*," 5.

4. Walsh, "Our 7-Mile Dive to Bottom," 114.
5. Piccard and Dietz, *Seven Miles Down*, 168.
6. Ibid.
7. Walsh, "Dive #70 by the Bathyscaph *Trieste,*" acoustic recording transcript (San Diego, Calif.: Navy Electronics Laboratory, 23 January 1960), transcript 11. This leak partially repaired itself about 1505 during ascent and slowed to a manageable inflow.
8. Piccard and Dietz, *Seven Miles Down*, 167–171.
9. Walsh "Our 7-Mile Dive to Bottom," 116.
10. Piccard and Dietz, *Seven Miles Down*, 171.
11. Walsh "Our 7-Mile Dive to Bottom," 116.
12. Conflict between primary sources: two sources state, identically, "without formal discussion" (Piccard and Dietz, *Seven Miles Down*, 172, and Walsh, "Our 7-Mile Dive to Bottom"). Piccard and Abercrombie, "Man's Deepest Dive," recounts the following exchange, Piccard says, "In my opinion, it isn't anything serious—we are not losing any gasoline. Let's go on and we'll see later." To that Walsh replies, "OK."
13. Walsh, "Our 7-Mile Dive to Bottom," 119.
14. Dietz, Dictabelts, 21.
15. Walsh, "Our 7-Mile Dive to Bottom," 110–121.
16. Walsh, "Dive #70 by the Bathyscaph *Trieste*," transcript 11. Also, Jacques Piccard, Dictabelts, cassette #1 transcript, 14.
17. Alan J. Jamison and Paul H. Yancey, "On the Validity of the *Trieste* Flatfish: Dispelling the Myth," *Biological Bulletin*, no. 3 (2012), 171, http://www.journals.uchicago.edu/doi/full/10.1086/BBLv222n3p171.
18. Voss to Mathers. Also Piccard, "Introduction to Krupp AG's *Pressnotiz,*" 5.
19. On 10 March 1960, Arthur E. Maxwell at the Office of Naval Research released corrected depth calculations: 35,803 feet plus or minus 16 feet. The *Guinness World Records* lists the dive at 35,797 feet.
20. Voss to Mathers. Also Piccard, "Introduction to Krupp AG's *Pressnotiz,*" 4. Piccard explained that sound takes 14 seconds to traverse 11,000 fathoms, up then back down; "in air it would have taken 5 times longer."
21. Don Walsh, "Dive #70 by the Bathyscaph *Trieste*," 11. Also, Voss to Mathers; and Piccard, "Introduction to Krupp AG's *Pressnotiz*," 4.
22. Piccard and Abercrombie, "Man's Deepest Dive," 238.
23. Piccard and Dietz, *Seven Miles Down,* 159. There were 100 U.S. flags and one Swiss flag folded and stowed in the bottom of the sphere. Piccard made certain that the bottom flag, *the deepest flag,* was Swiss.
24. Piccard, Dictabelts, 85.
25. Ibid., 16.
26. Christina Reed, *Marine Science: Decade by Decade* (New York: Facts on File, 2009), 132. Dräger cans contain a chemical that removed carbon dioxide from the sphere's atmosphere in an exothermic reaction; when activated the canisters generate warmth. See Walsh, "Dive #70 by the Bathyscaph *Trieste*," 14.
27. Walsh, "Our 7-Mile Dive to Bottom," 120.
28. Voss to Mathers. Also Piccard, "Introduction to Krupp AG's *Pressnotiz*," 5.

29. Walsh, "Our 7-Mile Dive to Bottom," 120. Also, Voss to Mathers; and Piccard, "Introduction to Krupp AG's *Pressnotiz*," 5.
30. Andreas B. Rechnitzer, "Supplement to Walsh's Dive #70 Recording" (San Diego, Calif.: Navy Electronics Laboratory, 23 January 1960), 15–17.
31. Piccard and Dietz, *Seven Miles Down*, 159.
32. Walsh, "Review of Bathyscaph *Trieste* Operations," 27.
33. Paolo Valenti, *Storia del Contiere Navale di Monfalcone* [History of the Monfalcone Naval Shipyard] (Trieste: Edizioni Luglio, 2007), http://www.betasom.it/forum/index.php?showtopic=37983.
34. Voss to Mathers; Krupp AG, "*Pressnotiz* for the Indian Industries Fair," 3.
35. Pietro Spirito, "Yolanda: The Woman Who Named the Piccard Bathyscaph *Trieste*," *Il Piccolo* (29 March 2009), 21.
36. Walsh, "Review of Bathyscaph *Trieste* Operations," 37.
37. Walsh, "Going the Last Seven Miles," 18.
38. Walsh, "In the Beginning," 12.
39. Piccard and Abercrombie, "Man's Deepest Dive," 239, photo caption.
40. "Bathyscaph's Team Honored," *San Diego Union* (5 February 1960).
41. Walsh e-mail to Lee Mathers, 25 September 2015: "We carried 100 American flags. We wanted to make sure we covered as many persons of influence as we could. A lot went to members of Congress and senior Navy officials. We also carried about 100 philately 'covers' each signed by Jacques and myself during the long flight back to D.C. after our deepest dive."
42. Piccard and Dietz, *Seven Miles Down,* 180.
43. "NEL Men to Pilot Bathyscaph Dive," *San Diego Union* (13 February 1960).
44. "NEL Team to Tell of Sea Research," *San Diego Union* (20 April 1960).
45. "Navy Plans Award for S.D. Scientist," *San Diego Union* (14 May 1960).
46. Bertrand Piccard notes that participants in the drift of the Gulf Stream by Jacques Piccard's mesoscaphe *Ben Franklin* in July–August 1969 included two representatives of the Naval Oceanographic Office and a former Navy submarine officer as pilot. Bertrand Piccard, e-mail to Lee Mathers, 11 October 2017.
47. "Swiss Inventor of Bathyscaph Going Home," *San Diego Union* (17 March 1960); "Piccard Plans Better Craft," *San Diego Union* (27 March 1960).
48. "Swiss Scientist Entertained," *San Diego Union* (8 November 1960).

Chapter 8. The Navy Electronics Laboratory Years

Epigraph: Don Walsh, "Review of Bathyscaph *Trieste* Operations 1958–1961" (San Diego, Calif.: Navy Electronics Laboratory, December 1961), 59.

1. Voss to Mathers; Krupp AG, "*Pressnotiz* for the Indian Industries Fair," 2.
2. Arthur E. Maxwell, quoted in Weir, *Ocean in Common*, 320.
3. Voss to Mathers; Piccard, "Introduction to Krupp AG's *Pressnotiz*," 5.
4. Walsh, "Our 7-Mile Dive to Bottom," 121.
5. Walsh, "Technological and Operational Aspects," 62–63.
6. Walsh, "Review of Bathyscaph *Trieste* Operations," 33.
7. Ibid., 32.
8. Ibid., 35.

9. Rechnitzer, "Summary of the Bathyscaph *Trieste* Research Program Results," 33.
10. The *Trieste*'s sphere was only one-half the volume of the Gemini space capsule and considerably less than that of an Apollo capsule.
11. Walsh, "Technological and Operational Aspects," 62–64, and "Review of Bathyscaph *Trieste* Operations," 36.
12. "Review of Bathyscaph *Trieste* Operations," 36, 38.
13. Richard D. Terry, *The Deep Submersible* (New York: Western, 1966), 193.
14. Rechnitzer, "Summary of the Bathyscaph *Trieste* Research Program Results," 30–31; Terry, *Deep Submersible*, 193.
15. Rechnitzer, "Summary of the Bathyscaph *Trieste* Research Program Results," 33.
16. Walsh, "Review of Bathyscaph *Trieste* Operations," 39.
17. Ibid., 40.
18. Walsh, "Technological and Operational Aspects," 62–66.
19. Ibid., 41.
20. Ibid.
21. George W. Martin, e-mail to authors, 23 July 2014.
22. Walsh, "Technological and Operational Aspects," 42.
23. Martin Klein, "Martin's Story: Bathyscapth *Trieste* Alumni Association," n.d., https://web.archive.org/web/20120112034423/www.bathyscaphtrieste.com/contents/story/marty.html.
24. Lawrence A. Shumaker, "Evaluation of External Lighting Systems for the Bathyscaph *Trieste*," NEL Report 1094 (San Diego, Calif.: Navy Electronics Laboratory), 21 December 1961.
25. Walsh, "Review of Bathyscaph *Trieste* Operations," 44–48.
26. "Bathyscaph Gets Key Navy Role," *San Diego Union* (14 May 1962).
27. Walsh, "Review of Bathyscaph *Trieste* Operations," 53, and "Technological and Operational Aspects," 52–53.
28. David W. Waltrop, Don Walsh, Richard Taylor, Beauford Myers, and Lee J. Mathers, "CIA's Underwater Space Mission Revealed" (Chantilly, Va.: Smithsonian National Air & Space Museum, 26 April 2013), video recording "CIA's Underwater Space Mission Revealed," US Stream TV, David Waltrop, 26 April 2013, http://www.ustream.tv/recorded/32000249. Walsh's comments at video time 01:30–01:35.
29. "Capsule Lost Forever, Expert Says," *San Diego Union* (22 July 1961).
30. "Repairs Delay Dive Series of *Trieste*," *San Diego Union* (21 September 1961).
31. "*Trieste* Finds Water Clear as Air at 3,840 Feet," *San Diego Union* (24 June 1962); "'Fireworks' Seen from Bathyscaph," *San Diego Union* (26 April 1962); "Bathyscaph in Dive of 4,000 Feet," *San Diego Union* (27 October 1961).
32. Eric G. Barham, "The Deep Scattering Layer as Observed From the Bathyscaph *Trieste*," *Proceedings of the 16th International Congress of Zoology* (Washington, D.C., 27 August 1963), 298–300.
33. Ibid.
34. Bryant Evans, "Apartments at Sea? *Trieste* Finds Fish Have 'Em," *San Diego Union* (1 July 1962); Eric G. Barham, N. J. Ayer Jr, and R. E. Boyce of the Navy Electronics Laboratory, "Megabenthic Fauna in the San Diego Trough: Photographic Study from Bathyscaph *Trieste*," Second International Oceanographic Congress, Moscow, USSR, May–June 1966, in *Deep Sea Research and Oceanographic Abstracts* (Oxford [U.K.]: Pergamon, 1967), 773–784.

35. Bryant Evans, "*Trieste*: Its Findings Tell New Sea Theory," *San Diego Union* (17 May 1962).
36. "Bathyscaph Gets Key Navy Role."
37. Hugh Bradner (1915–2008), obituary, https://scripps.ucsd.edu/news/2462 (accessed September 2015). Bradner was the inventor of the diving wetsuit and other diving-related innovations, as well as a noted geophysicist.
38. Walsh to Mathers, 27 September 2015.
39. "Sub Pay Denied Bathyscaph Diver," *San Diego Union* (28 May 1960); "Senate Votes *Trieste* Crew Hazard Pay," *San Diego Union* (30 June 1960); "Congress OKs *Trieste* Hazardous Duty Pay," *San Diego Union* (1 July 1960); and "Two San Diegans Win Bathyscaph Duty Pay," *San Diego Union* (21 September 1962).
40. Don Walsh, "Oceans: *Alvin* and *Trieste*," U.S. Naval Institute News (9 November 2012), http://news.usni.org/2012/11/09/oceans-alvin-and-trieste.
41. See https://www.whoi.edu/what-we-do/explore/underwater-vehicles/hov-alvin/dive-statistics/; and https://www.capecodtimes.com/news/20200329/whois-alvin-to-reach-new-depths-after-8m-upgrade.
42. "Position Filled on Bathyscaph," *San Diego Union* (23 May 1962).
43. "Deep Sea Pilot to Be Honored," *San Diego Union* (1 July 1962).
44. "Bathyscaph's Walsh Transferred to Sub," *San Diego Union* (14 July 1962).
45. The *Archimède* operated into the 1970s; she then was placed in operational reserve at Toulon.

Chapter 9. SubSunk

Epigraph: Wakelin, "*Thresher*: Lesson and Challenge," 761.

1. Frank A. Andrews, "Searching for the *Thresher*," Naval Institute *Proceedings* (May 1964), 69–77. Also see N. Polmar, *Death of the* Thresher (Philadelphia: Chilton Books, 1964); revised as *The Death of the USS* Thresher (Guilford, Conn.: Lyons, 2001). These books are based in large part on interviews with the first commanding officer of the *Thresher*, then-Cdr. Dean Axene, USN. Also, Bruce Rule, *Why the USS* Thresher *(SSN-593) Was Lost* (Ann Arbor, Mich.: Nimble Books, 2017).
2. James H. Wakelin Jr., "*Thresher*: Lesson and Challenge," *National Geographic Magazine* (June 1964), 761.
3. Bishop, "U.S. Navy Involvement with DSV *Trieste*," 15. Also Naomi Oreskes, "A Context of Motivation: US Navy Oceanographic Research and the Discovery of Sea-Floor Hydrothermal Vents," *Social Studies of Science* (1 October 2003), 710.
4. Kenneth V. Mackenzie, "Early History of Deep Submergence Navigation aboard *Trieste*," *Navigation: Journal of the Institute of Navigation* (Spring 1970).
5. George W. Martin, interviews with Lee Mathers, May–July 2014.
6. Herbert L. Graybeal, "Design of the Bathyscaph *Trieste II*," First U.S. Naval Symposium on Military Oceanography, Naval Oceanographic Office, Washington, D.C., June 1964, 2. Primary design criteria were: reserve buoyancy sufficient to allow for loss of any two gasoline tanks; towing speed of ten knots; and electrical underwater propulsion endurance of eight hours at two knots. Calculations indicated that mounting the Krupp sphere would reduce reserve buoyancy by about a ton, thus limiting payload and adding about 1.5 tons to the air weight of the *Trieste II*.
7. George W. Martin, "*Trieste*: The First Ten Years," Naval Institute *Proceedings* (August 1964), 58.

8. Donald L. Keach, "Oral History Interview, Donald Keach," by Gary Weir (Washington, D.C.: Naval History and Heritage Command, 1995).
9. "NEL Alerted: Bathyscaph May Aid Hunt," *San Diego Union* (11 April 1963).
10. "LSD Picked to Ferry *Trieste* East," *San Diego Union* (12 April 1963).
11. "Navy Restudies *Thresher* Design: Plans 22 Craft," *The New York Times* (13 April 1963).
12. "Piccard Willing to Help," *The New York Times* (13 April 1963); also "LSD Picked to Ferry *Trieste* East."
13. "LSD Picked to Ferry *Trieste* East."
14. Martin interviews.
15. "*Trieste* Commander Explains Deep Diver," Associated Press (19 April 1963). Also, Polmar, *Death of the Thresher,* 65–68.
16. "Anderson Sees Little Hope of Sub Recovery," *The New York Times* (12 April 1963).
17. Keach, "Oral History Interview."
18. Martin interviews.
19. Ibid.
20. Andrews, "Searching for the *Thresher.*"
21. "*Trieste* Due for Descent," *Utica* (N.Y.) *Observer-Dispatch* (23 June 1963).
22. "*Trieste* Readied for Test Dive," *Bridgeport* (Conn.) *Post* (1 May 1963).
23. "*Trieste* in 702-Foot Trial Plunge," *Lowell* (Mass.) *Sun* (5 May 1963).
24. Martin interviews.
25. Lawrence A. Shumaker, "*Trieste II,*" *Naval Engineers Journal* (August 1964), 515–517.
26. Martin, "First Ten Years," 60.
27. Frank A. Andrews, "Search Operations in the *Thresher* Area: 1964, Section II," *Naval Engineers Journal* (October 1965), 776.
28. Donald L. Keach, "Down to *Thresher* by Bathyscaph," *National Geographic* (June 1964), 774.
29. Andrews, "Searching for the *Thresher.*"
30. Ibid. Also, Edward E. Henifin, interviews with Lee Mathers, June–September 2014; also Kenneth V. Mackenzie, "Early History of Deep Submergence Navigation aboard *Trieste,*" 10–11. Mackenzie provides a complete description of the "cookie" system. Henifin suspected that the non-deployment occurred because the cookies had been in the direct sun on the fantail of the *Allegheny,* where the heat caused them to set permanently in a rolled configuration.
31. Andrews, "Searching for the *Thresher,*" 77.
32. R. D. Gerald, "Taut Line Navigation Buoys Used in the *Thresher* Search," Marine Technology Society Meeting #24 (1964), 57. Also, Mackenzie, "Early History of Deep Submergence Navigation aboard *Trieste,*" 5. Mackenzie also discusses attempts to add a transponder close to the buoy's anchor.
33. Frank A. Andrews, "Search Operations in the *Thresher* Area: 1964, Section I," *Naval Engineers Journal* (August 1965), 558–559.
34. Sydney T. Knott, "Navigational Techniques Used in the *Thresher* Search," Annual National Meeting of the Institute of Navigation, New York, 15 June 1964. Knott describes a similar "short line" acoustic navigation system fielded by WHOI on the *Atlantis II* near the end of the *Thresher* search in 1963. Knott received the Navy's Meritorious Public Service citation for his work during the 1963 *Thresher* search; Vicky Cullen, *Down to the Sea for Science* (Woods

Hole, Mass.: Woods Hole Oceanographic Institution, 2005), 117; and Kenneth V. Mackenzie, "Position Determination under the Sea," Annual Conference, Marine Technology Society, Washington, D.C., 29 June 1966; and Ramesh N. Vaishnav and Edward B. Magrab, "Exact Theory of Tracking of Moving Underwater Object by Short Base Navigation System Attached to an Imperfectly Stabilized Moving Ship," *The Journal of the Acoustical Society of America* (March 1970), 912. Online at: http://asa.scitation.org/doi/10.1121/1.1911978 (accessed 27 July 2017).

35. Andrews, "Searching for the *Thresher*," 77.
36. Ibid., 518.
37. Martin interviews.
38. Andrews, "Search Operations in the *Thresher* Area: 1964, Section II," 771.
39. Martin interviews.
40. Andrews, "Search Operations in the *Thresher* Area: 1964, Section II," 778.
41. Ibid.
42. The *Fort Snelling* had a crew of approximately 340 plus berthing for 325 troops.
43. Andrews, "Search Operations in the *Thresher* Area: 1964, Section II," 778.
44. Andrews, "Searching for the *Thresher*."

Chapter 10. The *Thresher* Search, 1963

Epigraph: Keach, "Down to *Thresher* by Bathyscaph," 775.

1. "*Trieste* Geared for Dive Today," *The New York Times* (23 June 1963).
2. Andrews, "Searching for the *Thresher*," 77.
3. Keach, "Down to *Thresher* by Bathyscaph," 770.
4. Keach, "Oral History Interview."
5. Martin interviews.
6. Jacques Piccard dumped both tubs for an emergency surfacing on Dive #7 in 1953.
7. Martin interviews. Newspaper articles at the time mentioned the *Trieste*'s bouncing up to 480 feet above the bottom; Martin remembered it more like 900 feet.
8. Keach, "Down to *Thresher* by Bathyscaph," 766.
9. Ibid., 772.
10. The Polish-built *Kuprin* was a stern-trawler; her gross tonnage was 2,900 tons, and she was 280 feet long.
11. Eugene J. Cash interviews with Lee Mathers, June–September 2014.
12. "Deadline Nearing in *Thresher* Search," *San Diego Union* (27 June 1963).
13. "*Trieste* Makes Fourth Dive in Search of Nuclear Sub *Thresher*," (Provo, Utah) *Sunday Herald* (30 June 1963).
14. Cash interviews.
15. Ibid.
16. "Hunt for *Thresher* Is Ended by *Trieste*," *The New York Times* (1 July 1963); "S.D. Grandfather Aids Sub Search," *San Diego Union* (1 July 1963); and "Hope of Finding *Thresher* Voiced," *San Diego Union* (19 July 1963).
17. Frank A. Andrews, "Underwater Navigation," in *Hydronautics,* ed. H. E. Sheets and V. T. Boatwright Jr. (New York: Academic, 1970), 233.
18. Ibid., 234–235.

19. Martin interviews.
20. Andrews, "Underwater Navigation," 234–235.
21. Keach, "Down to *Thresher* by Bathyscaph," 775–776.
22. Henifin interviews.
23. Polmar, *Death of the* Thresher, 78. Henifin, who would relieve Brad Mooney as officer-in-charge of the *Trieste*, would still later, in the Pentagon, become involved in the *Trieste*'s classified operations.
24. Andrews, "Underwater Navigation," 235.
25. Mackenzie, "Early History of Deep Submergence Navigation aboard *Trieste*," 5–6.
26. "*Trieste* Geared for Dive Today."
27. Fred Korth, "Statement of Secretary of the Navy Fred Korth on 5-Month-Long Search for Submarine *Thresher*," Press Release No. 1207-63 (5 September 1963).
28. Ibid.; John Maier, "USS *Thresher* Banners Installed on the USS *Albacore*," *Anchor Watch: The Journal of the Historic Naval Ships Association* (May–August 2013), 17, https://archive.hnsa.org/anchorwatch/pdf/2013summer.pdf (accessed 18 July 2017). Walter Brundage, "NRL's Deep Sea Floor Search Era: A Brief History of the NRL/*Mizar* Search System and Its Major Achievements," NRL Report 6208 (Washington, D.C.: Naval Research Laboratory, 1988), records the location of the *Thresher*'s wreck as 41°44 N / 64°-56 W.
29. Korth, "Statement of Secretary of the Navy."
30. "3 Leaders of *Thresher* Hunt Are Decorated by Navy," *The New York Times* (7 September 1963).
31. "Navy Honors *Trieste*, Crew," (Salt Lake City, Utah) *Deseret News* (9 October 1963).
32. Andrews, "Searching for the *Thresher*."
33. "*Trieste* Officer Takes Sub Job," *San Diego Union* (24 September 1963).
34. Martin interviews.
35. Shumaker, "*Trieste II*," 514.
36. Walsh to Mathers, 25 September 2015.
37. Martin interviews.
38. Frank A. Andrews, "My View: Submarine Development Group 2," 50th Anniversary of Submarine Development Group 2, New London, Conn., May 1999, 6.
39. "Bathyscaphe *Trieste* Is Left Rusting," *Somerset (Penna.) Daily American* (13 September 1967).
40. Acting Secretary of the Navy reply to Smithsonian Institution letter dated 30 March 1967 and Smithsonian reply dated 1 May 1967, copies in the coauthors' files; "Bathyscaph *Trieste* Going to Washington," *Long Beach Independent* (21 December 1967).

Chapter 11. The *Trieste II*

Epigraph: Graybeal, "Design of the Bathyscaph *Trieste II*," 5.

1. "MI-Built *Trieste* Float to San Diego: Bathyscaph Job Done Here Helps Navy in Depth Diving," *Mare Island Grapevine* (8 November 1963).
2. "Deep-Diving Craft to Be Built at Yard," *Mare Island Grapevine* (8 February 1963).
3. Graybeal, "Design of the Bathyscaph *Trieste II*," 7.
4. The Model Basin is located at Carderock, Md., a suburb of Washington, D.C.
5. Graybeal, "Design of the Bathyscaph *Trieste II*," 6.

6. "Rep. Bob Wilson," *San Diego Union* (3 November 1963).
7. "*Trieste* Relaunched," *San Diego Union* (18 January 1964).
8. Ibid.
9. Staehle interviews with Lee Mathers, July–August 2014.
10. Brad Mooney's flag-level career included service as the Oceanographer of the Navy and as the Chief of Naval Research; he passed away on 31 May 2014, having given several interviews for this book. Keach had three back-to-back command tours: in the hunter-killer submarine *K-1* (SSK 1), the *Trieste,* and the *Darter* (SS 576). His final active-duty Navy assignment was Deputy Director of Navy Laboratories, 1971–1973; he was relieved in that position by Don Walsh.
11. Adm. David L. McDonald, Chief of Naval Operations, letter to Ed Link, 2 May 1964.
12. "*Thresher*: The Navy Learns from Disaster," *San Diego Union* (5 April 1964).
13. "Hunt Renewed for *Thresher*," The New York Times News Service (10 June 1964), as appearing in the *Ottawa Journal* (11 June 1964).
14. "*Trieste II* Due in Boston," First Naval District News Release #83-64 (3 April 1964).
15. "Bathyscaph *Trieste II* Arrives in Boston," *San Diego Union* (4 May 1964).
16. "Navy Launches New Bathyscaph," *Newport Daily News* (25 May 1964).
17. George L. Hersch later became a licensed clinical psychologist, practicing in Berkeley, Calif. He received a PhD in psychology from the Wright Institute, University of Maine, and a second PhD, in zoology, from the University of California at Berkeley.
18. In 1966, Martin was with the Naval Ship Systems Command in Washington, where he participated in the underwater search for the nuclear weapon lost off Palomares, Spain. After retiring from active service as a captain he joined Lockheed, working on the Navy's Deep Submergence Rescue Vehicle (DSRV) program.
19. Andrews, "Search Operations in the *Thresher* Area: 1964, Section II," 774.
20. "Hydronaut Reveals Spooky Tales of *Trieste*'s Exploits," *San Diego Union* (14 April 1966).
21. There were later incidents of shot rusting in the ballast tubs. In such cases it was necessary for *Trieste* crewmen to enter the tubs via topside access hatches and chip them out entirely, using hammers and chisels.
22. Lt. Cdr. Mooney letter to Lt. George Martin, 28 August 1964. Mooney started the letter on 28 August, but did not complete and send it until 15 October.
23. "Bathyscaphe Is Taken Out for Deep Sea Tests," Associated Press (8 June 1964); also "Bathyscaphe Returns," *Bridgeport* (Conn.) *Telegram* (12 June 1964).
24. Mooney to Martin.
25. Rear Adm. Millard Firebaugh interview with Lee Mathers, 17 July 2015.
26. "*Trieste* to Renew *Thresher* Hunt," *San Diego Union* (12 June 1964).
27. In 1964 an advertising campaign for the boxed rice-and-pasta mix "Rice-A-Roni" hit television and radio with a catchy jingle that wormed its way into the national subconscious. Thus, "stoetzeroonie" was a play on it off Ray Stoetzer's name. Capt. Andrews established the definitive spelling of this "technical" term in "Search Operations in the *Thresher* Area: 1964, Section II," 776.
28. Andrews, "Underwater Navigation," 263–265.
29. Ray Stoetzer letter to the Lee Mathers, September 2015, enclosing a photo of the *Trieste II* in 1964 signed by Mooney and inscribed to Stoetzer with the words, "Your Stoetzeroonies helped!" See chapter 21 for a *Trieste* operation in 1977 in which the stoetzeroonies played a key role.

30. "*Trieste II* Commander Speaks to Rotary Club," *Portsmouth Herald* (10 July 1964).
31. Mooney to Martin.
32. Brundage, "NRL's Deep Sea Floor Search Era," 2.
33. Chester L. Buchanan, "A Love Affair," NRL Progress Report 1973, 82–83; reprinted in Brundage, "NRL's Deep Sea Floor Search Era," 14–15; also Brundage, "NRL's Deep Sea Floor Search Era," 2.
34. "In actual operation, standard deviations of 200-300 feet in position [of the camera, thus of photographs] occurred at depth of 10,000 feet, even under the most favorable environmental conditions: Naval Research Laboratory, *Mizar*'s Underwater Search System," NRL Progress Review 1971 (Washington, D.C., 1971), 206.
35. Andrews, "Search Operations in the *Thresher* Area: 1964, Section I," 552.
36. Firebaugh interview with Lee Mathers.
37. Brundage, "NRL's Deep Sea Floor Search Era," 3.
38. Andrews, "Search Operations in the *Thresher* Area: 1964, Section I," 553.
39. The tradition began when Dutch admiral Maarten Tromp, after a decisive victory at the battle of Dungeness in 1652, hung a broom from his mast to indicate that he had "swept the British from the seas."
40. Buchanan, "Love Affair," 82–83.
41. Andrews, "Search Operations in the *Thresher* Area: 1964," Section I, 553; Section II, 775 [emphasis added].
42. Mooney to Martin.
43. Ibid.
44. A reversal of the center of buoyancy and the center of gravity is a threat to most deep submersibles and particularly to a bathyscaph, with its large flotation tank over a small but heavy personnel sphere.
45. Mooney to Martin.
46. Andrews, "Search Operations in the *Thresher* Area: 1964, Section I," 553.
47. Martin Klein, Lab Book, Klein Associates, 4 August 1964; http://www.bathyscaphtrieste.org/contents/klein/image35.htm.
48. The authors of this volume have been unsuccessful in positively identifying "Fitzgerald." He may have been a representative from the Navy Electronics Laboratory, replacing Mackenzie for that dive.
49. Mooney to Martin.
50. Ibid.
51. Dennis Curtis interview with Lee Mathers, August 2014.
52. Buchanan, "Love Affair," 82–83. Also Mooney, interviews with Lee Mathers, April–October 2012; Howland interviews with Lee Mathers, July–August 2014.
53. Weir, *Ocean in Common*, 328–329.
54. Curtis soon returned to Yale University, earned a law degree, and made a career as a legal educator. In 2017 he was Clinical Professor Emeritus of Law at Yale's Law School.
55. Andrews, "Underwater Navigation," 235–236.
56. "*Trieste* Geared for Dive Today."
57. Andrews, "Search Operations in the *Thresher* Area: 1964, Section II," 773.
58. Naval Research Laboratory, "1966 NRL Abstracts" (Washington, D.C., 1966), 1–2.

59. "Bathyscaph Coming Home," *San Diego Union* (9 October 1964).
60. "Bathyscaph, Racing Yacht Arrive on Ship," *San Diego Union* (30 October 1964).

Chapter 12. Between Disasters

Epigraph: Paul M. Fry, "Submarine Which Found Missing H-Bomb Achieved Fame on Its First Assignment," *Albuquerque Journal* (20 March 1966), 3.

1. Fred Korth, "Deep Submergence Systems Review Group: Establishment Of," SecNav Notice 3100, Washington, D.C. 24 April 1963.
2. "Sub *Thresher* Sinking Made Improvements," (Corona, Calif.) *Daily Independent* (22 June 1966).
3. Korth, "Deep Submergence Systems Review Group."
4. "Navy Recalls Ex-*Nautilus* Chief to Head Rescue Study," *Corpus Christi* (Tex.) *Caller-Times* (26 April 1963).
5. General Dynamics, "DSSP Large Object Salvage Study," prepared for Navy Special Projects Office, Bureau of Naval Weapons Contract NOsp 65185-c (Groton, Conn., 1965).
6. Korth, "Deep Submergence Systems Review Group."
7. Ibid.
8. Ibid.; Rear Adm. Edward C. Stephan, "Report of the Deep Submergence Systems Review Group: Summary Report," NavExOS P-2452 (Washington, D.C.: Executive Office of the Secretary of the Navy, 1 March 1964), 15.
9. Rear Adm. Ignatius J. Galantin, "Deep Submergence Systems Project; Designation Of," NavMatInst 5430.24 (Washington, D.C.: Naval Material Command, 9 February 1966), 2; and John Craven, "The Navy's Deep Submergence Systems Project," *Undersea Technology* (December 1966), 6.
10. John Piña Craven, *The Silent War: The Cold War Battle beneath the Sea* (New York: Simon & Schuster, 2001), 109.
11. Craven, *Silent War*, 233.
12. Staehle interviews.
13. "How 'Scaph Will Look: Latest Dope," *Mare Island Grapevine* (21 May 1965).
14. The description of the CURV system can be found on the Wayback Machine: https://web.archive.org/web/20161008061956/http://www.public.navy.mil/spawar/Pacific/Robotics/Pages/CURV.aspx (accessed 1 August 2017).
15. Michel interviews with Lee Mathers, June–August 2014.
16. "Donald Saner Named Pilot of *Deep Quest*," *Van Nuys* (Calif.) *News* (6 February 1968).
17. E. W. Seabrooke Hull, "*Trieste II* Leads in DSR/V Electronics," *Geo-Marine Technology* (May 1966), 20–24.
18. Hull, "*Trieste II* Leads in DSR/V Electronics," 20–24. The periscope arrangement was little used, and another five years passed before it reappeared in a more acceptable form.
19. Ibid.
20. The best published account of the incident is Ted Szulc, *The Bombs of Palomares* (New York: Viking, 1967).
21. Craven, *Silent War*, 165.
22. John Van Voorhis interview with Lee Mathers, 31 August 2015.
23. The *NR-1* was completed in 1969, developed by Adm. H. G. Rickover as a test platform for a small nuclear power plant. Subsequently it was used as a deep-ocean research and recovery

vehicle. It displaced 372 tons submerged and carried a crew of five. Its publicly reported operating depth was 3,000 feet. The *NR-1* was taken out of service in 2008; proposals to build an *NR-2* as an HY-130 material test vehicle were abandoned.

24. "*Trieste II* Pilot Cited," *San Diego Union* (14 April 1966).
25. "Navy Unit Gets New CO," *San Diego Union* (20 May 1966). After a superior performance in command of the *Menhaden*, including two patrols in the Tonkin Gulf, Mooney was involved in locating the sunken Soviet missile submarine *K-129,* which was partially salvaged by the Central Intelligence Agency in 1974; see N. Polmar and Michael White, *Project Azorian: The CIA and the Raising of the K-129* (Annapolis, Md.: Naval Institute Press, 2010).
26. "*Trieste* Returns from Renovation," *San Diego Union* (25 May 1966).
27. "*Trieste* to Return to Ocean Depths," *San Diego Union* (19 June 1966).
28. "Navy's First Class of Hydronauts Delayed, Awaits Repairs on *Trieste,*" *San Diego Union* (14 July 1966).
29. Edward E. Henifin e-mail to Lee Mathers, 22 July 2014.
30. Hull, "*Trieste II* Leads in DSR/V Electronics," 24.
31. "*Pensate Profunde* Their Motto," *Indiana* (Penna.) *Gazette* (11 July 1966), 35.
32. "Navy in Dire Need of Hydronauts," (Fort Walton Beach, Fla.) *Playground Daily News* (11 July 1966), 10.
33. The bathyscaph simulator/trainer was relocated from Charlottesville, Va., to Ballast Point, Point Loma, San Diego, in 1967 but was not yet operational at the new location in July of that year. It involved a full-scale mockup sphere with a viewing window similar to that of the Terni sphere, a sandbox with a full-sized mechanical arm, and a TV system.
34. "Plans Sea Visits," *Tipton* (Ind.) *Tribune* (18 August 1966), 7.
35. National Academy of Sciences, *Report on the International Geophysical Year* (Washington, D.C.: National Science Foundation, 18 February 1959), 83.
36. "Capt. Lytle to Succeed Capt. Dornin at L.B. Base," (Long Beach, Calif.) *Independent Press-Telegram* (27 June 1965).
37. Michel interviews.
38. Shawn Cheadle, "Defense Priority Ratings: Know the Rules," *Microwave Journal* (June 2003). Except during the *Scorpion* operation, Priority-1 orders were used sparingly in the *Trieste* program to avoid calling attention to the covert aspects of the bathyscaph's workup for Winterwind.
39. "Navy Intensifies Undersea Research," (Long Beach, Calif.) *Independent Press-Telegram* (12 January 1966).
40. "Nuclear Navy Here," *Long Beach Independent Press-Telegram* (30 January 1966).
41. Craven, *Silent War,* 231–232.
42. Richard A. Landgraff, *A History of the Long Beach Naval Shipyard* (Seattle, Wash.: CreateSpace, 2009), 65, 367.
43. Floating dry docks, like other Navy service craft, are placed "in service," vice "in commission" as are *ships.*

Chapter 13. A "*Trieste III*"

Epigraph: *The Wizard of Oz*, movie directed by Victor Fleming et al. (MGM, 1939).

1. Craven, *Silent War*, 128. Craven claimed this call and meeting occurred in "early 1965"; however, DSSP footprints can be seen in both Mooney and Shumaker receiving orders as

officer-in-charge of *Trieste II* in January 1964. By October 1964 Mooney already was cognizant of the Air Force project and of the so-called *Trieste III* or DSV-1. It is almost certain that Craven's vetting into the Sand Dollar program occurred prior to October 1964.

One of the authors of this book, N. Polmar, as an employee of the Northrop Corporation, was assigned full time to DSSP for four years (1967–1970); at that time he worked only on "white" programs.

2. Sherry Sontag and Christopher Drew, *Blind Man's Bluff: The Untold Story of American Submarine Espionage* (New York: PublicAffairs, 1998), 52.
3. Ibid., 51. When there were conflicts between *Blind Man's Bluff* and Craven's *Silent War* we have resolved the conflict in favor of the former. Craven was interviewed extensively by Sontag and Drew; his memoirs are marked by purposeful misdirection as well as numerous inadvertent errors, e.g., referring to Don Walsh as *Donald* Walsh.
4. The *Halibut* was completed in 1960 as a Regulus cruise missile submarine (SSGN 587); with the end of the Regulus program she was extensively converted to a special mission submarine (SSN 587). She participated in numerous clandestine operations, including the location of the Soviet ballistic missile submarine *K-129,* which was partially salvaged in Project Azorian. For the *Halibut* conversion see Norman Polmar and Michael White, *Project Azorian: The CIA and the Raising of the K-129* (Annapolis, Md.: Naval Institute Press, 2010), 51, 54, 197–199.
5. Deputy Chief of Naval Operations (Fleet Operations and Readiness) (Op-03), "U.S. Navy Deep Submergence/Ocean Engineering Program: 1970–1980" (Washington, D.C.: Office of the Chief of Naval Operations, June 1968). See http://citeseerx.ist.psu.edu/viewdoc/download?doi=10.1.1.937.4330&rep=rep1&type=pdf (accessed 16 September 2017).
6. John E. Bennett, "Oral History Interview: John E. Bennett," interview by Donald R. Lennon, East Carolina University Digital Interview, interview #138, 6 February 1994. The *ARD 20* was given the name *White Sands* (hull number ARD 20) in 1968; the "name" *ARD 20* is used throughout this chapter. Also David S. Brandwein, "Telemetry Analysis," *Studies in Intelligence* (Fall 1964), https://www.cia.gov/library/center-for-the-study-of-intelligenc/kent-csi/vol8no4/html/v08i4a03p_0001.htm.
7. Sontag and Drew, *Blind Man's Bluff,* 136.
8. Staehle interviews with Lee Mathers.
9. Craven, *Silent War,* 136.
10. Staehle interviews with Lee Mathers.
11. Craven, *Silent War,* 233. Craven's narrative seems to place this event in 1968, but the content of the story suggests that it more likely occurred in 1964.
12. Fry, "Warhead Recovery System is Proposed" and "Underseas Study May Prove Boon," *Ogden* (Utah) *Standard-Examiner* (9 May 1964). Capt. John H. Dolan headed the DSSRG's small-object-recovery study group. "Increased Deep-Ocean Study Due," *Corpus Christi* (Tex.) *Caller-Times* (31 May 1964) records that Dolan would move with the project into the Special Project Office.
13. "Deep Sea Project Funds Face Slash," *San Diego Union* (3 February 1965).
14. Sontag and Drew, *Blind Man's Bluff,* 54, 61, 64.
15. "Navy Bares Undersea Development Project," (Long Beach, Calif.) *Independent Press-Telegram* (24 July 1965).

16. Graybeal, "Design of the Bathyscaph *Trieste II*," 8.
17. John R. Smyth, *One Life to Give: "Pensate Profunde"* (San Diego, Calif.: Helen L. Smith, 1990), 94.
18. Henifin e-mail to Lee Mathers, 20 August 2014.
19. The second sphere is referred to as the "MINSY sphere." It replaced the Hahn & Clay sphere on the *Trieste II* (DSV 1) during its 1974–1975 overhaul.
20. Staehle interviews with Lee Mathers. Lieutenants Staehle, Denis, and occasionally Forst were the Deep Submergence Systems Project staff representatives at vendor tests and trials.
21. Michel interviews with Lee Mathers, June–August 2014.
22. Staehle interviews with Lee Mathers.
23. Michel interviews with Lee Mathers.
24. Submarine Development Group 2 already existed in the Atlantic Fleet. In general, odd-numbered fleet organizations were in the Pacific and even-numbered ones in the Atlantic.
25. Sontag and Drew, *Blind Man's Bluff*, 77.
26. "Sub Admiral Calls Ocean Uses Vital," *San Diego Union* (13 August 1967).
27. James W. Watts was the first officer-in-charge of the *ARD 20* after her reactivation in 1966, taking command on 4 September that year. On 25 January 1968, Watts was relieved by Lt. Cdr. Charles E. Poarch.
28. *Trieste* Program Dive Logs, NMNW.2014.020.001, Naval Underwater Museum, Keyport Wash., http://www.navalunderseamuseum.org/wp-content/uploads/2016/01/Trieste-Dive-Log-2014_020_001-low-res.pdf. In 2017 the Naval Undersea Museum's dive logs did not include the dives between 1963 and 1967. The following address on the Wayback Machine documents those dives: https://web.archive.org/web/20120112034206/http://www.bathyscaphtrieste.com:80/contents/divelog/logt2.html. All cited deck logs of ships operating with the *Trieste* are at the Naval History and Heritage Command, Navy Yard, Washington, D.C.
29. Staehle interviews with Lee Mathers.
30. See Polmar and White, *Project Azorian*.

Chapter 14. Mission Lost

Epigraph: Henifin interviews with Lee Mathers, August 2014.

1. The *Dakar* was built as the British submarine *Totem*, completed in 1945. Purchased by Israel in 1965, she sailed from England to Israel on 9 January 1968. Her wreckage was located in 1999 between the islands of Cyprus and Crete at a depth of approximately 9,800 feet.
2. Henifin interviews with Lee Mathers.
3. "Obituary, John Reid 'Jack' Smyth," *Times-Herald* (Vallejo, Calif.) (10 February 2009). In addition to managing the *DSV-1* construction program and overseeing the bathyscaph's workup, Smyth would participate in the *Trieste*'s investigation of the *Scorpion* wreck in 1969.
4. Ibid.
5. Royal G. Fort (crewman), interview with Lee Mathers, 9 May 2015.
6. Staehle, Fort, and James F. Kuczkowski, interviews with Lee Mathers, 8–9 May 2015.
7. The authors have interviewed several participants in the *Trieste*'s classified operations of September–December 1967, with Mr. Polmar having worked for DSSP at that time. However, they were unable to reconstruct a dive log for the period, and efforts to get the surviving documentation declassified by the Navy have not been successful.

8. Fort interview.
9. Anthony J. Taromino interview with Lee Mathers, August 2013. The Naval Ship Systems Command, established in 1966 to replace the Bureau of Ships, would be folded into a new Naval Sea Systems Command on 1 July 1974.
10. Robert Nevin interview with Lee Mathers, 6 August 2014.
11. See Cdr. Richard A. Mobley, USN (Ret.), *Flash Point North Korea: The* Pueblo *and the EC-121 Crises* (Annapolis, Md.: Naval Institute Press, 2003).
12. See Polmar and White, *Project Azorian*.
13. The Israeli attack on the U.S. spy ship *Liberty* (AGTR 3) was made in the mistaken belief that it was an Egyptian ship, the American embassy having advised the Israeli government that there were no U.S. ships in the area. See Capt. A. Jay Cristol, USN (Ret.), *The* Liberty *Incident Revealed: The Definitive Account of the 1967 Israeli Attack on the U.S. Navy Spy Ship* (Annapolis, Md.: Naval Institute Press, 2013).
14. An administrative reorganization of CNO's staff in 1968 upgraded submarine activities to a Deputy Chief of Naval Operations position held by a vice admiral. In October 1969, the new Deep Submergence Systems Division was assigned to a rear admiral with an intelligence sub-specialty. Cdr. Brad Mooney, who was OpNav's first Deep Submergence Submarine Plans and Programs Officer in 1968, moved into the Deep Submergence Systems Division under Rear Adm. Maurice (Mike) Rindskopf upon its founding in 1969.

Chapter 15. *Scorpion* Preparations

Epigraph: Commander-in-Chief U.S. Atlantic Fleet, message DTG 082247Z August 1969.

1. Andreas B. Rechnitzer, W. W. Denner, and E. C. Estes, "An Assessment of Remotely Operated Vehicles to Support the AEAS Program in the Arctic" (Reston, Va.: Science Applications International, 15 September 1986), 1–26, http://www.dtic.mil/dtic/tr/fulltext/u2/a231730.pdf.
2. Stephen Johnson, *Silent Steel: The Mysterious Death of the Nuclear Attack Sub USS* Scorpion (New York: John Wiley, 2006), 113–142. Also Bruce Rule, *Why the USS* Scorpion *(SSN-589) Was Lost: Death of a Submarine in the North Atlantic* (Ann Arbor, Mich.: Nimble Books, 2011).
3. "Bathyscaphe May Retrieve Sub Sections," (Phoenix) *Arizona Republic* (1 November 1968).
4. "*Trieste II* Won't Search for *Scorpion*," (Long Beach, Calif.) *Independent Press-Telegram* (2 November 1968).
5. Smyth, *One Life to Give*, 97.
6. "Mooney Heads Navy Program," *Portsmouth* (N.H.) *Herald* (9 October 1968).
7. Sontag and Drew, *Blind Man's Bluff*, 206.
8. Staehle interviews with Lee Mathers, July 2014–September 2015.
9. On 20 August 1968, Lt. Werner F. Ruch was relieved by Lt. Myers as executive officer of the *White Sands*.
10. Staehle interviews with Lee Mathers.
11. Johnson, *Silent Steel*, 208.
12. Forman, "From Beebe and Barton to Piccard and *Trieste*," 34.
13. Smyth, *One Life to Give*, 99.
14. Beauford Myers, interviews with Lee Mathers, June 2013–September 2015.
15. Dunn interviews with Lee Mathers.

16. Smyth, *One Life to Give*, 106–107.
17. Troy Brown e-mail to Beau Myers, 23 August 2014.
18. David T. Byrnes, interview with Lee Mathers, 19 August 2014. Walsh was "stashed" at Submarine Development Group 1 for several months between permanent duty assignments and there was involved with the creation of a Deep Submergence pin, his own design. He sent his design to an Italian jeweler for "prototyping" at his own expense. Don Walsh memo "The U.S. Navy Deep Submersible Pilot's Badge."
19. Myers interviews with Lee Mathers.
20. Dunn interviews with Lee Mathers.
21. Michael D. Roberts, *Dictionary of American Naval Aviation Squadrons,* vol. 2, *The History of VP, VPB, VP(H) and VP(AM) Squadrons* (Washington, D.C.: Naval Historical Center, 2000), 89, 158.
22. Smyth, *One Life to Give*, 104. The *Calypso,* built as a minesweeper in the United States for the Royal Navy, was completed in 1943.

Chapter 16. At the *Scorpion* Site

Epigraph: Secretary of the Navy's award of the Navy Unit Commendation Ribbon to the *Trieste* Integral Operating Unit, undated.

1. Ross Saxon e-mail to Lee Mathers, 26 June 2015.
2. Beauford Myers, interviews with Lee Mathers, June 2013–September 2015.
3. Larry C. (Lorenzo) Hagerty interview with Lee Mathers, 1 July 2015.
4. Saxon e-mail to Lee Mathers. Also, Nevin interview. Nevin's recollection was that none of the *Mizar*'s deep-ocean transponders were successfully reactivated and that Capt. Gautier reseeded the area with new DOTs based solely on the *Mizar*'s latitude/longitude fix of the wreck site. Saxon's memory of the episode was more specific and more in conformance with preparations described prior to the first *Trieste* dive.
5. Dunn interviews with Lee Mathers.
6. Saxon e-mail to Lee Mathers.
7. From 1968, standard operating procedures for bottom navigation involved long-baseline acoustic navigation with three deep-ocean transponders tethered 300 feet above the bottom, spaced two to three miles apart. Called NAVNET, this system, as installed in the *Trieste,* was designed for independent onboard navigation from within the sphere.
8. Robert F. Nevin, "The Navigation of Scorpion Operations, Phase II," slide presentation: *Trieste* (DSV-1), San Diego, Calif., Commander Submarine Development Group 1, 1969. All discussion here of the NAVNET at the *Scorpion* site, the Mark 15 computer, etc., is based upon Nevin's slide presentation, the text of which was provided the authors by Nevin.
9. Hagerty interview with Lee Mathers.
10. Myers interviews, June 2013–September 2015.

Chapter 17. *Scorpion* Site Investigation

Epigraph: Dunn, e-mail to Lee Mathers, 12 August 2014.

1. Johnson, *Silent Steel*, 210–211.
2. Ibid., 211.
3. Dunn interviews with Lee Mathers, June 2013–August 2014.

4. Myers interviews with Lee Mathers, June 2013–September 2015.
5. Craven, *Silent War*, 236.
6. Dunn interviews with Lee Mathers, June 2013–August 2014. Dunn passed away on 14 September 2014.
7. Nevin interviews with Lee Mathers, June–July 2014. Nevin passed away on 26 March 2017.
8. Craven, *Silent War*, 237.
9. Ira B. McCarty, "About Tow'n," *Kansas City (Missouri) Times* (12 August 1969). The article introduced Hawks' letter with: "Larry E. Hawks, chief warrant officer in the Navy . . . is a hydronaut diving in the bathyscaph *Trieste II* off the Azores. He recently described some of his adventures in a letter to his mother."
10. Andreas Rechnitzer, foreword to *The Trieste: The Story of the United States Navy's First Inner Space Ship*, by Dick Snyder (San Carlos, Calif.: Golden Gate Junior Books, 1964), 6.
11. Craven, *Silent War*, 237.
12. Peter A. Tyrrash and Eric L. Swensen, Scorpion *Operations Phase II* [cruise book] (San Diego: *White Sands*, 1969).
13. Johnson, *Silent Steel*, 213. Field ascribed the colors blue and yellow to the clothing sighted. It is extremely difficult to describe color at great depths accurately due to the effects of the lighting and the total darkness surrounding the illuminated area; color cannot be discerned on a monochrome TV monitor.
14. Saxon to Lee Mathers.
15. Peter Palermo, *Evaluation of Data and Artifacts Related to USS* Scorpion *(SSN 589), Prepared for Presentation to the CNO Scorpion Technical Advisory Group* (Structural Analysis Group, 29 June 1970), exhibits 7.1, 7.2, and 7.3.
16. Saxon to Lee Mathers.
17. Ibid.
18. USS *Apache* (ATF 67) deck log, 1969. Exact times given for any activity at the *Scorpion* site are from the *Apache*'s deck log, a copy of which is held by Larry Hagerty, executive officer of *Apache* during the *Scorpion* operations.
19. This and other quotations from Byrnes, "Running Lights" (Leonardtown, Md.: unpublished manuscript, 2016).
20. This was the second sighting of an "umbrella," previously seen in a photo taken by the *Mizar*. A marine biologist later evaluated that the photo showed a deep-sea creature, not a human artifact. Perhaps both "umbrellas" were misidentifications, but Byrnes swore by his original evaluation.
21. Rule e-mail to Lee Mathers, 13 April 2017. See Rule, *Why the USS* Scorpion *(SSN-589) Was Lost*.
22. *Scorpion* Court of Inquiry, sections 7.1.3, 7.2, of the Special Advisory Group (SAG) report of 29 June 1970. Also Rule e-mails to Lee Mathers, 13 April and 24 March 2017. The late Daniel L. McMillin, an engineer at Bell Laboratory, was the source of the copies of the helicorder grams from both the Newfoundland SOSUS arrays and the Canary Island hydrophones, which are the only records of the detection of the battery explosion signals still in existence as the audiotapes did not cover that earlier time period. Also, Robert E. LaGassa, letter to Dr. Robert Ballard, "USS *Scorpion* Sinking Analysis," (20 October 2013), available at http://www.usna63.org/tradition/Bob-LasGassa-Insights.pdf.

23. Robert Fishback interview with Lee Mathers, 8 May 2015.
24. Hagerty interview.
25. Nevin interviews.
26. Hagerty interview. Lonnon would pass away on 13 February 2012; he had retired as a commander in 1985.
27. Ibid. Hagerty received a personal "well done" from the Commander-in-Chief, Pacific Fleet, for successfully completing as navigator the longest tow in U.S. Navy history.
28. Duane Tollaksen, e-mail to Lee Mathers, 17 August 2015.

Chapter 18. Neglect and Decline

Epigraph: Bartels e-mail to Lee Mathers, 29 July 2016.

1. The admiral in charge of the Chief of Naval Operations' deep submergence office held two titles: Director, Deep Submergence Systems Division; and Deep Submergence Program Coordinator. The first two admirals in that role, Rear Adm. Maurice Rindskopf and Rear Adm. Walter Dietzen, also held "black" roles in the national underwater reconnaissance program as staff director of the new National Underwater Reconnaissance Office.
2. Bertrand Piccard e-mail to Lee Mathers.
3. Nevin interviews.
4. Leonard B. Molaskey, memorandum: "Improved Viewing Capability for Bathyscaph *Trieste II*" (Danbury, Conn.: Perkin-Elmer Optical Technology Division, 20 December 1971), 5.
5. Commander Submarine Development Group 1, message: "Officer Undergoing Medical Treatment," 082137Z April 1970.
6. Beauford Myers, letter to Commander Submarine Development Group 1 relieving Nevin as officer-in-charge *White Sands* (ARD 20), 18 August 1970, Myers' files. Also Nevin interviews with Lee Mathers, June–August 2014. Nevin's next duty station was in the Office of the Oceanographer of the Navy. He served there for one year before moving to the National Atmospheric and Ocean; in 1974 he became the chief staff officer of the Naval Undersea Center in San Diego.
7. Malcolm G. Bartels, officer-in-charge of the *White Sands*, letter to the Commander, Submarine Development Group 1, "Aircraft Recovery Operation 9–10 October 1970, Report Of," 15 October 1970.
8. Myers e-mail to Lee Mathers, 15 July 2015.
9. Malcolm G. Bartels, "Back from the Bottom," U.S. Naval Institute *Proceedings* (November 1971), 101–104.
10. Myers e-mail to Lee Mathers; Malcolm Bartels e-mail to Lee Mathers, 16 July 2015. The Hellcat subsequently was placed on display at the Naval Aerospace Museum in Pensacola, Florida.
11. The USS *Pigeon* (ASR 21) and *Ortolan* (ASR 22), developed under the DSSP submarine rescue program, each was designed to carry and support two DSRVs and to support saturated-diving operations.
12. See Lee J. Mathers and Beauford E. Myers, "The Navy's Deep Ocean Grab," *Naval History* (February 2013), 40–48, https://www.usni.org/magazines/navalhistory/2013-01/navys-deep-ocean-grab.
13. Robert Perry, *A History of Satellite Reconnaissance,* vol. 3B, *Hexagon*, BYE 17017–74 (Washington, D.C.: National Reconnaissance Office, 1973), 104–105. The quotation is from Richard

J. Chester, *A History of the Hexagon Program*, BIF 007-0253-85 (Washington, D.C.: National Reconnaissance Office, 1985), 59, https://web.archive.org/web/20160310053325/http://nro.gov/foia/declass/GAMHEX/GAMBIT%20and%20HEXAGON%20Histories/1.pdf.

14. USS *Apache* deck logs, 1971–1972; Chester, *History of the Hexagon Program*, 59.
15. John L. McLucas, *Reflections of a Technocrat: Managing Defense, Air, and Space Programs during the Cold War* (Maxwell Air Force Base, Ala.: Air University Press, 2006), 74.
16. National Reconnaissance Office, "Sensor Subsystem Monthly Technical Report No. 55," Washington, D.C., August 1971, 2–7.
17. Robert A. Frosch, memorandum to John L. McLucas, Director, Naval Reconnaissance Office, "Deep Sea Recovery of Hexagon Reentry Vehicle," 18 August 1971. Frosch, as Assistant Secretary of the Navy for Research and Development, had no operational authority to commit or deploy Navy ships. However, as the director of the National Underwater Reconnaissance Program, reporting to David Packer, the Deputy Secretary of Defense, he did have authority to commit and deploy the *Apache*, *White Sands*, and *Trieste* for a mission that extended over ten months.
18. Shor, *Scripps Institution of Oceanography,* 107.
19. USNS *De Steiguer* (T-AGOR 12) deck logs, September–October 1971.
20. Dwight E. Boegeman interviews with Lee Mathers, August–October 2012. Boegeman was the Scripps Marine Physical Laboratory Deep-Tow engineer on board the *De Steiguer* in 1971.
21. Leonard Molaskey memorandum: "Trip Report: Recovery of RV #3" (Danbury, Conn.: Perkin-Elmer Optical Technology Division, 18 November 1971).
22. Brad Mooney telephone interviews with Lee Mathers, 12 April, 15 April, 15 September, 25 September, and 13 October 2012.
23. In preparation for Winterwind, Westinghouse constructed a kludge, the "sinking frame" that probably could have recovered the target re-entry vehicle without the construction of a purpose-built claw. The sinking frame remained at Ballast Point throughout this operation.
24. Leonard Molaskey memorandum: "Trip Report: Recovery Hook Test Program, San Diego, California, 20–21 September 1971" (Danbury, Conn.: Perkin-Elmer Optical Technology Division, 24 September 1971).
25. Leonard Molaskey, memoranda: "Trip Report: Recovery of #3—Activities of 27 September through 3 October 1971" (4 October 1971); ser. ME-64: "Status of RV-3 Recovery: Telecon to Cdr. Mooney, Sub Dev Group One" (8 October 1971); ser. ME-65: "Telecon with Lt. Cdr. Doyle" (12 October 1971); ser. ME-66: "Status of RV-3 Recovery" (13 October 1971). All Danbury, Conn.: Perkin-Elmer Optical Technology Division.

Chapter 19. The Deepest Recovery

Epigraph: Doyle to Mathers, e-mail, 17 August 2015.

1. See Lee Mathers and Beauford Myers, "The Navy's Deep Ocean Grab," *Naval History* (February 2013), 40–49.
2. Deck logs of USS *Apache* (ATF 67), *Quapaw* (ATF 110), and *Safeguard* (ARS 25), 1971-1972.
3. A "tip of the hat" to Tom Clancy for *Hunt for Red October* (Annapolis, Md.: Naval Institute Press, 1984) quotation from the screenplay.
4. Beauford Myers, interviews with Lee Mathers, 2008–2010; Myers, letter home, 10 November 1971.

5. Leonard Molaskey, memorandum: "Trip Report–Recovery of RV #3" (Danbury, Conn.: Perkin-Elmer Optical Technology Division, 10 December 1971).
6. Capt. Samuel Packer, Commander, Submarine Development Group 1, was then personally involved with the first mother submarine mating—that of the USS *Hawkbill* (SSN 666) with the rescue submersible *Mystic* (DSRV 1) in January–February 1972, off San Diego.
7. CIA On-Scene Senior Representative, memorandum: "Prognosis for RV-3 Recovery Operations" (February 1972).
8. Col. Frank S. Buzard, Hexagon Program Director, message to Brig. Gen. David. D. Bradburn, Director, NRO staff (7 January 1972).
9. David W. Waltrop, "An Underwater Ice Station Zebra: Recovering a KH-9 Hexagon Capsule from 16,400 feet below the Pacific Ocean," *Quest: The History of Spaceflight Quarterly* (August 2012), 4–17, https://www.cia.gov/library/publications/cold-war/underwater-ice-station-zebra/ice-station-zebra.pdf.
10. Ibid., 14.
11. Robin Salser e-mail to Lee Mathers, 2 January 2008. Salser provided contemporaneous documentation in the form of pages from his petty officer's record book ("wheelbook"). Three declassified NRO documents present slightly different versions. Interviews by Lee Mathers with individuals on-site at the time were used to reconcile them.
12. CIA On-Scene Senior Representative, "Memorandum for the Record," 24 May 1972, attachment 3, 3–4.
13. The Navy press release of 23 May 1972, was picked up by the Associated Press and the story was published in newspapers throughout the United States, e.g., "Bathyscaph Makes Record 3-Mile Dive," *Mexico* (Missouri) *Ledger* (24 May 1972). While the identity of the recovered object remained classified for decades, the *Guinness Book of World Records* (1973) recorded the operation as "the deepest manned salvage, 16,400 feet." Bartels received the Superior Achievement Award for Outstanding Performance as a Practicing Navigator 1972, from the Institute of Navigation in Manassas, Virginia.
14. "*Trieste* Recovers Electronics Package from 16,400-ft. Depth," *Undersea Technology* (July 1972), 25.
15. Richard Taylor video recording, "CIA's Underwater Space Mission Revealed," David Waltrop, moderator, Smithsonian National Air and Space Museum, US Stream TV, 26 April 2013, http://www.ustream.tv/recorded/32000249.
16. John L. McLucas, Director, NRO, memorandum to the Assistant Secretary of the Navy (Research and Development), "Deep Sea Recovery of Hexagon Reentry Vehicle" (15 May 1972).
17. CIA On-Scene Senior Representative, "Memorandum for the Record," 2.
18. Charles W. Hudiburgh, Memorandum to Commander, Submarine Development Group One, 3 May 1972, *White Sands* (ARD 20) Command History 1971, File Box 859A Navy Operational Archives, Washington, D.C.
19. Walsh e-mails to Lee Mathers, 25 August 2015, and 23 September 2016.
20. Bartels, e-mail to Lee Mathers, 15 July 2015; Bartels was executive officer of the *Catfish* for the sinking exercise.
21. Richard L. Abbott interview with Lee Mathers, 11 July 2015.
22. Frank G. Charlton interview with Lee Mathers, 10 September 2015.

23. Bartels e-mail to Lee Mathers 15 July 2015. In June 1973, Bartels received orders to the U.S. Pacific Command on Oahu as the oceanographic and hydrographic programs officer; booklet "Change of Command Ceremony *Trieste II* (DSV 1)," Commander, Submarine Development Group 1 (28 June 1973).
24. "Robots Explore for Sea Tasks at Great Depths," *Defiance* (Ohio) *Crescent News* (14 April 1979), 2; https://newspaperarchive.com/defiance-crescent-news-apr-14-1979-p-2.
25. "Teleprobe to Aid Navy in Seeking Sunken Ships," *Bridgeport* (Conn.) *Telegram* (27 November 1969). The article relates a sea trial off San Clemente earlier in 1969 conducted by F. M. Daugherty of the Office of the Oceanographer; it describes the Teleprobe as an "overweight guppy . . . mounted on a sled-like frame, 18 feet long and weighing 500 pounds," with side-scanning sonar, cameras, and a magnetometer.
26. "The Sea Pioneers Who Dive 10,000 Feet Deep," *Oakland* (Calif.) *Tribune* (23 June 1973). This was a "puff piece" with a photo showing "Hydronauts Malcolm Bartels, Richard Abbott, Roger Whitaker, and R. D. Baker" with the *Trieste* in the *White Sands*' docking well.
27. *Apache* deck logs, July–August 1973.
28. R. D. Baker e-mail to Lee Mathers, 27 July 2015.
29. Ibid.
30. Ibid.
31. Richard Abbott e-mail to Lee Mathers, 29 July 2015.
32. Enlisted hydronauts were being certified on the *Trieste*, *Turtle*, *Sea Cliff*, *Alvin*, and the DSRVs *Mystic* and *Avalon*.
33. Howe later served as a lieutenant in the Marine Corps.
34. "Oldest Tug in Navy to Retire," *San Diego Union* (24 March 1974).
35. Johnson, Willard E., John H. Howland, and Ronald J. Doyle, "Deep Submergence: The U.S. Navy's Operating Forces," Paper #OTC 2570, Offshore Technology 8th Annual Conference, Houston, Tex., 6 May 1976, table 1. The *Alvin*'s dive information is available at http://www.marine.whoi.edu/divelog.nsf/Dives%20in%201970s?OpenViewandStart=1.

Chapter 20. A New Support Ship

Epigraph: Robert D. Ballard and Will Hively, *The Eternal Darkness: A Personal History of Deep-Sea Exploration* (Princeton, N.J. Princeton University Press, 2000), 61.

1. At that time one of the coauthors of this book, Norman Polmar, was editor of the U.S. sections of the annual *Jane's Fighting Ships*.
2. In addition to Firebaugh's exposure to the bathyscaph on the Development Group 1 staff, he had been involved in the 1964 *Thresher* search on Capt. Frank Andrews' staff. Firebaugh was later promoted to rear admiral.
3. *Point Loma* (AGDS 2) "Commissioning Ceremony Booklet" (San Diego: Commander Submarine Development Group 1, 3 July 1976). The principal address was by Rear Adm. J.G. Williams Jr., Commander, Submarine Force Pacific Representative San Diego.
4. James Worthington, e-mail to Lee Mathers, 15 August 2015.
5. Thomas G. Vetter, e-mail to Lee Mathers, 25 July 2015. The original four small viewports in the Hahn & Clay sphere were there to allow visual monitoring of the forward portside and forward starboard ballast-release valves, the after ballast-release valve, and (through the entry hatch) the clearing of water from the access tube. The two viewports intended to have views

of the forward shot tubs were obstructed by equipment racks inside the sphere. The new MINSY sphere had only the main window forward, a viewport through the access hatch, and a small rear viewport.

6. The authors of this book have not been able to determine the ultimate disposition of the Hahn & Clay sphere.
7. Johnson, Howland, and Doyle, "Deep Submergence," 507.
8. R. D. Baker, interviews with Lee Mathers, July 2015. Baker recalled that the bathyscaph's activities at Alameda attracted the attentions of local reporters. In March 1975 *The Los Angeles Times* and other newspapers broke the story of the CIA's attempt to recover a Soviet submarine in the Central Pacific, reporting the effort's name (inaccurately) as Project Jennifer. The *Trieste*'s crew was cautioned not to speak of the newspaper reporting in any way.
9. The *Salmon* was built as a radar picket submarine (SSR); changed to an attack submarine (SS) in 1961.
10. Thomas G. Vetter, *30,000 Leagues Undersea: True Tales of a Submariner and Deep Submergence Pilot* (Dumphries, Va.: Tom Vetter Books, 2012), 328–334.
11. For a detailed summary of the SH-3 Sea King search and partial salvage by the *Trieste* see ibid., 55–60; for the sinking of the workboat *Bertram*, 93–101.
12. Ibid., 48–49.
13. George Ellis was a much-respected senior chief petty officer, a mentor to subordinates, and a rock for officers. He later was a pilot of the *Alvin* at Woods Hole as a civilian employee.
14. Vetter to Mathers, and a timeline Vetter developed from his contemporaneous records that was used to structure other participants' memories.
15. ANGUS was the first unmanned search and survey system developed by Woods Hole. It was designed to work primarily in rugged volcanic terrain to depths of 20,000 feet. Its various subsystems were mounted within a heavy-duty steel frame capable of withstanding a head-on collision with a vertical rock outcropping.
16. Robert Ballard, "The History of Woods Hole's Deep Submergence Program," in National Research Council, *50 Years of Ocean Discovery: National Science Foundation 1950–2000* (Washington, D.C.: National Academies Press for the Ocean Studies Board, 2000), 67–86, https://www.ncbi.nlm.nih.gov/books/NBK208815 (accessed 17 July 2017).
17. Larry J. Porter, interviews with Lee Mathers, 19 July 2014 and 9 May 2015. Also Richard Abbott, discussions with Lee Mathers, July–August 2015; and Robert Ballard, e-mails to Lee Mathers, 4–7 March 2018.
18. Vetter, *30,000 Leagues Undersea*, 345–351.
19. Vetter to Mathers.
20. Forman, "From Beebe and Barton to Piccard and *Trieste*," 35.
21. Vetter to Mathers; Vetter timeline.
22. Robert Ballard, e-mail to Lee Mathers, 6 March 2018.
23. Vetter, *30,000 Leagues Undersea*, 351.

Chapter 21. An Epiphany

Epigraph: John Van Voorhis, interviews with Lee Mathers, August 2015.

1. Glen Sjoblom interview with Lee Mathers, 29 September 2013. Sjoblom revealed that Naval Reactors was interested in conducting radiological surveys as part of an effort to

analyze whether or not the deep ocean was suitable for the disposal of de-fueled reactor plants from deactivated nuclear submarines. The Navy ultimately decided to place the de-fueled reactor compartments at the radioactive waste site operated by the Department of Energy at the Hanford Reservation in Washington State.

2. Sjoblom e-mail to Lee Mathers, 16 July 2015. Sjoblom could not recall which organization conducted the towed sonar search of the target area.
3. Martine Dreyfus Rawson and William B. F. Ryan, "Geologic Observations at the 2800 Meter Radioactive Waste Disposal Site and Associated Deepwater Dumpsite 10 (DWD-106) in the Atlantic Ocean" (Washington, D.C.: Environmental Protection Agency, September 1983).
4. See chapter 11.
5. Charles Copeland, interview with Lee Mathers, 9 May 2015.
6. Jon R. Losee, e-mail to Lee Mathers, 23 August 2015. Losee joined the *Trieste* again in 1979 for a re-examination of both the *Thresher* and the *Scorpion* wrecks. In 1979 the Navy had better sensors but obtained essentially the same results at both sites—low levels of Cobalt-60 from reactor piping but no danger from the reactor core or from the two Mark 45 nuclear torpedoes in the *Scorpion*. The 1979 "ops" also included Sherman Williams, involved with biological, sediment, and water sampling at the wreck sites. Les Parsons and Warrant Officer Burton (Burt) Tharp were the pilots.
7. Vetter e-mail to Lee Mathers; also Vetter, *30,000 Leagues Undersea*, 389–396.
8. "Sub Reactor Dumped in Sea Called Safe," *San Diego Union* (18 May 1980).
9. Dick Abbott e-mail to Lee Mathers, 7 August 2014.
10. Vetter e-mail to Lee Mathers, 27 August 2015.
11. Sperry Rand, "*Trieste II* Trainee's Guide," Publication No. GJ-13-1617 (Syosset, N.Y.: Sperry System Management Division, January 1968), 4-53 to 4-71.
12. Robert D. Ballard, "Improving the Usefulness of Deep-Sea Photographs with Precision Tracking," *Oceanus* (Spring 1975), 40–43. William M. (Skip) Marquet had put together Woods Hole's plotting system used by the *Alvin* and her mother ship *Lulu* on Project Famous in 1974, and it was Marquet's concept and suggestions that catapulted the early VanVoorhis/Vetter plotting work into an attempt to automate the *Trieste*'s tracking; Vetter e-mail to Lee Mathers, 26 July 2015. Marquet arrived on board the *Point Loma* with Ballard's team for the Cayman Trough dives and stayed through the *Thresher* dives. His main contribution was to assist VanVoorhis and Branchflower improve the accuracy of acoustic ranging. Marquet had a wealth of experience and deep technical expertise in these techniques.
13. John B. VanVoorhis interview with Lee Mathers, 31 August 2015. The Westinghouse deep ocean transponder was designed to be deployed through submarine torpedo tubes; for the *Trieste* operations, a few were hand-released over the side or stern of support ships. These Advanced Technology Navigation (ATNAV) deep-ocean transponders were in short supply owing to wear and tear, but also to occasional losses when the release signal failed to retrieve a DOT.
14. Ibid. The vital piece of the puzzle brought to the table by Marquet was to have each of the *Trieste*'s pings to the DOTs automatically slaved to an initiating ping, on a different frequency from the *Point Loma*. The latter then had both the distance to the *Trieste* and those between the *Trieste* and the DOTs, i.e., the location of the bathyscaph within the DOT

grid. Another ping on the DOT frequency directly from the ship established its own location within the DOT grid. Thus two pings from the ship, on different frequencies, provided complete "situational awareness" on the surface of the *Trieste*'s bottom navigation.

15. Norman Branchflower e-mail to Lee Mathers, 15 July 2015. VanVoorhis was the technical engineer while Branchflower dealt with the computer system design and programming, using the on board HP 9825 desktop computer, which was one of the first desktops with a real-time clock. The computer had 16k RAM, a 32-character visual display, and a cash register–type thermal paper printer. The thermal paper output had to be hand-transcribed within 24-hours, before the printing faded. It took a couple of months after the *Thresher* dives to put the whole system into operation.
16. Vetter e-mail to Lee Mathers, 23 August 2015: "My earliest plots (Cayman Trough) are hand-plotted on grid paper onto which I traced the Ballard-supplied bottom topography." Also, "The World's People," *Sperry World* (November 1978). Sperry's president, Salvatore A. Conigliaro, presented to VanVoorhis the firm's Meritorious Public Service Citation, "for outstanding performance," including "development of new tracking and navigation methods" and a "major contribution to the successful completion of the Seafloor Geophysical Research Operations for the period 31 May 1977 to 17 November 1977." To this award Sperry added a financial bonus.
17. VanVoorhis interviews with and e-mail to Lee Mathers, August 2015.
18. Georges L. Weatherly and Mark Wimbush, "Near-Bottom Speed and Temperature Observations on the Blake-Bahama Outer Ridge," *Journal of Geophysical Research* (20 July 1980), 3971–3981.
19. Roger D. Flood e-mail to Lee Mathers, 28 July 2015.
20. Bruce C. Heezen and Michael Rawson, "Visual Observations of the Sea Floor Subduction Line in the Middle-America Trench," *Science* (1977), 423–425.
21. Vetter, *30,000 Leagues Undersea*, 321–327.
22. Larry Porter interviews with Lee Mathers, 19 July 2014, and 9 May 2015.
23. Raymond Freeman-Lynde interview with Lee Mathers, 24 July 2015; Freeman-Lynde states that they overshot on the deepest dive, bottoming at 20,242 feet. Also R. P. Freeman-Lynde, M. B. Cita, F. Jadoul, E. L. Miller, and W. B. F. Ryan, "Marine Geology of the Bahama Escarpment," *Marine Geology* (1981), 119–156; http://eesc.columbia.edu/courses/w4937/Readings/Topic16_Readings/Lynde_et_al_1981.pdf.
24. See chapter 15.

Chapter 22. Maturity

Epigraph: Les Parsons, interview with Lee Mathers, 9 May 2015.

1. Martin Mandelberg, "An Oceanographic Acoustic Beacon and Data Telemetry System Powered by a SNAP-21 Radioisotope Thermoelectric Generator," Engineering in the Ocean Environment, IEEE 1971 Conference, San Diego, Calif., 21–24 September 1971, 220–223.
2. Abbott e-mail to Lee Mathers, 12 July 2015.
3. Vetter, *30,000 Leagues Undersea*, 375.
4. H.V. Weiss and J. F. Vogt, "Radioisotope Thermoelectric Generators Emplaced in Deep Ocean: Recover or Dispose in Situ?" Technical Report 1106 (San Diego, Calif.: Naval Ocean Systems Center, March 1986), 20. The study recommended abandonment in place as safer.

5. Thomas G. Vetter e-mail to Lee Mathers, 19 July 2015. After four years of operating commercial submersibles in the San Diego area, Vetter returned to active Navy duty in 1983. He rejoined Submarine Development Group 1 as the first deputy chief of staff, specifically to oversee the training and mission activities of all Navy manned and unmanned deep submergence vehicles. This position later became Commanding Officer, Deep Submergence Unit, later changed to the Undersea Rescue Command.
6. Abbott e-mail to Lee Mathers 12 July 2015. The *Point Loma*'s commanding officer, Cdr. Waer dismissed Hartman during the 1979 Atlantic operations; Lt. Cdr. Jon Anderson arrived to assume the duties of executive officer. Hartman was hospitalized and diagnosed with a rare brain disorder; he passed away shortly thereafter.
7. The original dive logs contain an error, reversing the divers and observers for Dives #5-79 and #6-79; the first dive was to DOS-2, and the second dive took Losee to the *Thresher.*
8. Jon R. Losee interview with Lee Mathers, 18 July 2014, and e-mail to Mathers, 23 August 2015.
9. Jon R. Losee, "Deep Ocean Gamma Ray Measurements, 1979 *Trieste* Operation," Technical Report 567 (San Diego, Calif.: Naval Ocean Systems Command, 1979).
10. Clark D. Sachse interview with Lee Mathers, 2 April 2017.
11. Burton Tharp e-mail to Lee Mathers, 10 August 2015.
12. Wilson Whitmire interviews with Lee Mathers, June–August 2015.
13. Les Parsons e-mail to Lee Mathers, 27 August 2015.
14. Forman, "From Beebe and Barton to Piccard and *Trieste*," 36. The Mk 70 Mobile Submarine Simulator was introduced into the fleet in 1976; it was a modified Mk 46 torpedo, 12.75 inches in diameter. with its warhead and guidance package removed and replaced by an electronics package and an additional fuel tank installed; it was 8½ feet long.

Chapter 23. Senescence and Retirement

Epigraph: George H. Verd, e-mail to Lee Mathers, 5 August 2015.

1. The Yucca Mountain Nuclear Waste Repository in Nevada was selected in 1987 for long-term storage of high-level radioactive waste in the United States.
2. "Raising Sub Studied," *San Diego Union* (29 January 1980), and "Navy to Attempt Recovery of Sunken Research Vessel," *San Diego Union* (6 March 1980), and "*Trieste* to Lead Hunt for Research Vessel," *San Diego Union* (19 July 1980).
3. Forman, "From Beebe and Barton to Piccard and *Trieste*," 36. Forman is in error in his statement, "During the next two dives in August, the vehicle was located and recovered."
4. Both the 6 August and 9 August 1980 dives are recorded in the unclassified dive log as three-man dives. This is inaccurate and purposefully misleading, as part of the security regime.
5. Burton Tharp e-mail to Lee Mathers, 27 August 2015. The manganese nodules are the same as used in the CIA's cover story for the salvage ship *Hughes Glomar Explorer*'s recovery of part of the Soviet ballistic missile submarine *K-129* from the North Pacific in 1974. That is, the ship was publicly reported to be mining manganese nodules.
6. Burton Tharp, e-mail to Lee Mathers, 1 September 2015. This dive also was the only known dive where a bathyscaph pilot voided his bowels at depth. Tharp, piloting, placed the *Trieste* on the bottom and asked Parsons to "take a short nap." Each seat in the bathyscaph had a flip-up lid, covering a catch basin designed for that purpose.

7. Parsons, Tharp, VanVoorhis, and Rean interviews with Lee Mathers revealed that the extant unclassified dive log contained errors, at least partially due to a hydraulic oil stain on the original that required the copying by hand of several pages. Their memories are that there were only three dives in the Hawaii area, not five: one unsuccessful dive to locate the lost RUWS vehicle and two two-man dives near an uncharted deep-sea volcano/seamount as part of a classified operation.
8. George H. Verd e-mail to Lee Mathers, 5 August 2015.
9. Ibid.
10. The *Mystic* (DSRV 1) became operational in 1971 and the *Avalon* (DSRV 2) in 1972.
11. Daniel Rean e-mail to Lee Mathers, 29 August 2016.
12. Daniel Rean e-mail to Lee Mathers, 27 July 2015.
13. "Navy Retires Research Sub *Trieste II*," *San Diego Tribune* (19 May 1984).

Chapter 24. Postscript

Epigraph: Don Walsh, e-mail to Lee Mathers, 18 May 2019. For hadal, see chapter 7 of this volume.

1. Howard Talkington, "Manned and Remotely Operated Submersible Systems: A Comparison" NUC TP 511 (San Diego, Calif.: Naval Undersea Center, June 1976), 26.
2. See Don Walsh, "*Nereus* Explores the Oceans' Greatest Depth," Naval Institute *Proceedings* (September 2009), 88.
3. "Losharik" was a Russian cartoon horse.
4. The *Pressure Drop* was the former U.S. Navy towed-array surveillance ship *Indomitable* (T-AGOS 7).
5. Victor Vescovo and Kelly Walsh, interviews and e-mails with Lee Mathers, June–July 2020.

Bibliography

Books

Ballard, Robert D. *Explorations: A Life of Underwater Adventure.* New York: Hyperion, 1995.

Ballard, Robert D., and Will Hively. *The Eternal Darkness: A Personal History of Deep-Sea Exploration.* Princeton, N.J.: Princeton University Press, 2009.

Broad, William J. *The Universe Below: Discovering the Secrets of the Deep Sea.* New York: Simon & Schuster, 1997.

Chun, Carl. *Aus den Tiefen des Weltmeeres* [From the Depths of the Ocean]. Jena [Germany]: Gustav Fischer, 1903.

Clancy, Tom. *The Hunt for Red October.* Annapolis, Md.: Naval Institute Press, 1984.

Craven, John Piña. *The Silent War: The Cold War Battle beneath the Sea.* New York: Simon & Schuster, 2001.

Cristol, A. Jay. *The* Liberty *Incident Revealed: The Definitive Account of the 1967 Israeli Attack on the U.S. Navy Spy Ship.* Annapolis, Md.: Naval Institute Press, 2013.

Cullen, Vicky. *Down to the Sea for Science.* Woods Hole, Mass.: Woods Hole Oceanographic Institution, 2005.

Forman, Will. *The History of American Deep Submersible Operations, 1775–1995.* Flagstaff, Ariz.: Best, 1999.

Houot, Georges S., and Pierre Henri Willm. *2000 Fathoms Down.* London: Hamilton & Hart-Davis, 1955.

Johnson, Stephen P. *Silent Steel: The Mysterious Death of the Nuclear Attack Sub USS* Scorpion. Hoboken, N.J.: John Wiley, 2006.

Kaharl, Victoria A. *Water Baby: The Story of* Alvin. New York: Oxford University Press, 1990.

Landgraff, Richard A. *A History of the Long Beach Naval Shipyard.* Seattle, Wash.: CreateSpace, 2009.

Matsen, Brad. *Descent: The Heroic Discovery of the Abyss.* New York: Pantheon, 2005.

McLucas, John L. *Reflections of a Technocrat: Managing Defense, Air, and Space Programs during the Cold War.* Maxwell Air Force Base, Ala.: Air University Press, 2006.

Mobley, Richard A. *Flash Point North Korea: The* Pueblo *and the EC-121 Crises.* Annapolis, Md.: Naval Institute Press, 2003.

Moran, Barbara. *The Day We Lost the H Bomb: Cold War, Hot Nukes and the Worst Nuclear Weapons Disaster in History.* New York: Ballantine, 2009.

National Research Council. *50 Years of Ocean Discovery: National Science Foundation 1950–2000.* Washington, D.C.: National Academies Press for the Ocean Studies Board, 2000.

Paccalet, Yves. *Auguste Piccard, Professeur de Reve* [Auguste Piccard, Professor of Dreams]. Grenoble [France]: Editions Glénat, 1997.

Piccard, Auguste. *Earth, Sky, and Sea*. New York: Oxford University Press, 1956.

Piccard, Jacques, and Robert S. Dietz. *Seven Miles Down: The Story of the Bathyscaphe* Trieste. New York: G. P. Putnam's Sons, 1961.

Polmar, Norman. *Death of the Thresher*. Philadelphia: Chilton Books, 1964; revised edition New York: Lyons, 2001.

Polmar, Norman, and Thomas B. Allen, *Rickover: Controversy and Genius*. New York: Simon & Schuster, 1982.

Polmar, Norman, and Michael White. *Project Azorian: The CIA and the Raising of the* K-129. Annapolis, Md.: Naval Institute Press, 2010.

Reed, Christina. *Marine Science: Decade by Decade*. New York: Facts on File, 2009.

Romney, Carl. *Recollections*. Bloomington, Ind.: AuthorHouse, 2012.

Rothenberg, Marc, ed. *The History of Science in the United States: An Encyclopedia*. New York: Garland, 2001.

Rule, Bruce. *Why the USS* Scorpion *(SSN 589) Was Lost: The Death of a Submarine in the North Atlantic*. Ann Arbor, Mich.: Nimble Books, 2011.

———. *Why the USS* Thresher *(SSN-593) Was Lost*. Ann Arbor, Mich.: Nimble Books, 2017.

Sadoul, Numa. *Tintin et Moi—Entretiens avec Hergé* [Tintin and Me: Interviews with Hergé]. Tournai, Belg.: Casterman, 1975.

Sagalevich, Anatoly M. *The Deep: Voyages to the* Titanic *and Beyond*. Redondo Beach, Calif.: Botanical, 2009.

Sapolsky, Harvey M. *Science and the Navy: The History of the Office of Naval Research*. Princeton, N.J.: Princeton University Press, 1990.

Sheets, Herman E., and Victor T. Boatwright Jr., eds. *Hydronautics*. New York: Academy, 1970.

Shor, Elizabeth Noble. *Scripps Institution of Oceanography: Probing the Oceans 1936 to 1976*. La Jolla, Calif.: Scripps Institution of Oceanography, 1977. Http://scilib.ucsd.edu/sio/hist/gc29s5.pdf.

Smyth, John R. *One Life to Give: "Pensate Profunde."* San Diego, Calif.: Helen L. Smith, 1990.

Snyder, Dick. *The Trieste: The Story of the United States Navy's First Inner Space Ship*. San Carlos, Calif.: Golden Gate Junior Books, 1964.

Sontag, Sherry, and Christopher Drew. *Blind Man's Bluff: The Untold Story of American Submarine Espionage*. New York: PublicAffairs, 1998.

Spirito, Pietro. *Trieste a Stelle Strisce* [*Trieste* Stars & Stripes]. *Trieste* [Italy]: MGS, 1994.

Szulc, Tad. *The Bombs of Palomares*. New York: Viking, 1967.

Terry, Richard D. *The Deep Submersible*. New York: Western, 1966.

Valenti, Paolo. *Storia del Contiere Navale di Monfalcone* [History of the Monfalcone Naval Shipyard]. Trieste [Italy]: Edizioni Luglio, 2007.

Vetter, Thomas G. *30,000 Leagues Undersea: True Tales of a Submariner and Deep Submergence Pilot*. Dumphries, Va.: Tom Vetter Books, 2012.

Vyborny, Lee, and Don Davis. *Dark Waters: An Insider's Account of the* NR-1, *the Cold War's Undercover Nuclear Sub*. New York: New American Library, 2003.

Weir, Gary E. *An Ocean in Common: American Naval Officers, Scientists, and the Ocean Environment*. College Station: Texas A&M University Press, 2001.

Woodward, Bob. *Veil: The Secret Wars of the CIA*. New York: Simon & Schuster, 1987.

Yount, Lisa. *Robert Ballard: Explorer and Undersea Archaeologist*. New York: Chelsea House, 2009.

Articles

Notes: The large number of articles from *The New York Times* and the *San Diego Union* are cited in the book's endnotes.

USNI = U.S. Naval Institute.

Andrews, Frank A. "Search Operations in the Thresher Area: 1964, Section I." *Naval Engineers-Journal* (August 1965).

———. "Search Operations in the Thresher Area: 1964, Section II." *Naval Engineers Journal* (October 1965).

———. "Searching for the Thresher." USNI *Proceedings* (May 1964).

———. "Underwater Navigation." In *Hydonautics,* edited by Herbert E. Sheets and Victor T. Boatwright Jr. New York: Academic, 1970.

Ballard, Robert. "Improving the Usefulness of Deep-Sea Photographs with Precision Tracking." *Oceanus* (Spring 1975).

Bartels, Malcolm G. "Back from the Bottom." USNI *Proceedings* (November 1971).

Bishop, Charles B. "U.S. Navy Involvement with DSV *Trieste*." *Marine Technology Society Journal* (Winter 2009).

Brandwein, David S. "Telemetry Analysis." *Studies in Intelligence* (Fall 1964). Https://www.cia.gov/library/center-for-the-study-of-intelligence/kent-csi/vol8no4/html/v08i4a03p_0001.htm.

Busby, R. Frank. "Design and Operational Performance of Manned Submersibles." *Proceedings of the American Society of Mechanical Engineers* (1968).

Busby, R. Frank, and William E. Hart. "Sea Floor Navigation for Deep Submergence Vehicles." *Navigation* (Summer 1966).

Cheadle, Shawn. "Defense Priority Ratings, Know the Rules." *Microwave Journal* (June 2003).

Dietz, Robert S. "Deep-Sea Research in the Bathyscaph *Trieste*." *The New Scientist* (April 1958).

———. "1000-Meter Dive in the Bathyscaph *Trieste*." *Limnology and Oceanography* (January 1959).

Fisher, Robert L. "Meanwhile, Back on the Surface: Further Notes on the Sounding of Trenches." *Marine Technology Society Journal* (Winter 2009).

Flood, Roger D., and Charles D. Hollister. "Submersible Studies of Deep-Sea Furrows and Transverse Ripples in Cohesive Sediments." *Marine Geology* (1980).

Forman, Will. "From Beebe and Barton to Piccard and *Trieste*." *Marine Technology Society Journal* (Winter 2009).

Freeman-Lynde, R. P., M. B. Cita, F. Jadoul, E. L. Miller, and W. B. F. Ryan. "Marine Geology of the Bahama Escarpment." *Marine Geology* (1981). Http://eesc.columbia.edu/courses/w4937/Readings/Topic16_Readings/Lynde_et_al_1981.pdf.

Fry, Paul M. "Submarine Which Found Missing H-Bomb Achieved Fame on Its First Assignment." *Albuquerque Journal* (20 March 1966).

Gardner, James V., Andrew A. Armstrong, Brian R. Calder, and Jonathan Beaudoins. "So, How Deep Is the Mariana Trench?" *Marine Geodesy* 37 (2014).

Goode, Sanchez. "Postscript to Palomares." USNI *Proceedings* (December 1968).

Hamilton, E. L. "Sediment Sound Velocity Measurements Made *in Situ* from Bathyscaph *Trieste.*" *Journal of Geophysical Research* 68, no. 21 (1963).

Heezen, Bruce C., and Michael Rawson. "Visual Observations of the Sea Floor Subduction Lines in the Middle-America Trench." *Science* 196 (1977).

Houot, Georges S. "Deep Diving by Bathyscaphe off Japan." *National Geographic* (January 1960).

Hull, E. W. Seabrook. "*Trieste II* Leads in DSR/V Electronics." *Geo-Marine Technology* (May 1966).

Jamieson, Alan J., and Paul H. Yancey. "On the Validity of the *Trieste* Flatfish: Dispelling the Myth." *Biological Bulletin* 222, no. 3 (2012), Http://www.journals.uchicago.edu/doi/full/10.1086/BBLv222n3p171.

Keach, Donald L. "Down to *Thresher* by Bathyscaph." *National Geographic* (June 1964).

Kohnen, William. "Human Exploration of the Deep Seas: Fifty Years and the Inspiration Continues." *Marine Technology Society Journal* (Winter 2009).

LaFond, E. "Dive Eighty-Four." *Sea Frontiers* 8 (1962).

Lill, Gordon G. "Oceanographic Activities in the Geophysics Branch, Office of Naval Research, 1950–1959." *Proceedings of the Royal Society of Edinburgh, Section B: Biology* (January 1972).

Mackenzie, Kenneth V. "Early History of Deep Submergence Navigation aboard *Trieste*." *Navigation: Journal of the Institute of Navigation* (Spring 1970).

———. "Sound-Speed Measurements Utilizing the Bathyscaph *Trieste*." *Journal of the Acoustical Society of America* 33, no. 8 (1961).

Marchand, Jean. *"Au Fond de Oceans avec de Professeur Piccard"* [To the Depths of the Oceans with Professor Piccard]. *La Science de la Vie* (January 1939).

Mare Island Naval Shipyard. "Deep-Diving Craft to Be Built at Yard." *Mare Island Grapevine* (8 February 1963).

———. "MI-Built *Trieste* Float to San Diego / Bathyscaph Job Done Here Helps Navy in Depth Diving." *Mare Island Grapevine* (8 November 1963).

Martin, George W. "*Trieste*: The First Ten Years." USNI *Proceedings* (August 1964).

Mathers, Lee J., and Beauford E. Meyers. "The Navy's Deep Ocean Grab." *Naval History* (February 2013).

Maxwell, Arthur E. "The Bathyscaph: A Deep-Water Oceanographic Vessel, Part 1—A Report on the 1957 Scientific Investigations with the Bathyscaph, *Trieste*." *U.S. Navy Journal of Underwater Acoustics* (April 1958).

Michel, E. John. "In the Trenches . . . Topside Remembrances by the Chief of the Boat, DSV *Trieste*." *Marine Technology Society Journal* (Winter 2009).

Moody, Dewitt H. "40th Anniversary of Palomares." *Faceplate* (September 2006).

Mooney, John B., Jr. "Marine Geodesy at the Office of Naval Research." *Marine Geodesy* 10, nos. 3–4 (1986).

Moore, D. G. "Erosional Channel Wall in the La Jolla Sea-Fan Valley Seen from Bathyscaph *Trieste II*." *Geological Society of America Bulletin* 76 (1965).

———. "Geological Observations from the Bathyscaph *Trieste* near the Edge of the Continental Shelf off San Diego." *GSA Bulletin* 74, no. 8 (1963).

Nicholson, William M., and Joseph A. Cestone. "The Effect of Electronics on the Design of Deep Submersibles." *Naval Engineers Journal* (June 1969).

Nickelsburg, Michael, and Stephen A Beckley. "A 'PLUSS' for Deep Ocean Search." *Naval Engineers Journal* (October 1975).

O'Connell, John F. "The Fatal *Thresher* Deep Dive: Different Ocean, Different Outcome?" *The Submarine Review* (Winter 2012).

Oreskes, Naomi. "A Context of Motivation: US Navy Oceanographic Research and the Discovery of Sea-Floor Hydrothermal Vents." *Social Studies of Science* (1 October 2003).

Piccard, Auguste. *"Le projet d'une exploration sous-marine belge"* [Belgium's Underwater Exploration Project]. *Actes de la Société Helvétique des Sciences Naturelles* 120 (1940).

Piccard, Jacques, and Thomas J. Abercrombie. "Man's Deepest Dive." *National Geographic* (August 1960).

Piccard, Jacques, and Robert S. Dietz. "Oceanographic Observations by the Bathyscaph *Trieste* (1953–1956)." *Deep-Sea Research* (1957).

Polmar, Norman. "U.S. Navy: The Deep Submergence Vehicle Fleet." USNI *Proceedings* (June 1986).

Polmar, Norman, and Lee J. Mathers. "The First Deepest Dive." USNI *Proceedings* (January 2020).

Polmar, Norman, and Don Walsh. "Tremendous Military Advantages Seen for Deep Diving Submarines and Submersibles." *Sea Power* (October 1970).

Rechnitzer, Andreas B. "*Trieste I* Deepest Manned Dive Passes 35th Anniversary." *Marine Technology Society Journal* (Winter 2009).

Reznick, Israel, and Aleksander Macander. "Syntactic Foams for Deep Sea Engineering Applications." *Naval Engineers Journal* (April 1968).

"Robots Explore for Sea Tasks at Great Depths." *Defiance* (Ohio) *Crescent News* (14 April 1979).

"The Sea Pioneers Who Dive 10,000 Feet Deep." *Oakland* (Calif.) *Tribune* (23 June 1973).

Shumaker, Lawrence A. *"Trieste II."* *Naval Engineers Journal* (August 1964).

Speiss, Fred N., and Arthur E. Maxwell. "Search for the *Thresher.*" *Science* (24 July 1965).

Spirito, Pietro. "La Madrina del Batiscafo '*Trieste*'" [The Godmother of the Bathyscaph *Trieste*]. Historical Diving Society Italy *HDS Notizie* (April 2012).

———. "Yolanda: The Woman Who Named the Piccard Bathyscaph *Trieste.*" *Il Piccolo,* 29 March 2009.

"Teleprobe to Aid Navy in Seeking Sunken Ships." *Bridgeport* (Conn.) *Telegram* (27 November 1969).

"*Trieste* Recovers Electronics Package from 16,400-ft. Depth." *Undersea Technology* (July 1972).

Vaishnav, Ramesh N., and Edward B. Magrab. "Exact Theory of Tracking of Moving Underwater Object by Short-Base Navigation System Attached to an Imperfectly Stabilized Moving Ship." *Journal of the Acoustical Society of America* (March 1970). Http://asa.scitation.org/doi/10.1121/1.1911978.

Wakelin, James H., Jr. "*Thresher*: Lesson and Challenge." *National Geographic* (June 1964).

Walsh, Don. "Conquering Inner Space." USNI *Proceedings* (January 1985).

———. "A Dive to the Bottom of the Sea . . . 50 Years Later," USNI *Proceedings* (January 2010).

———. "Flying Underwater." USNI *Proceedings* (July 2009).

———. "Going the Last Seven Miles." Naval Academy Alumni Association *Shipmate* (April 2000).

———. "In the Beginning . . . A Personal View." *Marine Technology Society Journal* (Winter 2009).

———. "Our 7-Mile Dive to Bottom." *Life* (15 February 1960).

———. "Race to the Bottom of the Sea: Seven Miles Deep." USNI *Proceedings* (January 2012).

Waltrop, David W. "An Underwater Ice Station Zebra: Recovering a KH-9 Hexagon Capsule from 16,400 Feet below the Pacific Ocean." *Quest* 19, no. 3 (August 2012). Https://www.cia.gov/library/publications/cold-war/underwater-ice-station-zebra/ice-station-zebra.pdf.

Weatherly, Georges L., and Mark Wimbush. "Near-Bottom Speed and Temperature Observations on the Blake-Bahama Outer Ridge." *Journal of Geophysical Research* (20 July 1980).

"The World's People." *Sperry World* (November 1978).

Papers and Conferences

Barham, Eric G. "The Deep Scattering Layer as Observed from the Bathyscaph *Trieste.*" *Proceedings of the 16th International Congress of Zoology,* Washington, D.C., 27 August 1963.

Barham, Eric G., N. J. Ayer Jr., and R. E. Boyce. "Macrobenthos of the San Diego Trough: Photographic Census and Observation from Bathyscaph *Trieste.*" *Deep-Sea Research* 14:773–784. This paper was presented at the Second International Oceanographic Congress, Moscow, USSR, May–June 1966, and published in the *Deep Sea Research and Oceanographic Abstracts,* vol. 14. Oxford, U.K.: Pergamon, 1967.

Busby, Roswell F. "Benefits to Military Oceanography from Manned Submersibles." In *Proceedings of the 7th U.S. Navy Symposium of Military Oceanography.* Annapolis, Md.: Naval Ship Research and Development Laboratory, 14 May 1970.

Committee on Undersea Warfare. "Proceedings of the Symposium on Aspects of Deep-Sea Research," Washington, D.C., 29 February–1 March 1956. *National Academy of Sciences Publication 473,* 1957.

Dill, R. F. "Military Significance of Deeply Submerged Sea Cliffs and Rocky Terraces on the Continental Slope." Fourth U.S. Navy Symposium on Military Oceanography (1967).

Fang, Bertrand T. "A Navigation Scheme for Deep Submergence Vehicles Using Combined Doppler and Position Measurements." Report 68-003. Space Science Department, Catholic University, Washington, D.C., 1968.

Gerald, R. D. "Taut Line Navigation Buoys Used in the *Thresher* Search." Marine Technology Society Meeting #24, 1964.

Green, Philip A. "Design Study of a Scientific Research Submersible System SRS-20,000." Seventh U.S. Navy Symposium of Military Oceanography, Naval Ship Research and Development Laboratory, Annapolis, Md., 14 May 1970.

Johnson, Willard E., John H. Howland, and Ronald J. Doyle. "Deep Submergence: The U.S. Navy's Operating Forces." Paper #OTC 2570. Offshore Technology 8th Annual Conference, Houston, Tex., 6 May 1976.

Knott, E. T. "Navigational Techniques Used in the *Thresher* Search." Annual National Meeting of the Institute of Navigation, New York, 15 June 1964.

Lowenstein, C. D., and J. D. Mudie. "On the Optimization of Transponder Spacing for Range-Range Navigation." Institute of Navigation, Manned Deep Submergence Vehicles Meeting, San Diego, Calif., 1966.

Mackenzie, Kenneth V. "Position Determination under the Sea." Annual Conference, Marine Technology Society, 29 June 1966.

Mandelberg, Martin. "An Oceanographic Acoustic Beacon and Data Telemetry System Powered by a SNAP-21 Radioisotope Thermoelectric Generator." Engineering in the Ocean Environment, IEEE 1971 Conference, San Diego, Calif., 21–24 September 1971.

Markel, Arthur L. "Adventures with *Aluminaut.*" Society of Automotive Engineers, February 1969.

Marquet, W. "Submersible Instrumentation Development." Woods Hole Oceanographic Institution Report 69-13, 1969.

Office of Naval Research. Bathyscaph Conference (Morning Session), 20 January 1958, Washington, D.C.

Rawson, Martine Dreyfus, and William B. F. Ryan. "Geologic Observations in the 2800 Meter Radioactive Waste Disposal Site and Associated Deepwater Dumpsite 10 (DWD-106) in the Atlantic Ocean." Washington, D.C.: U.S. Environmental Protection Agency, September 1983.

Schlosser, Arthur J. "Submersible Vehicle Machinery Operational Problems." Seventh Symposium of Military Oceanography, Naval Ship Research and Development Laboratory, Annapolis, Md., 14 May 1970.

Vaeth, J. Gordon. "Synthetic Training Equipment: Its Application to Oceanography." Environmental Science Services Administration, Rockville, Md., 1968.

Vine, Allyn C. "Vehicles as Instruments for Oceanographers." Symposium on Aspects of Deep Sea Research, 29 February–1 March 1956, Committee on Undersea Warfare. Washington, D.C.: National Academy of Sciences, 1957.

von Arx, W. S., ed. "Proceedings of the Symposium on Aspects of Deep Sea Research, Feb. 29–March 1, 1956." Pub. No. 473. Washington, D.C.: Committee on Undersea Warfare, National Academy of Sciences, 1957.

Waltrop, David W., Don Walsh, Richard Taylor, Beauford Myers, and Lee J. Mathers. "CIA's Underwater Space Mission Revealed." Chantilly, Va.: Smithsonian National Air & Space Museum, 26 April 2013.

Official Documents and Reports

Andrews, Frank A. "Meaning of Error in a Short Base Underwater Navigation System." Catholic University Report to the Office of Naval Research Code 466, Washington, D.C., 1966.

Bales, N. K, E. W. Foley, and R. M Watkins. "Dynamic Stability Characteristics of the USS *Point Loma* (AGDS-2)." Bethesda, Md.: David W. Taylor Naval Ship Research and Development Center, 1975.

Bartels, Malcolm G. "Aircraft Recovery Operations 9–10 October 1970, Report of ARD 20." San Diego, Calif., 1970.

———. "Change of Command Ceremony, *Trieste II* (DSV-1)." San Diego, Calif., 1973.

Bien, Albert, and Peter J. McDonough. "Naval Applications of the Man-in-the-Sea Concepts: Mission Definition." NWRC RM-50. Menlo Park, Calif.: Naval Warfare Research Center, December 1968.

Brundage, Walter. "NRL's Deep Sea Floor Search Era: A Brief History of the NRL/*Mizar* Search System and Its Major Achievements." NRL Report 6208. Washington, D.C.: Naval Research Laboratory, 29 November 1988.

Bureau of Ships. *Ship's Data U.S. Naval Vessels, Vol. III: Auxiliary, District Craft and Unclassified Vessels.* NAVSHIPS 250-012. Washington, D.C., 15 April 1945.

Busby, R. Frank. "Manned Submersibles." Washington, D.C.: Office of the Oceanographer of the Navy, 1976.

Brynes, David T. "Deep Submersible Logistic Support Design Concept." Master's thesis. Monterey, Calif.: Naval Postgraduate School, 1972.

Carroll, Jerry C., and James E. Walczak. "A Deep-Towed Magnetometer System." M-2-64. Washington, D.C.: Naval Oceanographic Office, May 1964.

Cawley, John D. "Review of Marine Navigation Systems and Techniques." Project Trident Report 1510165. Bureau of Ships Contract with Arthur D. Little, NObsr-81564 SS 050. Washington, D.C.: Bureau of Ships, 1965.

CIA Senior Representative. Memorandum: "Hexagon Recovery. Prognosis for RV-3 Recovery Operations." BYE-1071750-72. Washington, D.C.: Central Intelligence Agency, 3 February 1972.

Commander, Submarine Development Group 1. Booklet: *Change of Command Ceremony Trieste II (DSV 1).* San Diego, Calif., 28 June 1973.

———. "Officer Undergoing Medical Treatment." Radio message, 082137Z APR 70. San Diego, Calif., 8 April 1970.

Deputy Chief of Naval Operations, Fleet Operations and Readiness (OP-03). "U.S. Navy Deep Submergence/Ocean Engineering Programs: 1970–1980." Washington, D.C.: Office of the Chief of Naval Operations, June 1968.

Dietz, Robert S. "Bathyscaph *Trieste*." Technical Report 71-55. London: Office of Naval Research, 10 August 1955.

———. "Dives of the Bathyscaph *Trieste,* 1958–1963." Transcriptions of 61 Dictabelt recordings. Robert Sinclair Dietz Papers 1905–1994, Manuscript Collection 28. Archives of the Scripps Institution of Oceanography, San Diego, Calif. 1963.

First Naval District. "*Trieste II* Due in Boston." Press Release 83-64. Boston, 3 April 1963.

Frosch, Robert A., Deputy Secretary of the Navy (R&D). Memorandum to DNRO McLucas: "Deep Sea Recovery of Hexagon Reentry Vehicle." Washington, D.C., 18 August 1971.

Galantin, Vice Adm. Ignatius J., Chief of Naval Material. "Deep Submergence Systems Project; Designation Of." NavMatInst 5430.24. Washington, D.C.: Naval Material Command, 1966.

General Dynamics. "DSSP Large Object Salvage Study." Prepared for U.S. Navy Special Projects Office. Bureau of Naval Weapons Contract NOsp 65185-c. Groton, Conn., 1965.

Graybeal, Herbert L. "Design of the Bathyscaph *Trieste II.*" First U.S. Naval Symposium on Military Oceanography. Naval Oceanographic Office, Washington, D.C., June 1964.

Hightower, John D., George R. Beaman, George A. Wilkins, and Douglas W. Murphy. "Remote Unmanned Work System (RUWS) Electromechanical Cable System." U.S. Patent Office, Patent No. 4,010,619, 24 May 1976.

Hudiburgh, Charles W. Memorandum to Commander Submarine Development Group One, 3 May 1972. *White Sands* (ARD 20) Command History 1971. File Box 859A, Navy Operational Archives, Washington, D.C.

Keach, Donald L. "Oral History Interview, Donald Keach," by Gary Weir. Washington, D.C.: Naval History and Heritage Command, 1995.

Korth, Fred. "Notice: Deep Submergence Systems Review Group; Establishment Of." SecNav Notice 3100. Washington, D.C., 24 April 1963.

———. "Statement of the Secretary of the Navy Fred Korth on 5-Month-Long Search for Submarine *Thresher.*" Press Release No. 1207-63. Washington, D.C., 5 September 1963.

Larson, Rear Adm. Charles R., Commander, Submarine Development Group One. "Qualification and Certification Requirements for Operators of Untethered, Manned, Non-Combatant, Deep Submergence Vehicles and Deep Submergence Rescue Vehicles." ComSubDevGruOneInst 1540.4A. San Diego, Calif., 3 July 1978.

Lill, Gordon G. "Gordon G. Lill Oral History Interview." American Heritage Center, University of Wyoming, Laramie, 6 February 1978.

———. "An Interview with Dr. Gordon Lill," by David K. Van Keuren. Naval Research Laboratory, Washington, D.C., 20 March 1995.

Losee, Jon R. "Deep Ocean Gamma Ray Measurements, 1979 *Trieste* Operation." Technical Report 567. San Diego, Calif.: Naval Ocean Systems Command, 1979.

McLucas, John L. Memorandum: "Deep Sea Recovery of Hexagon Reentry Vehicle." Washington, D.C.: National Reconnaissance Office, 15 May 1972.

Molaskey, Leonard B. Memorandum: "Improved Viewing Capability for Bathyscaph *Trieste II*." Danbury, Conn.: Perkin-Elmer Optical Technology Division, 20 December 1971.

———. Memorandum: "Trip Report: Recovery Hook Test Program, San Diego, California, 20–21 September 1971." Danbury, Conn.: Perkin-Elmer Optical Technology Division, 24 September 1971.

———. Memorandum: "Trip Report: Recovery of RV #3." Danbury, Conn.: Perkin-Elmer Optical Technology Division, 18 November 1971.

National Academy of Sciences. *Report on the International Geophysical Year 1959*. Washington D.C.: National Science Foundation, 18 February 1959.

Naval Oceanographic Systems Center. "Underwater Vehicle History." Technical Report 1530. San Diego, Calif., 1989.

Naval Research Laboratory. "*Mizar*'s Underwater Search System." NRL Progress Review 1971. Washington, D.C., 1971.

———. "1966 NRL Abstracts." Washington, D.C., 1966.

Nevin, Robert F. "The Navigation of Scorpion Operations, Phase II." Slide presentation: *Trieste* (DSV-1), San Diego, Calif., Commander Submarine Development Group 1, 1969. Text of presentation forwarded to the authors by Nevin.

Newell, J. K. "*Trieste II* (DSV-1) Operator Qualification; Promulgation Of." *Trieste II* (DSV-1) Inst 1414.3A. San Diego, Calif., 23 April 1977.

Oak Ridge National Laboratory. "Isotopes and Radiation Technology: Quarterly Technical Progress Review." ORNL-TM-3382. Oak Ridge, Tenn., 1971.

Office of the Oceanographer of the Navy. "The Ocean Engineering Program of the U.S. Navy: Accomplishments and Prospects." Alexandria, Va., 1967.

Perry, Robert. *A History of Satellite Reconnaissance*. Vol. 3B, *Hexagon*. BYE 17017–74. Washington, D.C.: National Reconnaissance Office, 1973.

Peterson, Paul A. "Optimum Buoyancy Requirements for Weight-Limited Deep Submersibles." Master's thesis. Boston: Massachusetts Institute of Technology, June 1965.

Piccard, Jacques. Press Release: "Introduction to Krupp Sphere." Indian Industries Fair November/December 1961. New Delhi: Krupp AG, 1961.

Rechnitzer, Andreas B. "The 1957 Diving Program of the Bathyscaph *Trieste*." NEL Report 941. San Diego, Calif.: Navy Electronics Laboratory, 1959.

———. "Summary of the Bathyscaph *Trieste* Research Program Results (1958–1960)." NEL Report 1095. San Diego, Calif.: Navy Electronics Laboratory, 2 April 1962.

———. "Supplement to Walsh's Dive #70 Recording." San Diego, Calif.: Navy Electronics Laboratory, 23 January 1960.

Resnick, Israel. "Syntactic Foam Buoyancy Materials for Submerged Research Vehicles." Second U.S. Navy Symposium on Military Oceanography. Annapolis, Md.: Naval Ordnance Laboratory, 1965.

Shumaker, Lawrence A. "Evaluation of External Lighting Systems for the Bathyscaph *Trieste.*" NEL Report 1094. San Diego, Calif.: Navy Electronics Laboratory, 21 December 1961.

Sperry Rand. "*Trieste II* Trainee's Guide." Publication No. GJ-13-1617. Syosset, N.Y.: Sperry System Management Division, January 1968.

Stephan, Rear Adm. Edward C. "Report of the Deep Submergence Systems Review Group: Summary Report." NavExOS P-2452. Washington, D.C.: Executive Office of the Secretary of the Navy, 1 March 1964.

Submarine Support Facility, Ballast Point San Diego. "Lease Agreement with OIC, Deep Submergence Systems Project Technical Office." San Diego, Calif., 20 September 1966.

Talkington, Howard R. "Remotely Manned Undersea Work Systems at Naval Ocean Systems Center." Technical Report 237. San Diego, Calif.: Naval Oceanographic Systems Center, 15 April 1978.

Tryon, V. Paul. "Remote Unmanned Work System (RUWS) Mating Latch." U.S. Patent Office, Patent No. 3,987,741. 12 January 1976.

Tyrrash, Peter A., and Eric L. Swensen. Scorpion *Operations Phase II.* Cruise Book. San Diego, Calif.: USS *White Sands* (ARD 20), 1969.

Walsh, Don. "The Bathyscaph *Trieste*: Technological and Operational Aspects, 1958–1961." NEL Report 1096. San Diego, Calif.: Navy Electronics Laboratory, 27 July 1962.

———. "Dive #70 by the Bathyscaph *Trieste.*" Acoustic Recording Transcript. San Diego, Calif.: Navy Electronics Laboratory, 23 January 1960.

———. "Review of Bathyscaph *Trieste* Operations 1958–1961." San Diego, Calif.: Navy Electronics Laboratory, December 1961.

Waltrop, David W., Lee J. Mathers, Beauford Myers, and Richard Taylor. Panel Presentation: "An Underwater Ice Station Zebra: Recovering a Secret Spy Satellite Capsule from 16,400 ft. below the Pacific Ocean." Keyport, Wash.: Naval Undersea Museum, 3 November 2012.

Weiss, H.V., and J. F. Vogt. "Radioisotope Thermoelectric Generators Emplaced in Deep Ocean: Recover or Dispose in Situ?" Technical Report 1106. San Diego, Calif.: Naval Ocean Systems Center, March 1986.

Whitmire, Wilson R. Booklet: "USS *Point Loma* (AGDS 2) Commissioning Ceremony." San Diego, Calif., 3 July 1976.

Miscellaneous

Andrews, Frank A. "My View: Submarine Development Group 2." 50th Anniversary of Submarine Development Group 2. New London, Conn., May 1999.

Byrnes, David T. "Running Lights." Leonardtown, Md.: unpublished manuscript, 2016.

Mooney, John B., Jr. Letter to George Martin, 28 August 1964.

Myers, Beauford. Letter home, 10 November 1971.

Voss, Heinfield. E-mail correspondence with the authors, August 2013. Voss is the Krupp official historian and director of the Historisches Archiv Krupp, Essen, Germany.

Documents on the Internet

Ballard, Robert. "The History of Woods Hole's Deep Submergence Program." In National Research Council, *50 Years of Ocean Discovery: National Science Foundation 1950–2000.*

Washington, D.C.: National Academies Press for the Ocean Studies Board, 2000. Https://www.ncbi.nlm.nih.gov/books/NBK208815/ (accessed 17 July 2017).

Bennett, John E. "Oral History Interview: John E. Bennett." Interview by Donald R. Lennon, East Carolina University Digital Interview (Interview #138), 6 February 1994. Https://digital.lib.ecu.edu/11276/ (accessed 4 August 2017).

Brundage, Walter L., and Robert B. Patterson. "Ocean Bottom Photography from a Deep Towed Camera System." Naval Research Laboratory *Optical Engineering Journal,* 1 May 1966. Reprinted in Walter Brundage, "NRL's Deep Sea Floor Search Era: A Brief History of the NRL/*Mizar* Search System and Its Major Achievements." NRL Report 6208. Washington, D.C.: Naval Research Laboratory, 1988. Http://www.dtic.mil/dtic/tr/fulltext/u2/a204011.pdf (accessed 4 August 2017).

Buchanan, Chester L. "A Love Affair." NRL Progress Report 1973. Reprinted in Walter Brundage, "NRL's Deep Sea Floor Search Era: A Brief History of the NRL/*Mizar* Search System and Its Major Achievements." NRL Report 6208. Washington, D.C.: Naval Research Laboratory, 1988. Http://www.dtic.mil/dtic/tr/fulltext/u2/a204011.pdf.

Chester, Richard J. *A History of the Hexagon Program,* BIF 007-0253-85. Washington, D.C.: National Reconnaissance Office, 1985. Https://web.archive.org/web/20160310053325/http://nro.gov/foia/declass/GAMHEX/GAMBIT%20and%20HEXAGON%20Histories/1.pdf.

Chirini, Aldo, ed. *L'Esplorazione delle Profondita Marine e I Batiscafi* 'Trieste,' 'Archimède' & 'Alvin' [Deep Ocean Explorations by the Bathyscaphs "*Trieste*," "Archimède" & "Alvin,"'], AMA 82. Trieste [Italy]: Associazione Marinara Aldebaran, January 1999. Http://www.cherini.eu/pdf/Batisfera.pdf.

Department of the Navy, Press Release [no headline] dated 15 November 1959. 16 November 1959, Guam, Marianas. Http://www.onr.navy.mil/focus/ocean/vessels/submersibles8.htm.

———. Press Release: "Deep Sea Diving Craft Breaks Its Own World Record." Guam, Marianas, 8 January 1960. Https://web.archive.org/web/20021002062153/http://www.onr.navy.mil/focus/ocean/vessels/submersibles9.htm.

———. Press Release: "Navy Bathyscaph *Trieste* Makes 37,800-Foot Dive." Washington, D.C., 23 January 1960. Https://web.archive.org/web/20020418105908/http://www.onr.navy.mil/focus/ocean/vessels/submersibles11.htm.

Klein, Martin. "Martin's Story: Bathyscaph *Trieste* Alumni Association," n.d. https://web.archive.org/web/20120112034423/www.bathyscaphtrieste.com/contents/story/marty.html.

LaGassa, Robert. "USS *Scorpion* Sinking Analysis: Letter to Dr. Robert Ballard." Annapolis, Md.: U.S. Naval Academy Class of 1963. 20 October 2013. Http://www.usna63.org/tradition/Bob-LasGassa-Insights.pdf.

Library of Congress. Piccard Family Papers, circa 1470–1983. Https://lccn.loc.gov/mm77036145.

Maier, John. "USS *Thresher* Banners Installed on the USS *Albacore*." *US Naval Ships Association Journal Anchor Watch* (May/June/July/August 2013). Https://archive.hnsa.org/anchorwatch/pdf/2013summer.pdf.

Nicholson, William M. "Oral History Interview." Interview by Donald R. Lennon. OH Interview 177. East Carolina University, Greenville, N.C., 8 February 1999. Https://digital.lib.ecu.edu/text/11255.

Rechnitzer, Andreas B., W. W. Denner, and E. C. Estes. "An Assessment of Remotely Operated Vehicles to Support the AEAS Program in the Arctic." Reston, Va.: Science Applications International, 15 September 1986. Http://www.dtic.mil/dtic/tr/fulltext/u2/a231730.pdf.

Romero, Alberto. "Victor Aldo De Sanctis." Historical Diving Society Italy *HDS Notizie,* no. 22 (2002). Http://www.hdsitalia.org/sites/www.hdsitalia.org/files/documenti/HDSN22.pdf.

Scripps Institution of Oceanography. "Obituary: Hugh Bradner (1915–2008)." La Jolla, Calif., 8 May 2008. Http://scripps.ucsd.edu/news/2462.

Taylor, Richard. Video Recording: "CIA's Underwater Space Mission Revealed." David Waltrop, Moderator. Smithsonian National Air and Space Museum. US Stream TV, 26 April 2013. Http://www.ustream.tv/recorded/32000249_hearing_pt1.pdf (accessed 4 August 2017).

Toye, Sandra. "Deep Submergence: The Beginnings of *Alvin* as a Tool of Basic Research." In National Research Council, *50 Years of Ocean Discovery: National Science Foundation 1950–2000.* Washington, D.C.: National Academies Press for the Ocean Studies Board, 2000. Https://www.ncbi.nlm.nih.gov/books/NBK208813/.

Trieste Program Dive Logs. NMNW.2014.020.001. Naval Undersea Museum, Keyport, Wash., 1 February 2014. Http://www.navalunderseamuseum.org/wp-content/uploads/2016/01/*Trieste*-Dive-Log-2014_020_001-low-res.pdf.

U.S. House of Representatives. *A Report of the Committee on Oceanography of the National Academy of Sciences.* Special Subcommittee on Oceanography of the Committee on Merchant Marine and Fisheries, Eighty-Sixth Congress, First Session, Washington, D.C., 3 March 1959. Http://docs.lib.noaa.gov/noaa_documents/NOAA_related_docs/oceanography_inUS_1959

Walsh, Don. "Memorial Tribute: John B. Mooney Jr. 1931–2014." *National Academy of Engineering* (2016), vol. 20, pp. 180–185. Https://www.nap.edu/read/23394/chapter/30 (accessed 7 August 2017).

———. "Oceans: *Alvin* and *Trieste.*" U.S. Naval Institute *News,* 9 November 2012. Http://news.usni.org/2012/11/09/oceans-alvin-and-trieste.

Wilson, David A. "Personnel Implication of New Technological Developments: Undersea Technologies." SR70–3 [Project Argonaut]. San Diego, Calif.: U.S. Naval Personnel Research Activity, July 1969. Http://citeseerx.ist.psu.edu/viewdoc/download?doi=10.1.1.937.4330&rep=re p1&type=pdf.

Woods Hole Oceanographic Institution. "*Alvin* Dives 1970s." Marine: Dive Logs 1970s, Woods Hole Oceanographic Institution. Http://www.marine.whoi.edu/divelog.nsf/Dives%20in%201970s?OpenView&Start=1.

General Index

Ship/Submarine/Submersible Index

U.S. Navy ships are identified by their hull numbers. The prefix T- indicates a Navy ship operated by the Military Sea Transportation Service, renamed the Military Sealift Command in 1970; those ships have civilian crews.

About the Authors

Norman Polmar is an analyst, consultant, and author specializing in naval, aviation, and technology subjects. He has been a consultant or advisor on naval issues to three Senators, the Speaker of the House of Representatives, and three Secretaries of the Navy as well as to the director of the Los Alamos national laboratory, and to the leadership of the U.S., Australian, Chinese, and Israeli navies. He has written or coauthored more than fifty published books.

Lee J. Mathers is a former Surface Warfare Officer with an intelligence subspecialty. He attended the University of Utah, Central Michigan University, and the Defense Intelligence School, the last followed by intelligence duty in the Office of the Chief of Naval Operations. He has written for the magazines *Proceedings* and *Naval History* and was a key researcher for the Naval Institute book *Project Azorian* by Norman Polmar and Michael White. He now resides in Canada.

The Naval Institute Press is the book-publishing arm of the U.S. Naval Institute, a private, nonprofit, membership society for sea service professionals and others who share an interest in naval and maritime affairs. Established in 1873 at the U.S. Naval Academy in Annapolis, Maryland, where its offices remain today, the Naval Institute has members worldwide.

Members of the Naval Institute support the education programs of the society and receive the influential monthly magazine *Proceedings* or the colorful bimonthly magazine *Naval History* and discounts on fine nautical prints and on ship and aircraft photos. They also have access to the transcripts of the Institute's Oral History Program and get discounted admission to any of the Institute-sponsored seminars offered around the country.

The Naval Institute's book-publishing program, begun in 1898 with basic guides to naval practices, has broadened its scope to include books of more general interest. Now the Naval Institute Press publishes about seventy titles each year, ranging from how-to books on boating and navigation to battle histories, biographies, ship and aircraft guides, and novels. Institute members receive significant discounts on the Press' more than eight hundred books in print.

Full-time students are eligible for special half-price membership rates. Life memberships are also available.

For a free catalog describing Naval Institute Press books currently available, and for further information about joining the U.S. Naval Institute, please write to:

Member Services
U.S. Naval Institute
291 Wood Road
Annapolis, MD 21402-5034
Telephone: (800) 233-8764
Fax: (410) 571-1703
Web address: www.usni.org